Springer Remote Sensing/Photogrammetry

More information about this series at http://www.springer.com/series/10182

Venkat Lakshmi
Editor

Remote Sensing of Hydrological Extremes

Springer

Editor
Venkat Lakshmi
Department of Geophysics
Stanford University
Stanford, CA, USA

Professor
School of Earth, Ocean and Environment
University of South Carolina
Columbia, SC, USA

ISSN 2198-0721 ISSN 2198-073X (electronic)
Springer Remote Sensing/Photogrammetry
ISBN 978-3-319-43743-9 ISBN 978-3-319-43744-6 (eBook)
DOI 10.1007/978-3-319-43744-6

Library of Congress Control Number: 2016955245

Printed on acid-free paper

This Springer imprint is published by Springer Nature
The registered company is Springer International Publishing AG Switzerland

Preface

Water is the most important natural resource we possess on the planet. Too much water—floods—can cause tremendous damage by washing away roads and buildings, eroding land, and destroying crops and livestock. Floods often result in the loss of human lives, and can have huge impacts on national economies when occurring at large scales. At the other end of this spectrum are droughts. Droughts result in reduced crop yield when irrigation sources are unavailable and rainfall is the only source of water for crops.

The time scales, inception, and progress of floods and droughts are completely different. Whereas floods (specifically flash floods) can happen quickly, droughts take months and sometimes years to form. In the case of flooding, there is some advance warning in precipitation forecasts using atmospheric models. In cases of downstream floods, the flow time from the upper reaches of the catchment to the lower reaches and the outlet offers some advance lead time. In the case of droughts, the lack of precipitation coupled with high evapotranspiration is a prescription for disaster, and this could occur over long periods of months to years.

With rapid advances in computer modeling and observing systems, floods and droughts can be forecasted and assessed with greater precision today than ever before. Land surface models (especially over the entire Continental United States) can map the hydrological cycle at kilometer and sub-kilometer scales. In the case of smaller areas (not the entire Continental United States), there is even higher spatial resolution, and the only limiting factor is the availability of input data. In situ sensors are commonly automated and the data directly relayed to the Internet for many hydrological variables such as precipitation, soil moisture, surface temperature, and heat fluxes. In addition, satellite remote sensing has advanced to providing daily (or better) observations at kilometer to 10-km spatial scales.

We are at a critical juncture in studying hydrological extremes. The following features make this unique. Firstly, in recent years, floods (e.g., in China, Pakistan, Thailand, Laos, South Carolina, and Eastern China) and droughts (e.g., in California, Australia, and the Asian subcontinent) have centered on extremes involving water. Water plays an important role in the global economy and thus extreme events can have a large economic impact that is not limited to a specific region or country. Secondly, there have been major advances in the global monitoring of precipitation and measurement of soil moisture with the launch of the Global Precipitation Measurement (GPM) mission in February 2014. This follows the Tropical Rainfall Measurement Mission [TRMM] that provided data from 1997 to 2015 for the measurement of precipitation and the Soil Moisture Active Passive (SMAP) mission in January 2015 and the Soil Moisture and Ocean Salinity (SMOS) in November 2009. These have been complemented by sensors that monitor vegetation, surface temperature, and evapotranspiration (Moderate Resolution Imaging Spectroradiometer—MODIS), and the Gravity Recovery and Climate Experiment (GRACE) that estimates changes in surface and subsurface water storage and provides a clear picture of the land surface hydrological state. Lastly, we are at the end of an El-Nino event that alters the normal hydrological cycle. Thus, this is the perfect time to focus on methods to monitor hydrological extremes and provide support to emergency management using the latest technology.

This book—*Remote Sensing of Hydrological Extremes: Droughts and Floods*—provides an overview of the current state of the science on the monitoring of droughts and floods using remote sensing. The various chapters in the book span the diversity of geographical locations and satellite sensors and analyze methods for studying these extreme events. Droughts and floods are studied in Brazil, the Congo River Basin of Africa, the Mekong River Basin, the Magdalena River Basin in Columbia, South America, and various locations in United States. New tools and analyses have been developed to map floods, and these will prove valuable for recovery and rebuilding efforts. Analyses of the dependence of land surface variables in spatial and temporal propagation help to predict droughts. Both of these analyses (floods and droughts) are of value for future land-use planning.

This book will serve as a reference for students, teachers, practicing hydrologists, agriculture scientists, engineers, and those involved in water and emergency management and in the construction and insurance industry.

Water can be a blessing or a curse. . .predictability of water in a flood and lack of water in a drought is both a scientific problem and a societal necessity.

Stanford, CA
July 2016

Venkat Lakshmi

Acknowledgements

Venkat Lakshmi was awarded the Cox Visiting Fellowship for the 2015–2016 academic year by the School of Earth, Energy and Environmental Sciences, Stanford University.

Funding from the NASA Applied Sciences Program—Water Resources Applications, Program Manager Bradley D. Doorn is gratefully acknowledged.

Contents

Contributors

Aakash Ahamed, M.Sc. Hydrological Sciences Laboratory, NASA Goddard Space Flight Center, Greenbelt, MD, USA

A. AghaKouchak University of California, Irvine, CA, USA

Ali Ahmadalipour, M.S. Remote Sensing and Water Resources Lab, Department of Civil and Environmental Engineering, Portland State University, Portland, OR, USA

John Bolten Hydrological Sciences Laboratory, NASA Goddard Space Flight Center, Greenbelt, MD, USA

G. Robert Brakenridge Dartmouth Flood Observatory, Community Surface Dynamics Modelling System, University of Colorado, Boulder, CO, USA

Bob Brakenridge, Ph.D. CDSMS/INSTAAR, University of Colorado, Boulder, CO, USA

Rogger Escobar C. Department of Earth Sciences, School of Sciences, EAFIT University, Medellín, Colombia

Mark Carroll, B.S., M.P.S., Ph.D. SSAI and Biospheric Sciences Laboratory, NASA GSFC, Greenbelt, MD, USA

Eunsang Cho, M.S. Department of Civil and Environmental Engineering, University of New Hampshire, Durham, NH, USA

Michael H. Cosh, Ph.D. USDA Agricultural Research Service, Beltsville, MD, USA

Narendra Das Jet Propulsion Laboratory, NASA, Pasadena, CA, USA

X. Dong, Ph.D. China Three Gorges University, Yichang, Hubei, P.R. China

Colin Doyle Department of Geography and the Environment, University of Texas at Austin, Austin, TX, USA

A. Farahmand University of California, Irvine, CA, USA

Jessica Fayne, B.A., M.S. Department of Earth and Ocean Sciences, University of South Carolina, Columbia, SC, USA

Hydrological Sciences Laboratory, NASA Goddard Space Flight Center, Greenbelt, MD, USA

Otto Corrêa Rotunno Filho Laboratório de Recursos Hídricos e Meio Ambiente (LABH2O), Programa de Engenharia Civil, Instituto Alberto Luiz Coimbra de Pós-Graduação e Pesquisa em Engenharia—COPPE, Universidade Federal do Rio de Janeiro, Rio de Janeiro, RJ, Brazil

Nandita Gaur Department of Biological and Agricultural Engineering, Texas A&M University, College Station, TX, USA

Augusto Getirana Hydrological Sciences Laboratory, NASA Goddard Space Flight Center, Greenbelt, MD, USA

Earth System Science Interdisciplinary Center, University of Maryland, College Park, MD, USA

Y. He, Ph.D. National Meteorological Center, Beijing, P.R. China

Alfred Hubbard SSAI and Biospheric Sciences Laboratory, NASA GSFC, Greenbelt, MD, USA

Amor V.M. Ines Michigan State University, East Lansing, MI, USA

Jennifer M. Jacobs, Ph.D. Department of Civil and Environmental Engineering, University of New Hampshire, Durham, NH, USA

Raghavendra Jana King Abdullah University of Science and Technology, Thuwal, Saudi Arabia

Xinhua Jia, Ph.D. Department of Agricultural and Biosystems Engineering, North Dakota State University, Fargo, ND, USA

Hahn Chul Jung Hydrological Sciences Laboratory, NASA Goddard Space Flight Center, Greenbelt, MD, USA

Science Systems and Applications, Inc., Lanham, MD, USA

Hahn Jung, Ph.D. Hydrological Sciences Laboratory, NASA GSFC, Greenbelt, MD, USA

SSAI, Lanham, MD, USA

Albert J. Kettner Dartmouth Flood Observatory, Community Surface Dynamics Modelling System, University of Colorado, Boulder, CO, USA

Venkat Lakshmi Department of Geophysics, Stanford University, Stanford, CA, USA

School of Earth, Ocean and Environment, University of South Carolina, Columbia, SC, USA

Hyongki Lee Department of Civil and Environmental Engineering, University of Houston, Houston, TX, USA

National Center for Airborne Laser Mapping, University of Houston, Houston, TX, USA

C. Love University of California, Irvine, CA, USA

A. Mehran University of California, Irvine, CA, USA

Binayak P. Mohanty, Ph.D. Department of Biological and Agricultural Engineering, Texas A&M University, College Station, TX, USA

Hamid Moradkhani, Ph.D. Remote Sensing and Water Resources Lab, Department of Civil and Environmental Engineering, Portland State University, Portland, OR, USA

Joe Nigro, M.A. SSAI and Biospheric Sciences Laboratory, NASA GSFC, Greenbelt, MD, USA

F. Policelli, B.S., M.S. Hydrological Sciences Laboratory, NASA GSFC, Greenbelt, MD, USA

Department of Earth and Planetary Sciences, Johns Hopkins University, Baltimore, MD, USA

Vitor Paiva Alcoforado Rebello Laboratório de Recursos Hídricos e Meio Ambiente (LABH2O), Programa de Engenharia Civil, Instituto Alberto Luiz Coimbra de Pós-Graduação e Pesquisa em Engenharia—COPPE, Universidade Federal do Rio de Janeiro, Rio de Janeiro, RJ, Brazil

Juan D. Restrepo Department of Earth Sciences, School of Sciences, EAFIT University, Medellín, Colombia

Pedro J. Restrepo, Ph.D., M.Sc., Ingeniero Civil Consulting Engineer, Minneapolis, MN, USA

M. Sadegh University of California, Irvine, CA, USA

Yongchul Shin School of Agricultural Civil & Bio-Industry Engineering, College of Agriculture and Life Science, Kyungpook National University, Daegu, South Korea

Dan Slayback, Ph.D. SSAI and Biospheric Sciences Laboratory, NASA GSFC, Greenbelt, MD, USA

Z. Su, Ph.D. Faculty of Geo-Information Science and Earth Observation (ITC), University of Twente, Enschede, The Netherlands

M. J. Tourian University of Stuttgart, Stuttgart, Germany

Samuel E. Tuttle, M.A., Ph.D. Department of Civil and Environmental Engineering, University of New Hampshire, Durham, NH, USA

Carrie M. Vuyovich, M.S. Remote Sensing, GIS and Water Resources Branch, USACE Cold Regions Research and Engineering Laboratory, Hanover, NH, USA

L. Wang, M.Sc. Faculty of Geo-Information Science and Earth Observation (ITC), University of Twente, Enschede, The Netherlands

Hongxiang Yan, M.S. Remote Sensing and Water Resources Lab, Department of Civil and Environmental Engineering, Portland State University, Portland, OR, USA

Ting Yuan Department of Civil and Environmental Engineering, University of Houston, Houston, TX, USA

National Center for Airborne Laser Mapping, University of Houston, Houston, TX, USA

Ben Zaitchik, Ph.D. Department of Earth and Planetary Sciences, Johns Hopkins University, Baltimore, MD, USA

Mahkameh Zarekarizi, M.S. Remote Sensing and Water Resources Lab, Department of Civil and Environmental Engineering, Portland State University, Portland, OR, USA

Part I
Floods

Chapter 1
Satellite-Based Estimation of Water Discharge and Runoff in the Magdalena River, Northern Andes of Colombia

Rogger Escobar C., Juan D. Restrepo, G. Robert Brakenridge, and Albert J. Kettner

1.1 Introduction

Satellite-derived data provide important information about river characteristics such as channel geometry, water surface area, water levels, and floodplain extension (Rokni et al. 2015; Schumann and Moller 2015; Smith 1997). For example, variations in surface water area along river channels are well detected by various sensing instruments onboard satellites.

Although some studies suggested that the estimation of river discharge from satellite data could be possible (Brakenridge et al. 1994; Koblinsky et al. 1993), reliable applications of satellite-derived data in estimating stream flow have only slowly developed (e.g., Smith et al. 1995; Smith 1997; Bjerklie et al. 2003; Brakenridge et al. 2005). Main findings of these studies highlighted that: (1) for an accurate estimation of discharge from satellite, other data are needed, such as parameterization of morphological characteristics like sinuosity, slope, vegetation cover, and distribution of braided channels (Smith et al. 1996); (2) it is possible to calculate river discharge once velocities are estimated from radar or LIDAR and water surface elevations are used to estimate average depth (Bjerklie et al. 2003, 2005); and (3) river bank-full discharge can be estimated from remote sensing with an uncertainty of 24 % (Bjerklie 2007).

Such previous work indicates the capabilities of satellites data to monitor stream flow, but practical methods to monitor global runoff in real time were not available

R. Escobar C. (✉) • J.D. Restrepo
Department of Earth Sciences, School of Sciences, EAFIT University, Medellín AA 3300, Colombia
e-mail: rescobarcorrea@gmail.com; rescoba9@eafit.edu.co

G.R. Brakenridge • A.J. Kettner
Dartmouth Flood Observatory, Community Surface Dynamics Modelling System, University of Colorado, Boulder, CO 80309, USA

V. Lakshmi (ed.), *Remote Sensing of Hydrological Extremes*, Springer Remote Sensing/Photogrammetry, DOI 10.1007/978-3-319-43744-6_1

until recently. The first attempt to obtain near-real-time discharge from world rivers was developed by the Dartmouth Flood Observatory (DFO) in association with the Global Runoff Data Centre (GRDC) (Brakenridge et al. 2005). The method was based on measuring water surface changes through the Moderate Resolution Imaging Spectro-radiometer (MODIS) in rivers reaches between ~10 and 30 km. Latter, the Advanced Microwave Scanning Radiometer (AMSR-E) was utilized to estimate river discharge by capturing temperature changes in wet–dry surfaces. This study showed that microwave sensors are appropriate for measuring river discharge with an almost daily orbiting frequency (1–2 days) and low interference from clouds (Brakenridge et al. 2007).

More recently, the DFO has achieved important goals for more consistent calibration of discharge-responsive remote sensing signal to discharge values. The calibration is derived from estimates of the Global Water Balance Model (WMB) (Wisser et al. 2009), allowing calibration even in places without ground observations of discharge (Brakenridge et al. 2012; De Groeve et al. 2007). At present, the DFO (http://floodobservatory.colorado.edu), River Watch version 3 monitors river discharge in 343 sites along many important rivers around the world. River Watch sites are based on the passive microwave signals from AMRS-E, AMRS2, TRMM, and GPM, processed in collaboration with the Joint Research Centre (JRC) (Brakenridge et al. 2012). The satellite signal is calibrated site by site with discharge from the global Water Balance Model (WBM); only a few sites are calibrated with water discharge from ground-based gauging stations where information is available and consistent (Brakenridge et al. 2015).

For South America, numerous studies have used remote sensing to study river dynamics, especially in the Amazon River, including the analysis of hydrological regimes (Filizola et al. 2014; Vörösmarty et al. 1996), the estimation of river stage series (Birkett et al. 2002), the dynamics of floodplain water storage (Alsdorf et al. 2010; Frappart et al. 2005), flooding dynamics (Hall et al. 2011), and suspended sediment processes (Latrubesse et al. 2005; Restrepo et al. 2016). Other major studies include the analyses of flooding dynamics with MODIS imagery in the Paraná River (Handisyde et al. 2014) and the estimation of flood storage in the world largest wetland, the Pantanal, Brazil, by using the GRACE mission data and coupled modeling of hydro-climatic variables (Penatti Costa et al. 2015). The quite different and supplemental data provide by near-daily satellite microwave radiometry now can be incorporated on such topics as well.

Studies applying satellite data to estimate river discharge and flood dynamics in Colombia are scarce. A recent exercise of flood frequency mapping in the Magdalena basin employed Advanced Land Observing Satellite-Phased Array L-Band Synthetic Aperture Radar (ALOS-PALSAR) and developed maps of flooded areas in the lower catchment (Quiñones 2013). Other attempts obtained preliminary insights of suspended sediment dynamics in the Momposina floodplain (Fig. 1.1) by using MODIS imagery during La Niña floods in 2009 (Kettner et al. 2010; Syvitski et al. 2012) and flooding dynamics along the lower course of the Magdalena (Walschburger et al. 2015). To the best of our knowledge, no other studies employing satellite information have been conducted for Colombian fluvial systems.

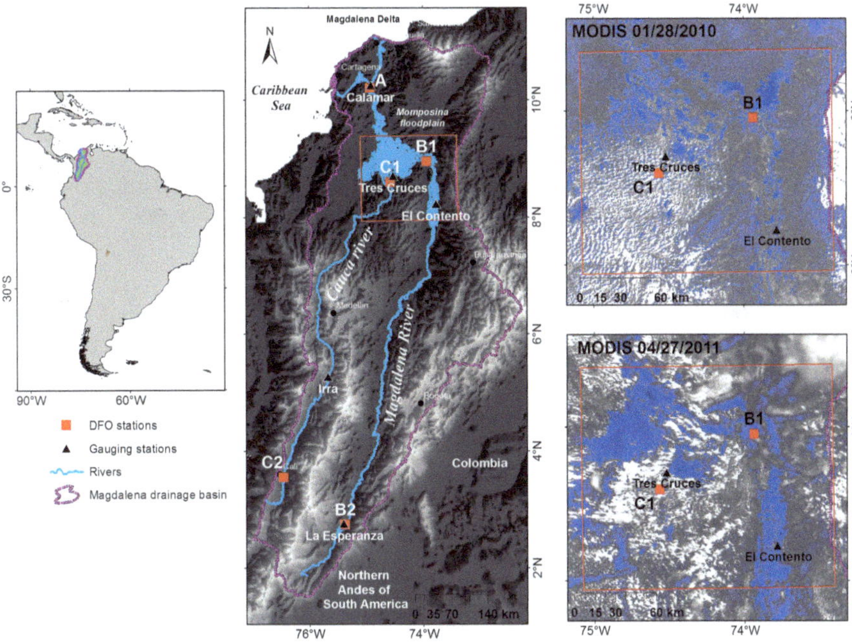

Fig. 1.1 Location of the Magdalena drainage basin showing the five DFO satellite-based gauging stations (*red squares*) and the corresponding hydrological ground-based stations (*black triangles*). We also show two different MODIS-Aqua images at the lower reaches of the Magdalena, the Momposina floodplain, during low discharge (*upper right*, Jan. 28, 2010) and high discharge conditions (*lower right*, Apr. 27, 2011)

During 2011, Colombia was considered a country under extreme risk in terms of natural disasters (e.g., floods, droughts, and landslides) (http://maplecroft.com/about/news/ccvi.html). Environmental decision-making and mitigation strategies for floods lack reliable scientific tools for drainage basin planning (Walschburger et al. 2015). Flood detection is mostly based on ground-based gauging stations records, with severe limitations in calibration, accuracy, and data availability. Unreliable rating curves between river discharge and river stage are constructed for gauging stations along river floodplains, resulting in the underestimation of river fluxes when extreme floods exceed the bank-full discharge or the maximum peak flow (Milliman and Farnsworth 2011). Implementing tools for near-real-time estimation of river discharge and flood magnitudes could allow better mitigation strategies by producing accessible data to scientists and decision makers.

It is our objective to measure the magnitude and variability of river discharge in the Magdalena drainage basin by utilizing satellite-derived data. We use the River Watch version 3 platform from the Dartmouth Flood Observatory (DFO) (Brakenridge et al. 2015) and provide runoff data in near real time through the DFO River Watch portal (http://floodobservatory.colorado.edu/DischargeAccess.html).

1.2 The Magdalena River Basin

Drainage basins in the Andes of Colombia, including the large Magdalena and its principal tributary, the Cauca (Fig. 1.1), are characterized by high precipitation rates and fast runoff response due to steep slopes (Restrepo et al. 2006). Regional climate is partly controlled by the occurrence of the El Niño Southern Oscillation (ENSO), which brings dry conditions during the El Niño and wet pulses during La Niña. Interannual variability of river discharge shows strong correlations with the ENSO at periods of 3–5 years (Poveda et al. 2001, 2006). Recently, strong La Niña events in 2008–2009 and 2010–2011 (Fig. 1.1) impacted the Magdalena drainage basin, causing infrastructural and agricultural damages of over $US 7.8 million (Hoyos et al. 2013). Also, the Magdalena catchment has become more vulnerable to strong climatic anomalies and associated floods due to severe changes in land cover and forest loss during the last three decades (Restrepo et al. 2015).

The Andes Mountains in Colombia consist of three nearly parallel, north–south-oriented ranges: the Western, Central, and Eastern cordilleras, which merge into a single range near the Ecuadorian border (Fig. 1.1). Between these ranges lie two river valleys: the high and narrow Cauca valley to the west and the lower and broader Magdalena valley to the east. The Magdalena River is the largest river system of the northern Andes of Colombia (Fig. 1.1), with a length of 1612 km. The drainage basin area covers 257,438 km^2 (24 % of Colombia), with headwaters located at an elevation of 3685 m (Restrepo et al. 2015). The Cauca River, the most important tributary of the Magdalena, has a drainage basin area of 66,750 km^2 or 26 % of the Magdalena basin (Restrepo et al. 2006). The mountainous section of the catchment is characterized by landslide activity over fossil sedimentary rocks. Most tributaries drain basins less than 6000 km^2 and are responsive to both natural and human-induced change (Restrepo and Syvitski 2006). The geomorphic setting of the Magdalena comprises subsiding foreland areas, an anastomosing river pattern, and tributary systems with high vertical aggradation (Latrubesse et al. 2005).

The Magdalena basin is characterized by monsoonal precipitation, with an average rainfall of 2050 mm y^{-1} for the whole basin. Precipitation patterns throughout the catchment vary greatly, from 500 mm y^{-1} in the eastern mountains of the middle and upper basin to more than 5000 mm y^{-1} in the western basin. The precipitation in the upper, middle, and lower reaches is 1535, 2185, and 1630 mm y^{-1}, respectively. The average runoff for the basin as a whole is 953 mm y^{-1}, with a runoff in the upper, middle, and lower reaches of 900, 1260, and 700 mm y^{-1}, respectively. The mean annual water discharge of the Magdalena (ground-based station A, Fig. 1.1) and Cauca rivers (ground-based station C1, Fig. 1.1) are 7200 and 2530 $m^3\ s^{-1}$, respectively (Restrepo et al. 2006).

1.3 Site Selection and Accuracy Statistics

We selected five locations along the main channels of the Magdalena and Cauca rivers for installation of "Space Gauging Stations" (Fig. 1.1). Each site exhibits either braided, anastomosing, or meandered fluvial morphology (relatively broad floodplains). Such morphologies favor significant variation of the site water surface area with changes in discharge. For remote sensing to discharge calibration, we used monthly estimated mean, peak, and low flows (daily values) calculated for 5 years (2003–2007) by the Global Water Balance Model (WBM), and daily river discharge data (1998–2015) from ground-based gauging stations, obtained from Instituto de Hidrología, Meteorología y Estudios Ambientales, IDEAM (Fig. 1.1) to calibrate each space gauging stations.

The River Watch microwave signal is itself a microwave brightness (radiance) ratio, computed by comparing the radiance from the site pixel to the brightest (driest) nearby pixel within a 9 × 9 pixel surrounding array (De Groeve et al. 2015). The time series for each site begins January 1, 1998 using the TRMM 37 GHz channel, and then the ratio ingests also data from AMSR-E, AMSR-2, and GPM (Fig. 1.2). The River Watch version 3 processor then computes a 7-day forward running mean from the daily time series of signal ratios, and in order to smooth some of the observed daily variability believed to be nonhydrological in origin.

Transformation of the remote sensing signal to river discharge values is performed using a rating equation. The calibrated discharge values were obtained

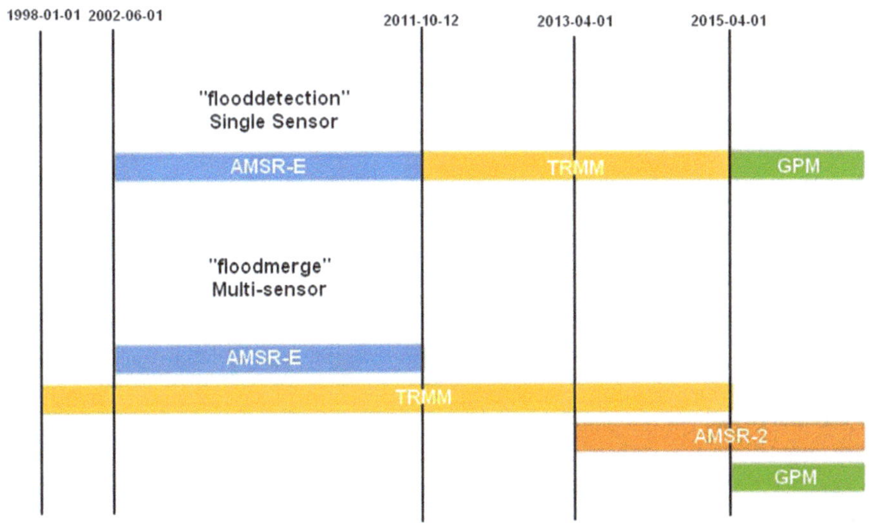

Fig. 1.2 The European Commission's (Joint Research Center) "global flood detection system" produces several microwave signal products. The present analysis uses the "floodmerge" product, with contributions by different sensors but all using the ground-sensing 36.5 or 37 GHz channels. The figure is from De Groeve et al. (2015)

via rating equations produced in two different ways: (a) via discharge values derived from WBM and (b) via ground-based observed discharges. For calibration purposes, we used 5 years (2003–2007) of model estimates and gauging station records. The WBM model produces daily discharge values for these years at each measurement site (its global grid resolution is 6 arc minutes). River discharge then is also obtained from the nearest gauging station. Comparisons were then established between discharge estimates from the WBM model and water discharge from gauging stations. Simple linear regression was based on 60 months, daily maximum, minimum, and mean pair-data points, 180 data pairs (Brakenridge et al. 2012). To improve the fitness of the models, and because along these highly seasonal rivers the model results exhibit obvious lag compared to observation, calibration vectors between ground-station discharge versus WBM runoff were also performed with different lag times. Once obtained, the rating equations are then used to transform the daily satellite signal data to daily river discharge during the 1998–2016 period.

To evaluate a priori the expected accuracy of the remote sensing, the rating curve R^2 and the observed signal/noise were used. The R^2 quality levels of the rating curve are classified as: $R^2 > 0.8 =$ "5," $R^2 > 0.7 =$ "4," $R^2 > 0.6 =$ "3," $R^2 > 0.44 =$ "2," and $R^2 < 0.44 =$ "1." A simple integer signal/noise (*S/N*) rating for only the satellite values is separately computed as follows:

1. The total observed range over 5 years in the signal, at individual River Watch array sites, varies between values as high as 0.3 (strong variance in signal) to very low values (0.01). The former indicates that a site is sensitive to and recording water surface area variability.
2. The average observed day-to-day departure over 5 years is the "noise" statistic. Most rivers monitored by River Watch exhibit long periods of relatively stable flow (even though floods do cause some periods of strong daily variation). Thus, the average daily variation is an indicator of the background level of variability that any hydrological signal must be visible above.
3. The *S/N* value compares the signal changes that are believed to be induced by the actual discharge variability to the daily changes that are produced by other variables (e.g., sensor detector noise). This ratio, which only refers to the remote sensing data, is classified into high and low values as follows: $S/N > 20 =$ "5," $S/N > 15 =$ "4," $S/N > 10 =$ "3," $S/N > 5 =$ "2," and $S/N < 5 =$ "1."

Finally, and overall accuracy assessment ranking is calculated as the average between the *S/N* ratio and the R^2 model/observation statistics. Sites rated as "fair" (2) or better are commonly very useful. Higher accuracy values indicate increasing goodness of fit of WBM modeling to remote sensing, and also increasing robustness of the remote sensing signal itself as a record of hydrological variability (Brakenridge et al. 2015).

1.4 Results

For station A, the signal ratio ranged between 0.84 and 0.99, while at B1 they varied between 0.74 and 0.98. Other stations, including C1, C2, and B2, witnessed ratio variation between 0.85 and 0.99 (Fig. 1.3). As expected, the sites A, B1, and C1, which are located along the lower floodplains of the Magdalena, exhibited larger variations in the signal ratio (from high variability in water surface area). Among these three sites, B1 shows the most accurate microwave signal. In contrast, B2 and C2, in the upper basin, with limited alluvial plains and less discharge, exhibited smaller variations in the microwave ratio information, indicating a possible weaker satellite detection of water surface variability; however, we later found that regardless of this relatively small range in the signal, the estimation of river discharge was still accurate.

As noted, river discharge values were obtained from the WBM model and also from five hydrological gauging stations. Determination coefficients (R^2) of the linear regression between observed and predicted water discharges were performed at different time lags (1–5 months). Positive lags indicate that the microwave signal was lagged with respect to river discharge and negative lags indicate that river discharge was lagged with respect to the microwave signal. Calibration improved in most of the stations once river discharge was lagged by some months (Table 1.1). For station A, paired correlation R^2 between satellite signal and water discharge-WBM increased from 0.11 at lag 0 to 0.40 at a lag of 2 months, indicating that the water routing in WBM for that station is too quick. However, results at station B1

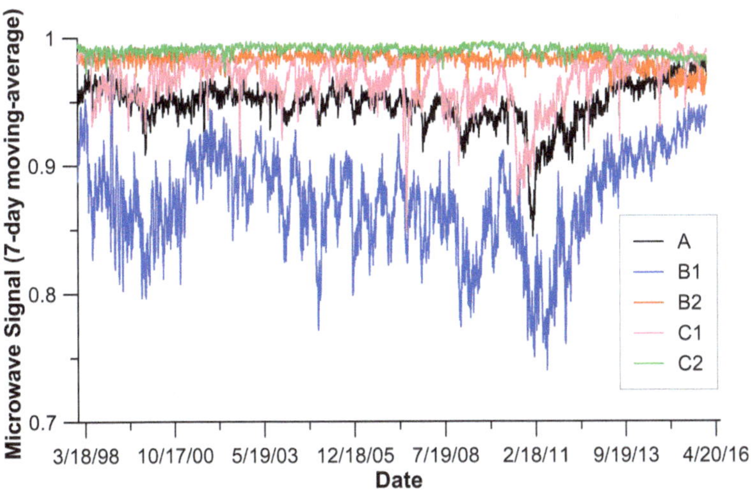

Fig. 1.3 Daily time series of microwave satellite signal processed by the JRC, Italy at five selected sites in the Magdalena River (Fig. 1.1). The microwave signal corresponds to the 7-day moving average of the "*C*/*M*" ratio (*M* being the radiance over the measurement site, and *C* being the brightest/driest "background" radiance from the 9 × 9 surrounding array)

Table 1.1 Determination coefficients (R^2) of regressions between the microwave signal and river discharges obtained from WBM model estimates and ground stations

Station:	A		B1		B2		C1		C2	
Lag	WBM	Ground	WBM	Ground	WBM	Ground	WBM	Ground	WBM	Ground
-5	0.13	0.03	0.10	0.00	0.32	0.45	0.01	0.00	0.37	0.30
-4	0.27	0.01	0.01	0.01	0.30	0.49	0.01	0.00	0.20	0.19
-3	0.37	0.06	0.03	0.01	0.21	0.34	0.06	0.02	0.13	0.10
-2	0.40	0.22	0.20	0.24	0.24	0.35	0.29	0.20	0.12	0.14
-1	0.30	0.32	0.52	0.57	0.22	0.38	0.47	0.45	0.22	0.29
0	0.11	0.20	0.64	0.32	0.37	0.43	0.41	0.39	0.48	0.57
1	0.03	0.00	0.24	0.02	0.34	0.45	0.25	0.13	0.37	0.43
2	0.03	0.05	0.04	0.00	0.40	0.41	0.19	0.06	0.26	0.28
3	0.08	0.02	0.03	0.02	0.26	0.34	0.13	0.04	0.11	0.08
4	0.15	0.01	0.12	0.18	0.22	0.36	0.12	0.04	0.11	0.09
5	0.12	0.06	0.30	0.34	0.20	0.36	0.08	0.06	0.19	0.16

R^2 values are computed by lagging 1–5 months the microwave signal with respect to discharge or vice versa. Positive lags indicate that the microwave signal was lagged with respect to river discharge while negative lags indicate that discharge was lagged with respect to the microwave signal. The highest obtained R^2 are highlighted in gray

indicate that the microwave signal and river discharge from WBM are well correlated ($R^2 = 0.64$) without incorporating a time lag. Also, the satellite signal and stream flow from ground-based observations are well correlated with $R^2 = 0.57$, but with a lag of 1 month. Other regressions between satellite signal and river discharge at stations C1, B2, and C2 showed R^2 determination coefficients between 0.40 and 0.64 (Table 1.1). The scatter plots of the best fitting regressions are presented in Fig. 1.4.

We transformed the satellite signal to river discharge (Fig. 1.5) by using the best fitting rating curves (Fig. 1.4). The equations used are summarized in Table 1.2. In general, satellite-observed river discharges follow well the interannual variability of measured stream flow at ground-based gauging stations (Fig. 1.5). However, the WBM-based rating curve overestimates the magnitudes of river discharge for station B1 during peak discharge conditions (Fig. 1.5).

The short wave variability in the upper basin stations B2 and C2 is well captured by the satellite signal, and the long wave variation is observed in the middle (B1 and C1) and lower sections of the Cauca and Magdalena rivers. Satellite-observed discharge matches well with observed discharge values from the ground, especially during peak discharge conditions (Fig. 1.5). It is worth noting that short-term fluctuations of stream flow are well captured by the satellite signal in station B2, a site located along the Magdalena below stream two major dams, Betania and El Quimbo. The river discharges at site C2 agree well with observed stream flow values despite the site's location in the upper basin, with a relatively narrow valley and limited alluvial plains. As we expected, observed versus observed discharges match quite well in the most downstream stations, A, B1, and C1 (Fig. 1.5).

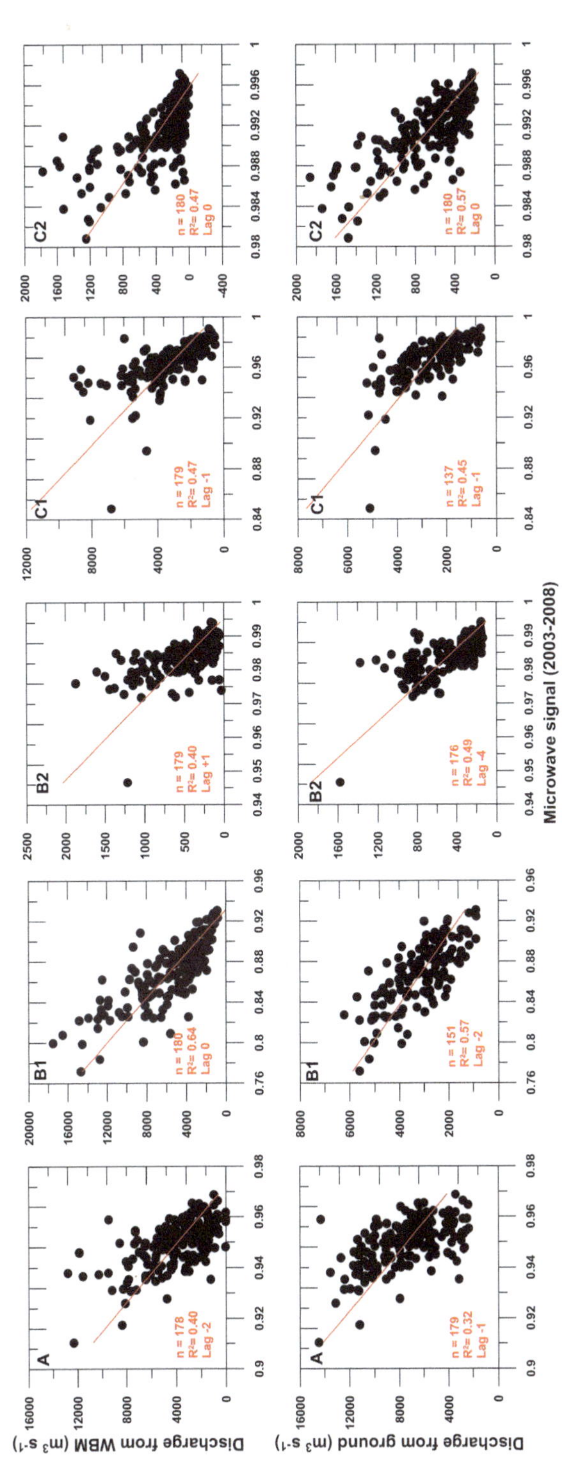

Fig. 1.4 Linear regressions between the microwave signal and river discharge estimates from WBM (*top*) and stream flow from ground-based gauging stations (*bottom*) along the Magdalena River (Fig. 1.1)

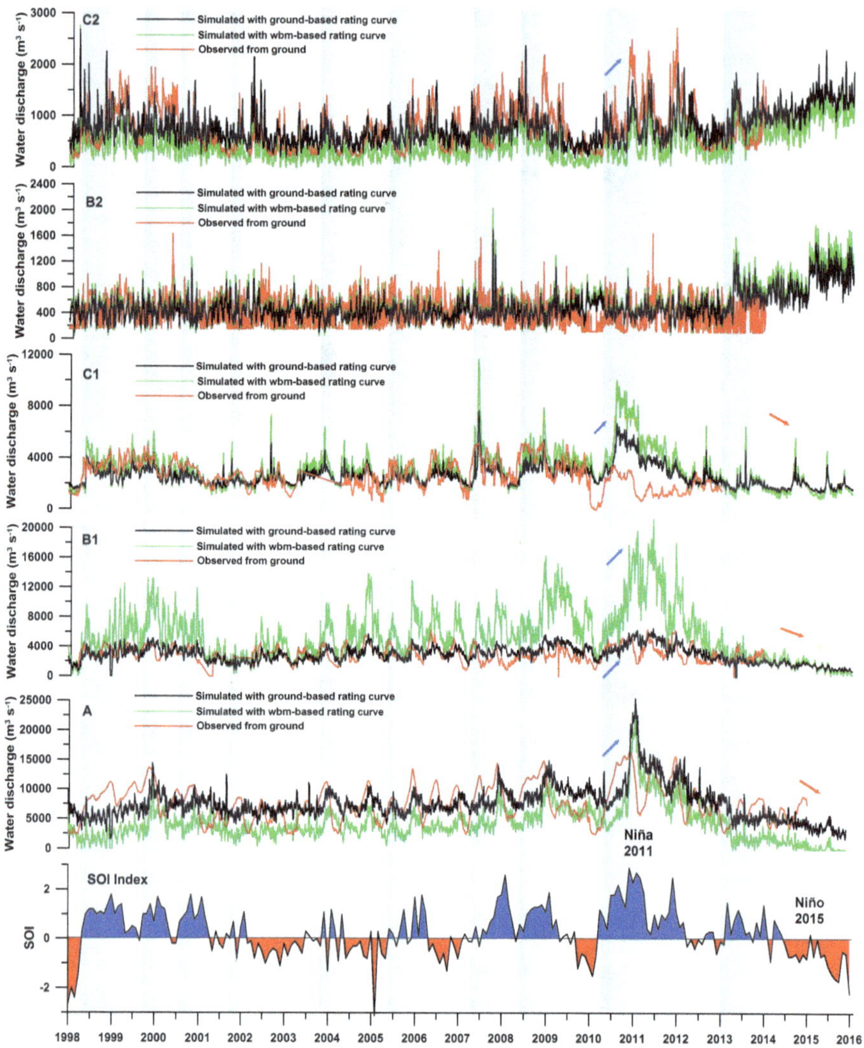

Fig. 1.5 Time series of river discharge for the Magdalena River obtained from rating curves using the WBM model estimates (*green line*) and discharge from ground-based gauging stations (*black line*). We also show ground-based observations of water discharge at each gauging station (*red line*). The Southern Oscillation Index (SOI) is shown at the *bottom*

Satellite-measured discharge variability shows a robust relation with the ENSO cycle. During the strong La Niña 2010–2011 that caused extended flooding in the Cauca and Magdalena rivers (Fig. 1.1), most of the gauging sites measured high discharges that are in close agreement with the observed positive values of the Southern Oscillation Index. Station B2 is not affected by the ENSO because its stream flow is largely human controlled by the operation of dams. In contrast, water

Table 1.2 Descriptive statistics of the correlation of model, satellite-derived, and ground-based river discharge

Site ID	Mean discharge WBM-based model ($m^3 s^{-1}$)	Mean discharge ground-based model ($m^3 s^{-1}$)	Mean discharge gauging station ($m^3 s^{-1}$)	Gauging station-based equation	R^2	R^2 quality	*S/N* rating	Accuracy
A	4420.82	7650.67	7667.71	$y = -170{,}746.49x + 169{,}553.97$	0.32	1	1	1
B1	5518.63	2963.19	3120.00	$y = -28{,}722.65x + 28{,}047.70$	0.57	2	1.5	1.75
B2	549.16	505.87	395.27	$y = -34{,}041.30x + 33{,}968.17$	0.49	2	1	1.5
C1	3051.16	2633.93	2642.99	$y = -42{,}939.55x + 44{,}086.00$	0.45	2	2	2
C2	485.68	794.04	748.72	$y = -89{,}021.55x + 88{,}924.22$	0.57	2	1	1.5

Mean water discharge at each gauging station is shown. We also include the values of signal/noise ratio (*S/N*) and the accuracy level according to the River Watch version 3 (Brakenridge et al. 2015)

discharge shows decreasing trends during the current El Niño phase 2015–2016 (Fig. 1.4).

The accuracy of obtained river discharge utilizing satellite-derived data and ground-based rating curves was evaluated by following the River Watch procedure (Brakenridge et al. 2015). Overall, mean observed discharges obtained from the ground-based rating curves are very similar to the stream flow values measured at gauging stations. In addition, most of the determination coefficients are close to 0.5 (Table 1.2). Consequently, long-term trends are well captured by observed values, while short-term variability might be more influenced by the microwave signal noise (Fig. 1.4). Data can be accessed from the Dartmouth Flood Observatory at http://floodobservatory.colorado.edu/DischargeAccess.html.

1.5 Discussion

Obtained values of water discharge via satellite data follow quite well the observed interannual variability of stream flow measured at ground-based gauging stations. Observed discharges at the most downstream stations, A, B1, and C2, represent appropriately the natural seasonality of water discharge along the Magdalena floodplains (Figs. 1.1 and 1.5).

In this study good satellite signal correlations to ground-based results were obtained once water discharges were lagged by 1–2 months with respect to the satellite signal. Satellite measurements of water discharge are based on variations of width of the water surface area while stream flow estimations at gauging stations are based on variations in stage height. Thus, as water level rises, it passes through a more width complex lagoon system. The phase lag observed between discharge and satellite signal is possibly related to the hydrological dynamics of channel–floodplain connections.

Gauging stations on the ground are usually placed on stable river cross-sections with homogenous channel geomorphology. Thus, many ground-based stations are located on confined channels without alluvial plains. In such cases, the satellite signal is limited. In contrast, water surface area in braided, meandered, or anastomosing channel reaches responds very sensitively to variations in width and to discharge. In addition, large discharge events may exceed the discharge capacity of the river channel cross-section. In these cases, satellite data capture more properly the stream flow magnitude in channels with lateral connectivity of floodplains, where effective width or water surface area can keep constant over longer periods of time (Smith et al. 1995, 1996). Our results clearly demonstrate the quality of satellite-derived water discharge along stations in the downstream reaches of the Magdalena and Cauca rivers (Fig. 1.4 and Table 1.2).

It is interesting to note that rating equations based on the WBM model have similar or even higher R^2 coefficients in comparison to rating curves based on river discharge from ground-based observations (Fig. 1.4). For all stations, observed time series of river discharge using the WBM-based rating curves follow closely the measured discharge variability of the ground-based stations. The only exception is station B1, where discharge is overestimated, especially during peak flows. The WBM-based rating curves can calculate river discharge appropriately in most of the sites; this approach thus can measure reasonably well discharge on entirely ungauged river reaches.

Accuracy of the Colombian sites was evaluated by using the same measures applied to the entire River Watch 3 array (Brakenridge et al. 2015). High *S/N* ratios indicate a strong signal compared to background noise. For the Magdalena River, low values were obtained for all stations (Table 1.2). However, the satellite signals still appear to monitor appropriately river discharge (e.g., the case of station C2, where microwave signal ranges between 0.97 and 0.99), but observed satellite river discharge matches very well with the observed stream flow at ground-based gauging stations (Fig. 1.5).

All South American rivers, independently of size, display a strong seasonal discharge and sediment load variability, typically by a factor of 5–10, when comparing low to high monthly discharge. The interannual variation of discharge and sediment load associated with the ENSO or El Niño-La Niña cycle is almost equally great, typically by a factor of 2–4, comparing low to high annual discharges (Richey et al. 1986; Depetris et al. 1996; Vörösmartry et al. 1996; Restrepo and Kjerfve 2000). For Colombia, relationships between river discharge anomalies and the ENSO have been found for the Magdalena (Restrepo and Kjerfve 2000; Restrepo 2013) and other Caribbean rivers such as the Sucio, Sinú, and Canal del Dique, which exhibit higher water discharge during La Niña and lower flows during El Niño (Restrepo et al. 2014). Further statistical analysis on annual stream flow data and ENSO anomalies show that the ENSO may be responsible for up to 65 % of the interannual stream flow variability in rivers such as the Magdalena, Cauca, Cesar, Rancheria, and Sinú (Gutiérrez and Dracup 2001; Restrepo and Kjerfve 2000; Restrepo et al. 2014, 2015). The River Watch data clearly record both types of flow variability.

In this study we have shown the capabilities of River Watch version 3 in estimating magnitudes of water discharge during flood events. Peak water discharges along the Magdalena River were clearly observed during La Niña events in 2008–2009 and 2010–2011 (Fig. 1.5). Only one station in the upper basin (B2), with a hydrological regime largely controlled by dams, did not show any seasonality associated with the La Niña cycle. Nevertheless, satellite-based discharge estimates reflect the magnitudes of stream flow during high discharge conditions, perhaps even more accurately than those observed at ground-based gauging stations along floodplains. The satellite-derived data for water discharge appear to be a reliable method for drainage basin planning and decision making toward flood mitigation. Many Latin American countries like Colombia lack real-time monitoring systems on river channels. Even along the Magdalena, the best gauged Colombian river, all stations along the main channel are commonly not recalibrated following major flood events, even though their actual rating curves are known to alter after such events. Because of such circumstances, satellite-derived water discharge offers an important complementarity, and for many river reaches the only reliable information concerning river discharges and including flood events.

1.6 Conclusions

For Andean rivers of Colombia, characterized by high rainfall rates, steep relief, and limited alluvial plains, microwave satellite radiometry is a robust tool for estimating near-real-time water discharge. In this study, stream flow series during the 1998–2016 were observed at five sites along the Magdalena River and its main tributary, the Cauca. The remote sensing is sensitive to measurement site water surface area; calibration to discharge is accomplished by comparison to either WBM modeled discharges or to ground station values. Time lags are an important factor affecting the statistical correlations among these three data time series. The best correlations were obtained once river discharge was lagged with respect to the satellite signal; a phase lag indicating that satellite detects slower variations in water surface area compared to the immediate variations in discharge-related river stage detected by ground-based stations. Regarding absolute values, river discharge estimates derived from the WBM model can greatly underestimate or overestimate stream flow values (as judged by comparison to the ground stations). In contrast, water discharges obtained from ground-based rating curves preserve the observed river discharge magnitudes.

Observed series of river discharge at the upper basin stations exhibited low seasonality and smaller variations in the satellite signal. In the middle and lower reaches of the Magdalena River, the accuracy of the satellite results is improved at longer time scales. Floods were well recognized by the observed series of river discharge, especially during La Niña event in 2010–2011, the strongest flooding event on record in Colombia. Four out of five studied stations witnessed high peak discharges during this period, with stream flow magnitudes being more accurate than the water discharge values gauged at ground-based stations.

Regions in Colombia such as the lower course of the Magdalena have been strongly affected by floods and associated sedimentary fluxes during the last decades. They do not have reliable ground-based data on river discharge and magnitudes of floods. This study is an innovative applied research generating near-real-time satellite river data for the region by deriving time series of stream flow and flooding data back to 1998.

Acknowledgments This study was carried out with support from the US National Academy of Sciences under the funded PEER project (grant number 70) "Satellite-based estimations of River discharge into the Cartagena Bay, Caribbean Colombia: Capacity building to mitigate sources of upstream runoff and associated risks of pollution." This project is implemented by the interaction between the Dartmouth Flood Observatory (DFO) at University of Colorado and EAFIT University-Colombia. Special thanks to the U.S. National Science Foundation, the Joint Research Centre (JRC), and NASA through the project Near Real Time Flood Inundation Prediction and Mapping (NNX14AQ44G-NASA).

References

Alsdorf D, Han SC, Bates P, Melack J (2010) Seasonal water storage on the Amazon floodplain measured from satellites. Remote Sens Environ 114:2448–2456. doi:10.1016/j.rse.2010.05.020

Birkett CM, Mertes LAK, Dunne T, Costa MH, Jasinski MJ (2002) Surface water dynamics in the Amazon Basin: application of satellite radar altimetry. J Geophys Res Atmos 107:D20. doi:10.1029/2001JD000609

Bjerklie DM (2007) Estimating the bankfull velocity and discharge for rivers using remotely sensed river morphology information. J Hydrol 341:144–155. doi:10.1016/j.jhydrol.2007.04.011

Bjerklie DM, Dingman SL, Vorosmarty CJ, Bolster CH, Congalton RG (2003) Evaluating the potential for measuring river discharge from space. J Hydrol 278:17–38. doi:10.1016/S0022-1694(03)00129-X

Bjerklie DM, Moller D, Smith LC, Dingman SL (2005) Estimating discharge in rivers using remotely sensed hydraulic information. J Hydrol 309:191–209. doi:10.1016/j.jhydrol.2004.11.022

Brakenridge GR, Knox JC, Paylor ED, Magilligan FJ (1994) Radar remote sensing aids study of the Great Flood of 1993. Eos Trans Am Geophys Union 75:521–527. doi:10.1029/EO075i045p00521

Brakenridge GR, Nghiem SV, Anderson E, Chien S (2005) Space-based measurement of river runoff. Eos Trans Am Geophys Union 86:185–192

Brakenridge GR, Nghiem SV, Anderson E, Mic R (2007) Orbital microwave measurement of river discharge and ice status. Water Resour Res 43:1–16. doi:10.1029/2006WR005238

Brakenridge GR, Cohen S, Kettner AJ, De Groeve T, Nghiem SV, Syvitski JPM, Fekete BM (2012) Calibration of satellite measurements of river discharge using a global hydrology model. J Hydrol 475:123–136. doi:10.1016/j.jhydrol.2012.09.035

Brakenridge GR, Kettner AJ, Syvitski JPM, Overeem I, De Groeve T, Cohen S, Nghiem SV (2015) Experimental satellite-based river discharge measurements: technical summary

De Groeve T, Kugler Z, Brakenridge GR (2007) Near real-time flood alerting for the global disaster alert and coordination system. In: Proceedings of the ISCRAM, pp 33–40

De Groeve T, Brakenridge GR, Paris S (2015) Global flood detection system data product specifications. JRC Technical Report. http://www.gdacs.org/flooddetection/Download/Technical_Note_GFDS_Data_Products_v1.pdf

Depetris PJ, Kempe S, Latif M, Mook WG (1996) ENSO controlled flooding in the Paraná River (1904–1991). Naturwissenschaften 83:127–129

Filizola N, Latrubesse EM, Fraizy P, Souza R, Guimaraes V, Guyot JL (2014) Was the 2009 flood the most hazardous or the largest ever recorded in the Amazon? Geomorphology 215:99–105. doi:10.1016/j.geomorph.2013.05.028

Frappart F, Seyler F, Martinez JM, León JG, Cazenave A (2005) Floodplain water storage in the Negro river basin estimated from microwave remote sensing of inundation area and water levels. Remote Sens Environ 99:387–399. doi:10.1016/j.rse.2005.08.016

Gutiérrez F, Dracup J (2001) An analysis of the feasibility of long-range streamflow forecasting for Colombia using El Niño—Southern Oscillation indicators. J Hydrol 246:181–196

Hall AC, Schumann GJP, Bamber JL, Bates PD (2011) Tracking water level changes of the Amazon basin with space-borne remote sensing and integration with large scale hydrodynamic modelling: a review. Phys Chem Earth 36:223–231. doi:10.1016/j.pce.2010.12.010

Handisyde N, Lacalle DS, Arranz S, Ross LG (2014) Modelling the flood cycle, aquaculture development potential and risk using MODIS data: a case study for the floodplain of the Rio Paraná, Argentina. Aquaculture 422–423:18–24. doi:10.1016/j.aquaculture.2013.10.043

Hoyos N, Escobar J, Restrepo JC, Arango AM, Ortiz JC (2013) Impact of the 2010-2011 La Niña phenomenon in Colombia, South America: the human toll of an extreme weather event. Appl Geogr 39:16–25. doi:10.1016/j.apgeog.2012.11.018

Kettner AJ, Restrepo JD, Syvitski JPM (2010) A spatial simulation experiment to replicate fluvial sediment fluxes within the Magdalena River Basin, Colombia. J Geol 118:363–379. doi:10.1086/652659

Koblinsky CJ, Clarke RT, Brenner AC, Frey H (1993) Measurement of river level variations with satellite altimetry. Water Resour Res 29:1839–1848. doi:10.1029/93WR00542
Latrubesse EM, Stevaux JC, Sinha R (2005) Tropical rivers. Geomorphology 70:187–206. doi:10.1016/j.geomorph.2005.02.005
Milliman JD, Farnsworth K (2011) River discharge to the Coastal Ocean—a global synthesis. Cambridge University Press, Cambridge
Penatti Costa N, Ribeiro de Almeida TI, Guimaraes Ferreira L, Arantes AE, Coe MT (2015) Satellite-based hydrological dynamics of the world's largest continuous wetland. Remote Sens Environ 170:1–13. doi:10.1016/j.rse.2015.08.031
Poveda G, Jaramillo A, Gil M, Quiceno N, Mantilla R (2001) Seasonality in ENSO related precipitation, river discharges, soil moisture, and vegetation index (NDVI) in Colombia. Water Resour Res 37:2169–2178. doi:10.1029/2000WR900395
Poveda G, Waylen PR, Pulwarty RS (2006) Annual and inter-annual variability of the present climate in northern South America and southern Mesoamerica. Palaeogeogr Palaeoclimatol Palaeoecol 234:3–27. doi:10.1016/j.palaeo.2005.10.031
Quiñones M (2013) Mapas de Frecuencias de Inundación y Tipos de Vegetación para las Cuencas del Río Magdalena y Cauca en Colombia. Technical report
Restrepo JD, Kjerfve B, Hermelin M, Restrepo JC (2006) Factors controlling sediment yield in a major South American drainage basin: the Magdalena River, Colombia. J Hydrol 316:213–232. doi:10.1016/j.jhydrol.2005.05.002
Restrepo JD, Kettner AJ, Syvitski JPM (2015) Recent deforestation causes rapid increase in river sediment load in the Colombian Andes. Anthropocene 10:13–28. doi:10.1016/j.ancene.2015.09.001
Restrepo JD, Syvitski JPM (2006) Assessing the effect of natural controls and land use chane on sediment yield in a major Andean River: the Magdalena drainage basin, Colombia. Ambio 35:44–53
Restrepo JD, Kjerfve B (2000) Magdalena river: interannual variability (1975–1995) and revised water discharge and sediment load estimates. J Hydrol 235:137–149
Restrepo JD (2013) The perils of human activity on South American deltas: lessons from Colombia's experience with soil erosion. Deltas: Landforms, Ecosystems and Human Activities, 358. IAHS Publ., p 143–152
Restrepo J, Ortíz JC, Pierini J, Schrottke K, Maza M, Otero L, Aguirre J (2014) Freshwater discharge into the Caribbean Sea from the rivers of Northwestern South America (Colombia): magnitude, variability and recent changes. J Hydrol 509:266–281
Restrepo JD, Park E, Aquino S, Latrubesse EM (2016) Coral reefs chronically exposed to river sediment plumes in the southwestern Caribbean: Rosario Islands, Colombia. Sci Total Environ 553:316–329. doi:10.1016/j.scitotenv.2016.02.140
Richey JE, Meade RH, Salati E, Devol AH, Nordin CF, dos Santos U (1986) Water discharge and suspended sediment concentrations in the Amazon River: 1982–1984. Water Resour Res 22:756–764
Rokni K, Ahmad A, Solaimani K, Hazini S (2015) A new approach for surface water change detection: integration of pixel level image fusion and image classification techniques. Int J Appl Earth Obs Geoinf 34:226–234. doi:10.1016/j.jag.2014.08.014
Schumann GJ-P, Moller DK (2015) Microwave remote sensing of flood inundation. Phys Chem Earth Parts A/B/C 83–84:84–95. doi:10.1016/j.pce.2015.05.002
Smith LC (1997) Satellite remote sensing of river inundation area, stage, and discharge: a review. Hydrol Process 11:1427–1439. doi:10.1002/(sici)1099-1085(199708)11:10<1427::aid-hyp473>3.0.co;2-s
Smith LC, Isacks B, Forster RR, Bloom A, Preuss I (1995) Estimation of discharge from braided glacial rivers using ERS1 synthetic aperture radar: first results. Water Resour Res 31:1325–1329

Smith LC, Isacks B, Bloom A (1996) Estimation of discharge from three braided rivers using synthetic aperture radar satellite imagery: potential application to ungauged basins. Water Resour Res 32:2021–2034

Syvitski JPM, Overeem I, Brakenridge GR, Hannon M (2012) Floods, floodplains, delta plains—a satellite imaging approach. Sediment Geol 267–268:1–14. doi:10.1016/j.sedgeo.2012.05.014

Vörösmarty CJ, Willmott CJ, Choudhury BJ, Schloss AL, Stearns TK, Robeson SM, Dorman TJ (1996) Analyzing the discharge regime of a large tropical river through remote sensing, ground-based climatic data, and modeling. Water Resour Res 32:3137–3150. doi:10.1029/96WR01333

Walschburger T, Angarita H, Delgado J (2015) Hacia una gestión integral de las planicies inundables en la cuenca del Magdalena-Cauca. In: Rodríguez Becerra M (ed) ¿Para dónde va el río Magdalena? Riesgos sociales, ambientales y económicos del proyecto de navegabilidad. Foro Nacional Ambiental-FESCOL, Bogotá, pp 145–168

Wisser D, Fekete BM, Vörösmarty CJ, Schumann AH (2009) Reconstructing 20th century global hydrography: a contribution to the Global Terrestrial Network-Hydrology (GTN-H). Hydrol Earth Syst Sci Discuss 6:2679–2732. doi:10.5194/hessd-6-2679-2009

Chapter 2
Remote Sensing of Drivers of Spring Snowmelt Flooding in the North Central U.S.

Samuel E. Tuttle, Eunsang Cho, Pedro J. Restrepo, Xinhua Jia, Carrie M. Vuyovich, Michael H. Cosh, and Jennifer M. Jacobs

2.1 Introduction

Melting of accumulated snow contributes significantly to runoff in northern North America and mountainous areas (up to 50–95 %) (WMO 2009). In many regions, snowmelt associated with the spring thaw creates an annual risk of flooding, which can endanger lives and damage private property and infrastructure. For instance, rapid snowpack melting and rainfall in the northeast U.S. in January 1996 led to 30 fatalities and over $1.5 billion in damages (Anderson and Larson 1996). Snowmelt flooding along the Red River of the North in 1997 damaged 85 % of all structures in Grand Forks, North Dakota, and caused a total of $4 billion in damages in the U.S. (Todhunter 2001). Weeks later, flood waters nearly overtopped levees in

S.E. Tuttle, M.A., Ph.D. (✉) • E. Cho, M.S. • J.M. Jacobs, Ph.D.
Department of Civil and Environmental Engineering, University of New Hampshire, Gregg Hall, 35 Colovos Road, Durham, NH 03824, USA
e-mail: samuel.tuttle@unh.edu; ec1072@wildcats.unh.edu; Jennifer.jacobs@unh.edu

P.J. Restrepo, Ph.D., M.Sc., Ingeniero Civil
Consulting Engineer, Minneapolis, MN, USA
e-mail: pr.engconsult@gmail.com

X. Jia, Ph.D.
Department of Agricultural and Biosystems Engineering, North Dakota State University, NDSU Dept 7620, PO Box 6050, Fargo, ND 58108, USA
e-mail: Xinhua.jia@ndsu.edu

C.M. Vuyovich, M.S.
Remote Sensing, GIS and Water Resources Branch, USACE Cold Regions Research and Engineering Laboratory, 72 Lyme Road, Hanover, NH 03755, USA
e-mail: carrie.m.vuyovich@usace.army.mil

M.H. Cosh, Ph.D.
USDA Agricultural Research Service, 10300 Baltimore Ave, Beltsville, MD 21032, USA
e-mail: Michael.Cosh@ars.usda.gov

V. Lakshmi (ed.), *Remote Sensing of Hydrological Extremes*, Springer Remote Sensing/Photogrammetry, DOI 10.1007/978-3-319-43744-6_2

Winnipeg, Manitoba, Canada, but still prompted $500–750 million in expenditures, including construction of a levee that is one of the largest peacetime efforts in Canadian history (Rannie 2016). In 2011, spring snowmelt in the north central U.S. partially contributed to major flooding in the Missouri, Souris, Ohio, and Mississippi river basins, responsible for up to $3.8 billion in direct damages (NWS-HIC 2011). Accurate river flow forecasting allows communities and governments to better plan for such flooding emergencies and mitigate the impacts of damaging floods.

Snowmelt flooding occurs when the primary source of floodwater is melting snow. Accumulated snowpack on the ground surface stores precipitated water that can be released days, weeks, or months later upon melting. Larger snowpacks store greater amounts of frozen water and have a larger risk and magnitude of spring floods. Soil moisture and soil frost status determine how much snowmelt can infiltrate into soils, because wetter soils and deeper frost (especially paired with high soil moisture) reduce infiltration, leading to more runoff (Todhunter 2001). Soil moisture status is often carried over from the late fall due to soil freezing, while the degree of soil freezing depends on winter air temperatures and the presence and depth of snow, which insulates the soil (Shanley and Chalmers 1999).

Other factors can exacerbate spring flooding. Precipitation during the spring melt adds water to the already elevated basin water content. Rainfall can increase the rate of snow melt by adding heat energy (Harr 1981). Rapid spring thaws deliver meltwater in shorter time windows and thus result in greater streamflow magnitudes (Todhunter 2001). Ice jams in river channels can impede river flows, raising the river height upstream of the blockages (Beltaos 2008). Additionally, changes in river grade may cause backwater effects (such as just north of Grand Forks, ND; Todhunter 2001), increasing flood heights. Finally, the spatial pattern of melting snow can interact with the river network configuration to deliver meltwater from individual tributaries simultaneously (increasing instantaneous flow volumes) or asynchronously (Miller and Frink 1984).

Accurate prediction of river flows due to snowmelt flooding requires knowledge of these factors, both temporally and spatially, and the ability to synthesize their interactions. However, many of these state variables, such as snow water equivalent, soil moisture, soil frost, and melt phase, are poorly spatially constrained. Traditionally, observations of these variables are provided by ground stations. However, ground stations provide only "point" observations from individual locations, and far too few stations exist in most regions to obtain a reliable spatial understanding of these hydrological states.

Fortunately, remote sensing can estimate many important drivers of snowmelt floods, including snow water equivalent (SWE) and soil moisture (among others). While many methods exist, passive microwave remote sensing has a robust history for snow and soil moisture, and thus is best prepared for operational use in flood forecasting. Significant snowmelt floods have occurred in the northern Plains of the United States. This area is well suited for passive microwave remote sensing (Vuyovich et al. 2014), so the potential for enhancing flood forecasting using remote sensing observations is high in this region.

In this chapter, we review the use of remote sensing observations to constrain snow water equivalent and soil moisture (two primary drivers of spring snowmelt floods), using the north central United States as a case study. We provide examples from previous studies where these methods have shown value, as well as specific examples from the Red River of the North basin. Attention is given to the potential for passive microwave instruments to contribute to operational flood forecasting.

2.2 Background on the Red River of the North Basin

The Red River of the North basin (RRB) provides an excellent location to showcase the challenges from snowmelt flooding in nonmountainous regions, as well as opportunities for remote sensing to contribute to scientific advancement and reduce the loss of human life and property. The Red River marks the border between the states of North Dakota and Minnesota and drains parts of eastern North Dakota, western Minnesota, and a small area of northeastern South Dakota (see Figs. 2.1 and 2.2). The river flows north from its headwaters in Wahpeton, ND to the United States–Canada border, and then on through Winnipeg, Manitoba, where it is joined by the Assiniboine River, and finally into Lake Winnipeg. In its entirety, the Red

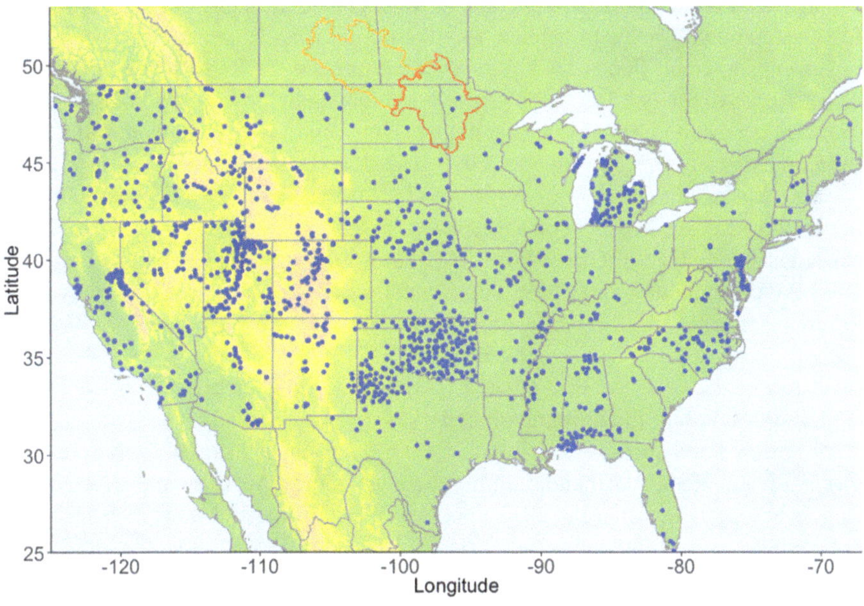

Fig. 2.1 Location of the Red River of the North basin (*red*) and observation sites that have contributed to the North American Soil Moisture Database (NASMD; *blue dots*) (Quiring et al. in press). The *red* polygon is the area of the RRB that we focus on in this chapter, and the *orange* polygon shows the tributary Assiniboine River basin. Notice the relative lack of observations available in the northern Great Plains, despite its agricultural importance

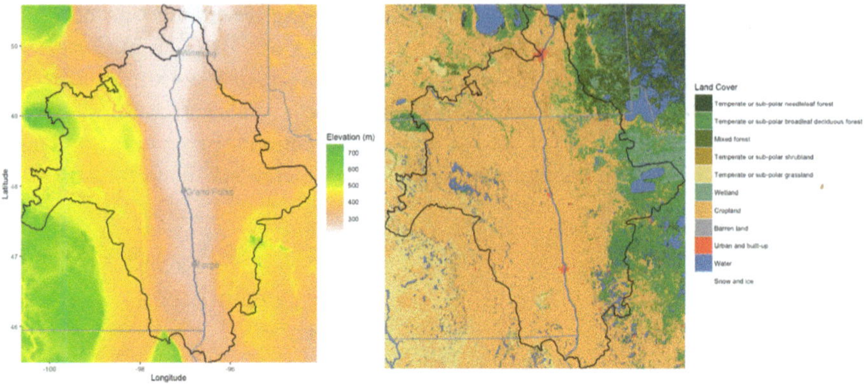

Fig. 2.2 Terrain (*left*) (from the ETOPO1 global relief model; Amante and Eakins 2009) and land cover (*right*) (NALCMS 2015) of the Red River of the North basin (excluding the Assiniboine River basin). The basin has very low relief and is dominated by cropland

River is approximately 885 km long and with a drainage area of 285,000 km^2 (including the 162,000 km^2 Assiniboine drainage), about 635 km and 101,500 km^2 of which are in the United States, respectively (Rannie 2016; Miller and Frink 1984). Along this distance, the main stem of the Red River drops only 72 m, for an average grade of 0.08 m/km. This lack of relief (Fig. 2.2) is one of the river's main features and partly helps to explain why the region is so vulnerable to flooding.

Spring snowmelt floods are a regular occurrence in the Red River of the North basin. Approximately 85 % of the yearly maximum peak flows in the RRB over the past century resulted from the spring melt, as opposed to only 15 % from summer or fall rain-driven events (Rasmussen 2016, Fig. 2.3). Despite the low average SWE compared to mountain snowpacks (Sturm et al. 1995; Brasnett 1999), spring floods in the Red River basin persist on a timescale of weeks to months (e.g., the Red River exceeded flood stage in Grand Forks, ND for 46 days during the record-breaking 1997 flood; Todhunter 2001) and floodwaters can extend for large distances from the main channel (e.g., flood width of up to 80–100 km near the U.S.–Canada border; Schwert 2003; Miller and Frink 1984).

In recent decades, the Red River experienced a number of large flood events (Fig. 2.3), including the aforementioned 1997 snowmelt flood. More recently, the 2009 spring flood included the highest ever river stage recorded at Fargo, ND, forcing evacuations and inundating hundreds of homes (Wazney and Clark 2016; Rogers et al. 2013), while the 2011 flood was the third-highest stage ever recorded at Grand Forks, ND (Stadnyk et al. 2016). It has been suggested that the increase in Red River flooding could be due to human-induced climate change, but the region is subject to multidecadal oscillations from very wet to dry periods (Knox 2000; Miller and Frink 1984; see Fig. 2.3), making causal attribution challenging (Hirsch and Ryberg 2012; Rasmussen 2016). Another possible explanation for the increase in flooding is change in agricultural practices, including extensive ditching and installation of subsurface drainage (Miller and Frink 1984) (e.g., Bois de Sioux

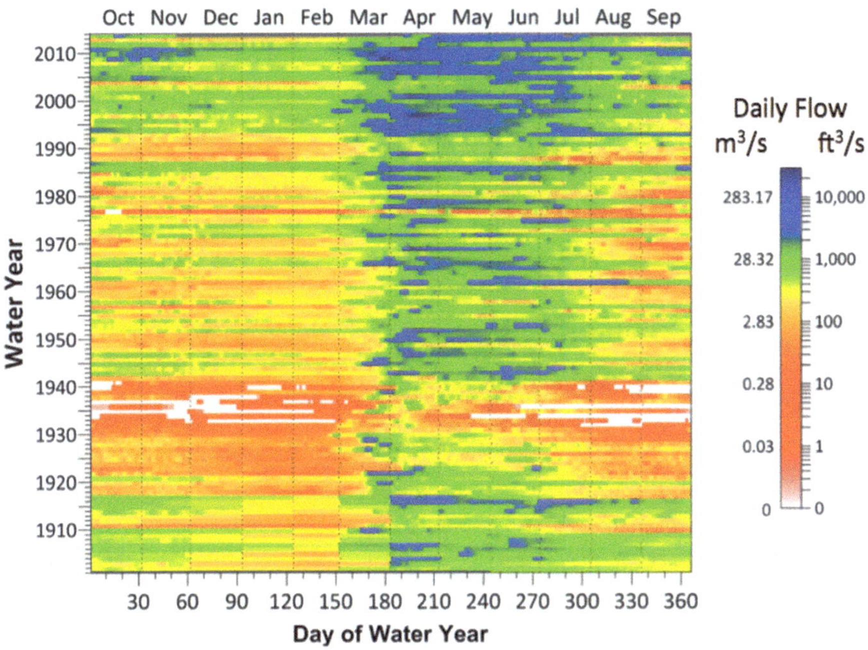

Fig. 2.3 Plot of daily Red River discharge at Fargo, ND (USGS gage 05054000) as a function of water year and day of water year (courtesy of Richard Koehler, NOAA NWS). Blue colors indicate high flows and red colors indicate low flows. Most of the highest flows occur from March to July, as a result of the spring snowmelt. The 1930s–1940s was a very dry period, while flows have increased in recent decades

Watershed District tile permit growth map: http://www.frontiernet.net/~bdswd/Files/Tile_Permit_Growth_Map_Reduced.pdf). However, it is difficult to isolate the effect of the man-made changes to peak river flows (Miller and Frink 1984; Rahman et al. 2014).

Despite the frequency of flooding in the region, ground observations of hydrological state variables are relatively sparse compared to other regions of the United States. Currently, frequent, high quality ground-based measurements of SWE and soil moisture are available from less than 20 and 10 locations, respectively, in the U.S. portion of the basin. This scarcity of SWE and soil moisture makes flood forecasting difficult in the Red River, but remote sensing can increase the spatial and temporal coverage of observations. Two issues that complicate satellite estimates of snow and soil moisture—rugged, steep terrain, and dense vegetation—are almost entirely absent from this region. The area is classified as 79 % cropland, 8 % grassland, shrubland, or wetland, and only 9 % is any type of forest (NALCMS 2015; Fig. 2.2). The minimal topographical relief is evident in Fig. 2.2. Due to these factors, remote sensing can be helpful to constrain hydrologic states that are important for accurate flood forecasting.

2.3 Snow Water Equivalent

SWE is the depth of liquid water that would result if the entire snowpack (at any given time and location) melted. Accurate flood forecasting requires knowledge of the magnitude and spatial distribution of SWE within a given basin. Remote sensing, especially passive microwave methods, can provide estimates of SWE and thus help to constrain this primary driver of spring flooding in areas such as the Red River of the North basin.

2.3.1 Passive Microwave

Most research in satellite remote sensing of SWE has focused on microwave frequencies, and specifically passive microwave radiometers. Passive microwave sensors detect the natural thermal emission of microwave radiation from the earth's surface. The primary instruments used for this purpose are listed in Table 2.1.

Multiple algorithms were developed over the past three decades to convert passive microwave brightness temperatures to depth and water equivalent of dry snow, beginning with Chang et al. (1982, 1987). The widely used "Chang algorithm" approach (and its later variants, e.g., Chang and Rango 2000; Kelly 2009) exploits the fact that higher frequency microwave radiation emitted from the earth's surface is scattered by snow particles, while snow is more transparent to microwaves at lower frequencies. Generally, the greater the snow depth, the greater the scattering, and the lower the detected brightness temperature, for a given snowpack temperature (Ulaby and Stiles 1980). Thus, passive microwave methods technically observe the effect of snow depth on thermally emitted radiation, so conversion of inferred snow depth estimates to SWE requires assumption of grain size and density (e.g., from Sturm et al. 1995, 2010). Chang et al. (1987) used the difference between brightness temperatures at 37 and 18 GHz frequencies (i.e., the "spectral gradient") of the SMMR satellite instrument and calibrated a simple linear coefficient to convert the brightness temperature difference to SWE, using an assumed snow density and grain size (Davenport et al. 2012). This approach was later modified for the SSM/I series of satellite instruments, and expanded to include more frequencies, and forest fraction and grain size corrections, for the current operational AMSR-E and AMSR2 algorithms (Kelly 2009). Other authors developed a modified "Chang" algorithm that further accounts for snow morphology over the course of the winter (Josberger and Mognard 2002).

In the past 15 years, new algorithms branched out beyond "Chang" methods, such as an algorithm based on the dense media transfer model (DMRT) (Kelly et al. 2003), which may soon be implemented as the operational AMSR2 algorithm (Richard Kelly, University of Waterloo, personal communication). Tedesco et al. (2004) developed an algorithm based on artificial neural networks, which compared well to other methods, especially when trained using experimental data.

Table 2.1 Passive microwave satellite sensors for SWE and soil moisture

Instrument	Satellite platform	Agency	Relevant frequencies (GHz)	Product spatial resolution	Temporal resolution	Operational period
Snow water equivalent						
AMSR2	GCOM-W1	JAXA	10.65, 18.7, 36.5, 89	10, 25 km	1–2 day	2012–present
AMSR-E	Aqua	NASA	10.65, 18.7, 36.5, 89	25 km	1–2 day	2002–2011
SSM/I and SSMIS	DMSP (multiple)	DoD	19.35, 37.0	25 km	1–2 day	1987–present
SMMR	Nimbus-7	NASA	18.0, 37.0	25 km	2–6 day	1978–1987
Soil moisture						
SMAP	SMAP	NASA	1.4	36 km	2–3-day	2015–present
AMSR2	GCOM-W1	JAXA	6.9, 10.65	25 km	1–2 day	2012–present
Aquarius	SAC-D	NASA	1.4	1 deg.	0.5–1 week	2011–2015
SMOS/ MIRAS	Proteus	ESA	1.4	25 km	2 day	2009–present
AMSR-E	Aqua	NASA	6.9, 10.65	25 km	1–2 day	2002–2012
SSM/I and SSMIS	DMSP (multiple)	DoD	19.35	25 km	1–2 day	1987–present
SMMR	Nimbus-7	NASA	6.6, 10.7	25 km	2–6 day	1978–1987

AMSR2 Advanced Microwave Scanning Radiometer 2, *GCOM-W1* Global Change Observation Mission—Water "Shizuku," *JAXA* Japanese Aerospace Exploration Agency, *AMSR-E* Advanced Microwave Scanning Radiometer for EOS, *NASA* National Aeronautics and Space Administration, *SSM/I* Special Sensor Microwave Imager, *SSMIS* Special Sensor Microwave Imager/Sounder, *DMSP* Defense Meteorological Satellite Program, *DoD* U.S. Department of Defense, *SMMR* Scanning Multichannel Microwave Radiometer, *SMAP* Soil Moisture Active Passive, *SAC-D* Satélite de Aplicaciones Cientificas-D, *SMOS* Soil Moisture and Ocean Salinity, *MIRAS* Microwave Imaging Radiometer with Aperture Synthesis, *ESA* European Space Agency

Other authors tested algorithms that use artificial neural networks, projection pursuit regression, general linear models, and nonlinear regression (Josberger et al. 1998; Singh and Gan 2000; Gan et al. 2009), but Chang-type methods remain in use as the primary algorithms for AMSR-E and AMSR2. Recent SWE products assimilate remote sensing and ground observations into computer models, including the Finnish Meteorological Institute (FMI)-led Global Snow Monitoring for Climate Research (GlobSnow) database (Luojus et al. 2013; Takala et al. 2011; Pulliainen 2006). The combination of models and observations may help to mitigate some of the shortcomings of passive microwave methods. Fundamental physics of microwave remote sensing of snow were summarized by Ulaby et al. (1981), Hallikainen et al. (1986), Mätzler (1987), Hallikainen (1989), and Chang et al. (1975). Rango (1993), Hall et al. (2006), and Dietz et al. (2012) provided reviews of snow remote sensing methods, while Clifford (2010) and Frei et al. (2012) reviewed global snow products.

Some of the earliest studies to examine the potential of microwave radiometers to detect snow took place in the northern Great Plains (e.g., Foster et al. 1980). Josberger et al. (1998) explored alternative methods (multiple linear regression, neural networks, and general linear models) to extract SWE from SSM/I microwave observations. Singh and Gan (2000) tested multiple "Chang-type" algorithms for use with SSM/I observations, also in the Red River basin. Foster et al. (2001) tested Chang et al. (1987) algorithm estimates of snow depth from SSM/I, modified for snow grain size and forest fraction, against ground observations of snow and gamma-derived snow depth estimates in the Roseau River basin, finding high correlation. Josberger and Mognard (2002) developed a Chang-type snow depth algorithm that uses air temperature to calculate a temperature gradient index as a proxy for snow grain metamorphism and applied it to SSM/I observations in the northern Great Plains. Mognard and Josberger (2002) examined the performance of the temperature gradient index algorithm in the region during the 1996/1997 snow season, finding that their dynamic algorithm agreed well with ground observations but algorithms that assume a constant grain size led to underestimation of snow depth (presumably due to underestimation of grain size). Mote et al. (2003) compared SWE from the SNTHERM model, the SSM/I SWE algorithm, and in situ measurements from five stations in the northern Great Plains, finding that the model underestimated SWE and the microwave algorithm overestimated SWE in late winter, likely due to metamorphism and grain growth. Dong et al. (2005) examined sources of uncertainty in SMMR Chang algorithm SWE across Canada using SWE derived from in situ snow depth observations, which indicated that the dominant errors in remotely sensed SWE are due to saturation of the microwave signal at high snow depth, high air temperature, and proximity to large water bodies. Chang et al. (2005) explored the uncertainties in point-based ground data as compared to larger scale SSM/I satellite estimates in the northern Great Plains. The authors found some significant differences between the two estimates in yearly snow depths but insignificant differences using the 10-year mean, and emphasized that approximately ten points measurements are necessary to produce a sampling error of 50 mm in 1° by 1° grid cells. Using airborne gamma radiation SWE estimates as "truth," Gan et al. (2009) compared an artificial neural network method of estimating SWE from SSM/I brightness temperature observations to projection pursuit regression and nonlinear regression methods in the Red River basin, finding that the neural network method was superior to the other two methods. Clearly, the northern Great Plains has been a test bed for developing and evaluating satellite SWE algorithms.

SWE remote sensing studies specific to the Red River basin decreased in recent years, but some other larger-scope analyses are worth noting, given their relevance to the use of passive microwave SWE estimates for flood forecasting. Vuyovich and Jacobs (2011) compared SSM/I and AMSR-E SWE estimates to a temperature index snow model driven by Tropical Rainfall Measuring Mission (TRMM) precipitation estimates in central Afghanistan and found that the satellite estimates were comparable but of higher magnitude than the modeled SWE for some winters. This analysis indicated that updating the snow model with satellite SWE estimates considerably improved modeled runoff and reservoir storage predictions and noted that rain on snow events is detectable in microwave observations as sharp decreases in SWE. Vuyovich et al. (2014) compared SSM/I and AMSR-E SWE estimates to NOAA NWS Snow Data Assimilation System (SNODAS) estimates in watersheds across continental United States, finding that the best agreement was for basins with less than 200 mm maximum annual SWE and forest fraction less than 20 %. Hancock et al. (2013) compared SSM/I, AMSR-E, and GlobSnow SWE estimates to remotely sensed snow covered area (SCA), in situ ground observations, and meteorological data across the northern hemisphere. The authors found that SSM/I and AMSR-E SWE estimates saturated in high snow covers, showed spurious spikes at the onset of spring melt, and overestimated under cold air temperatures and in forested areas, indicating that GlobSnow more accurately determines peak SWE accumulation and the seasonal appearance and disappearance of snowpack.

Analyses by the authors of this chapter suggest that passive microwave SWE estimates compare reasonably well to other observations in the northern Great Plains (e.g., Fig. 2.4). The root mean square error (RMSE) between AMSR-E SWE and 1172 U.S. Army Corps of Engineers (USACE) St. Paul District ground-snow-survey SWE observations taken between 2002 and 2011 is 34.7 mm with a coefficient of determination (R^2) of 0.36. This error is moderate, but the bias was only −3.8 mm, indicating that most of the error is likely due to the scale difference between point surveys and 25 km satellite pixels, along with the considerable variability of snow depth in the region (Chang et al. 2005; Cork and Loijens 1980).

The aforementioned research articles and further observations by the authors of this chapter emphasize important limitations of passive microwave SWE estimates (also discussed by Tedesco and Narvekar 2010):

1. For the wavelengths used by current satellites, the lower frequency microwave channels cannot penetrate deep snow covers, leading the SWE estimates to saturate at high snow depths. This is not a frequent concern for the Northern Great Plains due to its limited snowfall compared to mountainous regions.
2. The physics of wet snow are different than for dry snow. Wet snow favors absorption of microwave radiation rather than scattering, and thus, SWE retrieval is not feasible in the presence of liquid water. While microwave observations from wet snow change somewhat with increasing liquid water content (Kang et al. 2014), generally wet snow causes the brightness temperatures at different microwave frequencies to converge and thus leads the Chang-type SWE algorithms to estimate zero snow depth. Vuyovich et al. (2014) used the weekly maximum SWE to minimize impacts of wet snow on SWE estimates.
3. Accurate characterization of grain size and snow density throughout the winter is crucial for accurate SWE estimates. Snow metamorphosis over the course of the winter results in grain size and density increases (Bader et al. 1939; Colbeck 1982; Josberger and Mognard 2002), which can disproportionately decrease emitted microwave radiation. Increased attenuation at higher frequencies increases the temperature brightness differences and, without correction, makes the snow seem deeper than it is in reality (e.g., Chang et al. 1987,

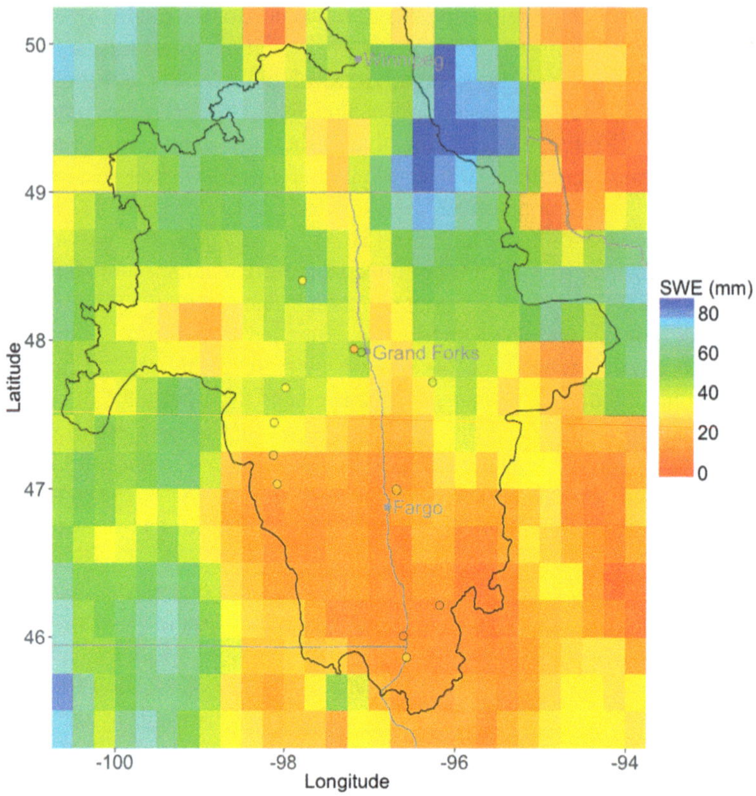

Fig. 2.4 Example comparison of AMSR2 passive microwave satellite SWE estimates to in situ ground observations, shown using the same color scale. The ground observations were collected between January 21 and 27, 2016, and the AMSR2 SWE (raster grid) is averaged over that period. A majority of the ground measurements are from manual snow tube surveys, but one point is from a snow pillow. The in situ and satellite estimates agree quite well, given the large disparity in observation scale, but the satellite may slightly overestimate SWE in the northwest RRB

Fig. 2.1). Thus, an algorithm that assumes a constant snow grain size and density will often underpredict SWE in early winter (when the grain size is small and density is low), and overpredict in the late winter (when the grain size is large and density is high). Changing snow properties and ice formation during melt-refreeze in early spring may also explain the spurious spikes in SWE observed by Hancock et al. (2013). In prairie environments such as the Red River basin, snowpacks are often shallow and air temperatures well below freezing, which can lead to the formation of large, platy crystals within the snowpack called depth hoar (Bader et al. 1939; Giddings and LaChapelle 1962), so accurate grain size estimates are crucial for accurate SWE estimates in this region.

4. Dense vegetation can inhibit accurate retrieval of SWE, but this is not a major concern for the cropland-dominated Red River basin.

Atmospheric effects also attenuate the microwave signal, but seasonal variation in this effect is negligible, and it may only be important during cloudy conditions (Wang and Tedesco 2007). For low relief, low vegetation watersheds such as the Red River basin, passive microwave SWE estimates provide the most value from early winter until snowmelt onset. Awareness of when and how snowpack physical conditions impact retrievals can optimize the useful information on SWE depth and spatial variability provided by passive microwave SWE observations. While there is still room for improvement in the global SWE algorithms, specific regional products or QA/QC analyses are an alternative approach to improving flood forecasting skill using microwave remote sensing.

2.3.2 Airborne Gamma Radiation Surveys

The potential use of natural terrestrial gamma radiation to estimate SWE for flood forecasting applications was recognized by the National Weather Service (NWS) in the late 1960s (Peck et al. 1971). Gamma radiation is naturally emitted from radionuclides in soil, including uranium, potassium, and thorium, and radiation from approximately the top 20 cm of the soil can be detected above the land surface (Carroll and Schaake 1983; Carroll 2001). However, water in any phase will attenuate gamma radiation, so the amount of gamma radiation reaching a detector from a bare soil is greater than that from the same soil when it is snow covered. The difference in radiation is proportional to the amount of SWE (i.e., mass of water; Peck et al. 1971).

Much of the North American-based research on airborne measurement of SWE using gamma radiation was conducted in the 1970s–1980s, and focused on "ground truthing" the data using ground measurements of SWE and snow depth, along with the correction for the effects of soil moisture, background radiation (e.g., from the aircraft and cosmic sources), atmospheric radon, Compton scattering, and air density on the gamma radiation signal (Peck et al. 1971, 1980; Carroll and Schaake 1983). Additional studies attempted to quantify the effect of uneven snow cover (e.g., from drifting; Cork and Loijens 1980; Carroll and Carroll 1989b) and vegetation (Carroll and Carroll 1989a) on the gamma radiation SWE estimates. During this time, researchers developed approaches to use the gamma radiation counts to estimate SWE, including single-flight methods that used the shape of the gamma spectra to infer SWE (Grasty 1982), and two-flight methods that include a calibration flight over bare soil in the fall (Carroll and Schaake 1983; Carroll 2001).

In 1980, the NWS set up an operational network of gamma radiation flight lines in the north central region on the United States, which now consists of over 2400 flight lines in 29 states and seven Canadian provinces. Each flight line is approximately 15–20 km long with a width of approximately 330 m. Thus, the areal coverage of each flight line is approximately 5–7 km^2. Currently, NOAA's National Water Center (NWC) maintains the flight line network and measures it each winter in support of river forecasting centers such as the North Central River Forecast

Center (NCRFC) (data freely available on http://www.nohrsc.noaa.gov/snowsurvey/).[1] The NWS opted for the two-flight method, utilizing the gamma radiation counts for the ^{40}K and ^{208}Tl photopeaks, and the total count between 0.41 and 3.0 MeV to obtain a single weighted estimate of SWE (Carroll and Schaake 1983; Carroll 2001). During the fall, gamma radiation for each flight line is measured over snow-free soil at a height of 150 m, in order to obtain a baseline estimate of soil moisture (Carroll 2001). The difference in gamma radiation measured between the fall flight and any subsequent identical flights over winter snowpack reflects the amount of SWE on the ground (as well as any potential changes in soil moisture status). Each year, the NWC obtains SWE estimates from zero to four times per line, depending on the presence of snow and flooding potential of the region.

Previous studies indicate that airborne gamma radiation techniques estimate SWE with an RMSE of 8.8 mm and an average bias of +54 mm (Carroll and Schaake 1983). NWC gamma radiation SWE estimates have become an important supplemental source of information for flood forecasting in the Red River basin (Brian Connelly, NCRFC, personal communication). A strength of the gamma estimates is that they are unaffected by the phase of water on the ground; the attenuation of gamma radiation is only dependent on the mass of water on the ground surface (e.g., dry snow, wet snow, ice, ponded water, etc.) and in the top 20 cm of soil. However, spatial coverage (only 5–7 km^2 per flight line) and logistics (limited aircraft flight time over the course of the winter season) limit the usefulness of the gamma flights. Also changes in below-ground water content during the winter can lead to inaccurate SWE estimates, as the gamma radiation method is affected by all water above approximately 20 cm depth.

In summary, it is challenging to capture the magnitude and variability of SWE at the spatial and temporal scales necessary for operational flood forecasting. Even the most trusted SWE observations from in situ surveys can be subject to undersampling errors (Carroll 2001) and lack the sampling density to characterize the considerable spatial variability of SWE in the region (Chang et al. 2005). Passive microwave and airborne gamma radiation methods provide complementary SWE estimates (Fig. 2.5). While gamma radiation observations have a proven track record for operational flood forecasting, microwave products appear to be of good enough quality to be informative. For these methods, validation from in situ observations, and even an intercomparison between the two remote sensing observations, is difficult because the spatial properties of the different measurement techniques are not the same. Microwave SWE algorithms continue to evolve to improve global retrieval performance. However, operational users of these products would be well served by an understanding of how products perform under the different snow conditions within their forecast region.

[1] Until recently, the flight line network was maintained by the NOAA NWS National Operational Hydrologic Remote Sensing Center (NOHRSC) in Chanhassen, Minnesota, before it became a part of the National Water Center (NWC).

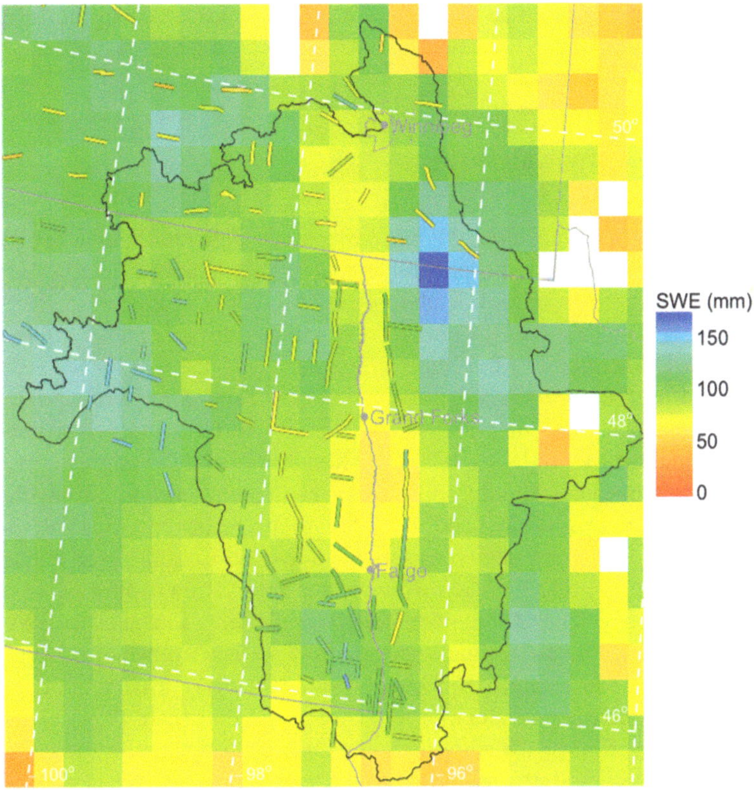

Fig. 2.5 Comparison of SWE estimated from NOHRSC airborne gamma radiation surveys and AMSR-E descending overpasses, shown on the same color scale. The gamma radiation surveys (*colored lines*) were conducted between March 11 and 14, 2009, and the AMSR-E SWE (raster grid) is averaged over that period. The gamma radiation and passive microwave estimates agree quite well, given the difference in measurement depths and footprint size

2.4 Soil Moisture

Fall soil moisture is important for spring flooding because it determines the partitioning of snowmelt between infiltration and runoff. Drier soil has a greater available water storage capacity. In areas with winters where air temperatures often remain well below freezing, such as the north central United States, soils regularly freeze in late fall and do not thaw until the spring (Todhunter 2001). This means that the soil water content in the fall becomes locked in the frozen soil and remains relatively constant until the following spring thaw. Thus, the soil wetness in late fall plays a role in determining the severity of spring flooding because drier soils are able to store more meltwater the following spring, as compared to wetter soils, resulting in less runoff. While the broader Great Plains region has a rich history as a test bed for microwave remote sensing of soil moisture, relatively few studies have been conducted in the Red River basin. These are summarized in the following sections, along with relevant studies from other nearby regions.

2.4.1 Passive Microwave

Remote sensing of soil moisture provides value for drought monitoring, agriculture, climate change, and land–atmosphere interactions. While active microwave satellite sensors (e.g., the Advanced Scatterometer (ASCAT); Wagner et al. 2013) can also observe soil moisture, passive microwave sensors are more numerous and have a more robust validation history. Many of the same sensors used for SWE estimation are also used for soil moisture estimation (see Table 2.1). While SWE algorithms utilize microwave measurements in the 10–89 GHz range, soil moisture remote sensing focuses on lower 1–10 GHz frequencies due to the greater penetration of those frequencies into the soil (1–5 cm) and greater transmissivity through vegetation. Older sensors operated at higher frequencies (e.g., SMMR, SSM/I, AMSR-E: 6–11 GHz), but newer sensors have trended toward lower frequencies (e.g., SMOS, SMAP: 1.4 GHz), allowing greater observation depths.

While some snow algorithms are relatively simple (e.g., Chang et al. 1987), soil moisture algorithms must account for the effects of surface temperature, vegetation, soil roughness, and soil dielectric properties on the thermal emission of microwave radiation in order to estimate soil moisture (Njoku and Entekhabi 1996). Generally, soil moisture algorithms use empirical relationships to determine the emissivity from a soil–water mixture, and modify the emissivity for the effects of soil roughness and vegetation. Due to this complexity, many algorithms employ an inverse modeling approach (as compared to the forward approach for SWE) to relate brightness temperature observations to soil moisture estimates (e.g., Jackson 1993; Owe et al. 2001; Njoku et al. 2003; Jones et al. 2009; Kerr et al. 2012; Santi et al. 2012).

Numerous remote sensing field validation campaigns have been conducted across the United States including Oklahoma (SGP97, SGP99, SMEX03, CLASIC07), Iowa (SMEX02, SMEX05), Georgia (SMEX03), Alabama (SMEX03), Arizona (SMEX04, SMAPVEX15), and Maryland (SMAPVEX08) (e.g., Colliander et al. 2012). Despite the relatively flat, low vegetation landscape of the northern Great Plains, presumably ideal for passive microwave remote sensing, no campaigns have taken place in North or South Dakota, or Minnesota (however, SMAPVEX12 was conducted west of Winnipeg, Canada in the Assiniboine River basin, and CanEX-SM10 was in the neighboring Saskatchewan). Most campaigns have coincided with experimental watersheds or other well-instrumented areas. The Red River basin (and the northern Great Plains in general; see Fig. 2.1) notably lacks in situ soil moisture measurements that could assist in validation and flood forecasting.

While soil moisture in the Red River basin appears to be understudied, passive microwave remote sensing has proven successful in more southern areas of the

Great Plains. Jacobs et al. (2003) provided an early demonstration of the potential value of remotely sensed soil moisture to improve rainfall-runoff predictions, using observations from the Oklahoma SGP97 experiment. Jackson et al. (2010) validated estimates from multiple AMSR-E soil moisture algorithms against in situ measurements in four research watersheds (Little Washita, OK; Reynolds Creek, ID; Little River, GA; and Walnut Gulch, AZ), finding a wide range of performance across algorithms. Two of the four tested algorithms met the NASA AMSR-E mission's 0.06 $m^3 m^{-3}$ RMSE requirement at the Oklahoma site, and simple bias and regression corrections yielded standard errors of 0.04–05 $m^3 m^{-3}$ for all four algorithms. A similar study (Jackson et al. 2012) validated SMOS soil moisture estimates, indicating an RMSE of approximately 0.043 $m^3 m^{-3}$ and near-zero bias for the Oklahoma site. Leroux et al. (2016) examined the performance of a prototype SMAP algorithm over a challenging, wide range of vegetation and soil conditions using airborne observations from the PALS instrument during the Manitoba, Canada SMAPVEX12 campaign, finding that soil moisture errors ranged from 0.05 to 0.1 $m^3 m^{-3}$.

Additional comparisons by the authors of this chapter show that L-band (1.2 GHz) passive microwave soil moisture products are promising for flood forecasting applications. Preliminary SMAP Level 3 results in the Red River basin show that the RMSE is 0.059 $m^3 m^{-3}$, with a bias of −0.011 $m^3 m^{-3}$ (and Kendall's τ rank correlation coefficient of 0.578), when evaluated against daily-averaged in situ soil moisture from 5 cm depth at the Glacial Ridge, MN SCAN site (Fig. 2.6; Schaefer et al. 2007). This comparison ignores the considerable spatial difference between the "point" measurements and gridded 36 km resolution satellite estimates, which is likely a significant source of error. SMOS estimates from

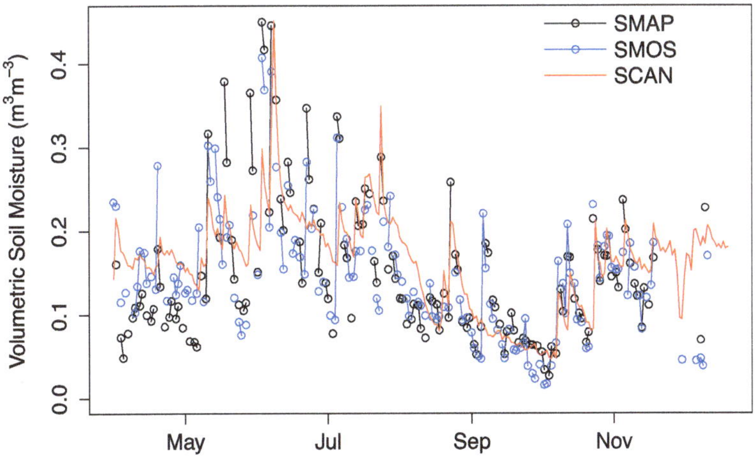

Fig. 2.6 Time series of 5 cm depth soil moisture at the Glacial Ridge, MN SCAN site (*red*) from April to December, 2015, along with passive microwave satellite estimates from SMAP (*black*), SMOS (*blue*). The SMAP and SMOS estimates agree quite well with the in situ observations, despite the large difference in observation scales

2015 also compare favorably to in situ SCAN data, with an RMSE of 0.054 $m^3\ m^{-3}$ and bias of $-0.013\ m^3\ m^{-3}$ (and Kendall's τ of 0.513). However, the comparison is not as encouraging for other products from older instruments that operate at higher frequencies (e.g., AMSR2 with LPRM algorithm: RMSE = 0.188 $m^3\ m^{-3}$, bias = 0.124 $m^3\ m^{-3}$, Kendall's $\tau = -0.017$).

Similar to SWE estimates, remotely sensed soil moisture has the ability to resolve the spatiotemporal evolution of the ground state. Figure 2.7 shows an example of a late fall soil moisture map from SMAP. The map clearly shows the relatively dry conditions prior to the fall-winter transition, especially in the southern portion of the Red River basin, although in situ observations suggest that SMAP estimates may be biased low for finer RRB soils. The Red River clay soils can develop large cracks under drought conditions, which serve as preferential drainage pathways. SMAP soil moisture estimates can potentially identify when cracking is likely to occur. Thus, the remote sensed soil moisture product has value to flood

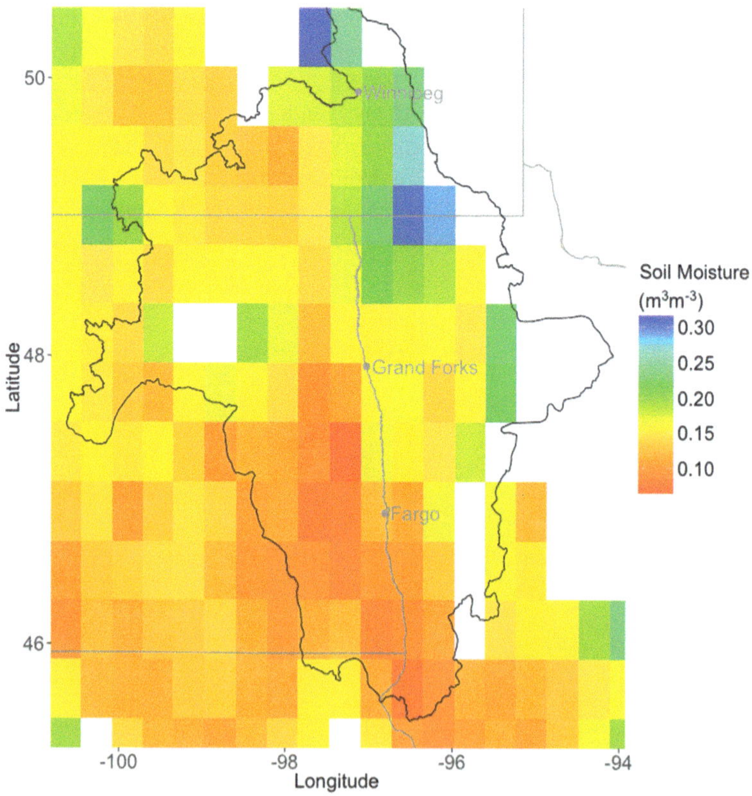

Fig. 2.7 Late fall soil moisture from SMAP (Level 3 product). The values shown are the mean soil moisture from November 15 to 28, 2015. The data were screened for the presence of open water (>10 % of pixel area), dense vegetation, precipitation, and snow or ice using SMAP data quality flags

forecasting centers for initializing operational models before the winter freeze-up and may provide insight into infiltration rates. In summary, comparisons between in situ and passive microwave estimates suggest that current satellite sensors can provide soil moisture estimates sufficient for operational use in flood forecasting, although some calibration may be necessary for finer soils.

2.4.2 Additional Soil Moisture Remote Sensing Methods

While passive microwave remote sensing has a strong history of performance in the Great Plains, the large spatial scale and relatively shallow depth of the soil moisture observations makes operational use of the products challenging in smaller basins. There are a number of additional airborne and satellite instruments which can provide observations to support the characterization of antecedent soil moisture conditions.

In addition to SWE estimations, airborne gamma radiation surveys produce estimates of soil moisture in the upper 20 cm of soil. Jones and Carroll (1983) and Carroll and Schaake (1983) summarize the method used to estimate soil moisture from gamma radiation observations, which is based on the difference in gamma radiation flux measured over dry and wet soils. The NWC estimates soil moisture for each flight line by obtaining gamma radiation observations in coincidence with gravimetric soil moisture samples (approximately 25 per 16–20 km flight line), yielding an empirical historical relationship. An early study of 155 flights over 42 flight lines found that the RMSE between in situ and airborne gamma radiation soil moisture is 0.039 $m^3\ m^{-3}$ soil moisture with an average bias of 0.001 $m^3\ m^{-3}$ (Jones and Carroll 1983). The NWC surveys the flight line network over bare soil approximately 1–2 times per fall (http://www.nohrsc.noaa.gov/snowsurvey/soil.html).

Rijal et al. (2013) explored the potential to use Landsat 5 TM data to detect soil moisture in the laboratory and field, at a much higher resolution (30 m) than passive microwave estimates (25–36 km). However, optical reflectance only observes soil moisture at skin depth, and thus relies on correlation with depth in order to estimate deeper soil moisture. For this reason, the authors used an empirical relationship to relate optical reflectance to soil moisture. The authors found an encouraging average difference of 0.02 $cm^3\ cm^{-3}$ between 15 cm depth estimated and field-measured soil moisture, but cautioned that their model should not be used elsewhere without validation. This method could show promise for providing soil moisture at the property scale with fair temporal resolution, which is not currently feasible using other methods.

As mentioned earlier, active microwave instruments also have a strong potential to provide high quality soil moisture estimates (e.g., ASCAT; Wagner et al. 2013). Additionally, Liu et al. (2011) created a multisensor soil moisture climatology that incorporates observations from both passive and active instruments starting in 1978 and currently extends through 2014, which could be useful for retrospective model testing.

2.5 Remote Sensing Data and Operational Flood Forecasting Models

While remote sensing observations have tremendous potential to inform antecedent moisture and snowpack states, translation to improved flood forecasting is not necessarily straightforward. In order for remote sensing data to be useful for operational flood forecasting, the data must (1) be trusted by the forecasters, and (2) successfully integrate into the flood modeling workflow. These points may seem trivial, but flood forecasting centers have established “tried and true” procedures. New datasets must demonstrate additional value beyond the current capability of the modeling system before they will be adopted into the forecasting workflow.

For example, the NOAA National Weather Service North Central River Forecast Center is responsible for forecasting river flows at 447 forecast points across the Upper Mississippi, Great Lakes, and Hudson Bay drainages at a 6-hourly time step. Briefly, the NWS hydrologic operational forecasting modeling system, the Community Hydrologic Prediction System (CHPS),[2] links the SNOW-17 snow model (Anderson 2006) to the Sacramento Soil Moisture Accounting (SAC-SMA) model (Burnash 1995) to a number of hydrologic and hydraulic flow routing models. The primary inputs to the model are mean areal temperature and precipitation from observations and NWS forecasts, and the main output is river discharge (and stage). The modeling system is lumped, with the watershed divided into many subbasins, and the NCRFC calibrates and runs the model at the subbasin scale. During operational forecasting, the NCRFC forecaster runs the model and obtains a river flow prediction, which is compared to observed streamflow at U.S. Geological Survey (USGS) streamgage locations. The forecaster then uses professional judgment to either accept the forecast, or to modify one or more model states and rerun the model, which will then change the predicted river flows. For instance, if the model has predicted a higher river discharge than the observed USGS streamflow for the past few time steps, the forecaster may decide to increase the model SWE to effectively reduce the meltwater routed to the stream, in order to better match the observations. In this case, the passive-microwave-derived SWE product might be well suited to help the forecaster understand variations across the watershed.

The current operational procedure does not have an established approach to assimilate observational data into the model. Furthermore, modeled streamflows depend on both model processes and model states, so streamflow prediction errors can arise from either source. Forecasters can compensate for deficiencies of the model by making manual changes, and most changes are based on observed streamflow rather than observed snow or soil moisture. These modifications may sometimes move the modeled states closer to “reality,” but in other cases, forecasters may make adjustments to model states that appear to contradict observations, but lead to better river forecasts. Due to the latitude given to the forecasters in

[2] http://www.nws.noaa.gov/oh/hrl/general/indexdoc.htm

model modification, remote sensing data will not be incorporated into the modeling procedure unless the forecasters find that it provides value to improving the streamflow predictions. For this reason, an early step toward incorporation of remotely sensed data in operational modeling is to make the data readily available, so that the forecasters can visualize it during their normal modeling procedure and gain familiarity with its strengths and weaknesses.

A second consideration for the use of remotely sensed data in operational flood forecasting is how the remote sensing data maps to the model variables. The physical dimensions and units of model states may not match those from available remotely sensed observations. For instance, SAC-SMA is a "bucket" model. Its soil moisture domains have a defined maximum depth of water (determined separately for each watershed via calibration), but no specified depth into the ground. Remote sensing estimates of soil moisture, on the other hand, provide a volumetric measure of soil moisture which reflects all soil moisture in a given soil depth range. Thus, input of remote sensing estimates into operational models requires a keen understanding of the model states, the remotely sensed data, and the assumptions that underlie each dataset. Some methods have been developed to address translation of soil moisture fields into and between models (Koster et al. 2009), which include cumulative distribution function (CDF) matching (Reichle and Koster 2004), and data assimilation (Reichle 2008), but to date, there is no standard approach. In order to enable enhanced operational flood forecasting with remote sensing observations, facilitation of mutual understanding through communication between forecasters and remote sensing scientists is likely to be as important as high quality data.

2.6 Conclusion

In the northern Great Plains, great potential exists for remote sensing to contribute to operational flood forecasting by constraining drivers of spring snowmelt flooding. The region is generally flat with low vegetation cover, which makes it suitable for most remote sensing techniques. Many remote sensing methods have been tested in the region and proven useful in nearby areas (e.g., airborne gamma radiation surveys, passive microwave, optical, and radar satellite instruments), and current platforms provide readily available near-real-time data (e.g., SMAP, SMOS, and AMSR2 soil moisture, AMSR2 and SSM/I SWE). Remote sensing soil moisture estimates in the Red River appear to be sufficiently accurate for flood forecasting applications. SWE estimates show promise but users must be aware of the limitations of the data.

The northern Great Plains, especially Red River of the North basin, lack the necessary ground observations for accurate spatial characterization of snow water equivalent and soil moisture at the subbasin scale, which can be provided by remote sensing observations. However, additional ground instrumentation is needed in order to validate remote sensing products in the region and to inform flood forecasting. While additional SWE and soil moisture measurement sites are surely necessary, observations of spring snow melting and ripening (which cannot be fully observed using passive microwave methods) are almost absent in the region, despite the importance of this time period for spring flooding. Additionally, snow grain size measurements, and more extensive snow density measurements, could help to improve SWE algorithms.

Flood forecasting and remote sensing in the northern Great Plains would also benefit from future field campaigns aimed to better understand snow and soil moisture processes in the region. Soil moisture field campaigns greatly outnumber snow campaigns over the past two decades. Snow campaigns hold great potential to advance the understanding of the region's hydrology. Such campaigns might benefit from an opportunistic, as opposed to scheduled, timeframe. Due to the yearly variability in weather and climate, it is difficult to determine when precipitation will occur, when snow will melt, etc., so it would be more effective to have a campaign ready for execution based on weather rather than planned dates.

While this chapter focuses on remote sensing methods used to estimate snow water equivalent and soil moisture, other exciting possibilities exist to remotely sense the drivers of spring snowmelt floods for use in operational flood forecasting. For instance, Ramage and Isacks (2002), Walker and Goodison (1993), and Grenfell and Putkonen (2008) developed methods to detect the presence of wet snow using passive microwave sensors. Snow covered area (SCA) from optical and infrared remote sensing (e.g., from the Moderate Resolution Imaging Spectroradiometer; MODIS) is considered to be a reliable method to observe snow cover extent and is already used operationally by some agencies (e.g., for hydropower generation in Norway; Winther and Hall 1999). The SMAP mission has produced a soil freeze–thaw product, while satellite instruments such as the Global Precipitation Measurement (GPM) mission (or ground-based radars such as the Next-Generation Radar (NEXRAD) network) can be used to detect spring precipitation. Synthetic aperture radar (SAR) and optical sensors (e.g., Landsat, MODIS, VIIRS) are useful for detecting river ice jams (Kreller et al. 2016). Additionally, during active flooding, remote sensing of flood heights may be obtainable with instruments such as the Surface Water Ocean Topography (SWOT) satellite mission, scheduled for launch in 2020. Finally, the extent of broad, slow-moving floods, such as those seen on the Red River, can be monitored using optical and infrared remote sensing (e.g., VIIRS, MODIS, Landsat; Wang et al. 2002; Zhan et al. 2002) or synthetic aperture radar (SAR; e.g., Wilson and Rashid 2005).

Improved snowmelt flood forecasts would immediately benefit large, vulnerable communities. Existing remote sensing tools provide a source of critically necessary information that can be made readily available to the forecasting community. Emerging remote sensing tools offer promise and, with further development, can complete the understanding of the significant changes in water and energy cycles throughout the entire winter and spring seasons.

Acknowledgements The authors gratefully acknowledge support from NASA Applied Sciences grant NNX15AC47G and NASA EPSCoR grant NNX11AQ34A. Additionally, thanks to Richard Koehler and Steven Quiring for allowing reproduction of their figures and data.

References

Amante C, Eakins BW (2009) ETOPO1 1 arc-minute global relief model: procedures, data sources, and analysis. NOAA Technical Memorandum NESDIS NGDC-24. National Geophysical Data Center, NOAA, Boulder
Anderson E (2006) Snow accumulation and ablation model–SNOW-17. US National Weather Service, Silver Spring
Anderson E, Larson L (1996) The role of snowmelt in the January 1996 floods in the northeastern United States. In: Proceedings of the 53rd Eastern Snow Conference, pp 141–149
Bader H, Haefeli R, Bucher E, Neher J, Eckel O, Thams C (1939) Snow and its metamorphism. In: U.S. Army Corps of Engineers (USACE) Cold Regions Research and Engineering Laboratory (CRREL) Snow, Ice, and Permafrost Research Establishment (SIPRE) Translation, vol 14. p 313
Beltaos S (2008) Progress in the study and management of river ice jams. Cold Reg Sci Technol 51 (1):2–19. doi:10.1016/j.coldregions.2007.09.001
Brasnett B (1999) A global analysis of snow depth for numerical weather prediction. J Appl Meteorol 38(6):726–740
Burnash RJC (1995) The NWS river forecast system—catchment modeling. In: Singh VP (ed) Computer models of watershed hydrology. Water Resources, Highlands Ranch, pp 311–366
Carroll T (2001) Airborne Gamma Radiation Snow Survey Program: a User's Guide, Version 5.0. National Operational Hydrologic Remote Sensing Center (NOHRSC), Chanhassen, p 14
Carroll SS, Carroll TR (1989a) Effect of forest biomass on airborne snow water equivalent estimates obtained by measuring terrestrial gamma radiation. Remote Sens Environ 27 (3):313–319
Carroll SS, Carroll TR (1989b) Effect of uneven snow cover on airborne snow water equivalent estimates obtained by measuring terrestrial gamma radiation. Water Resour Res 25 (7):1505–1510
Carroll TR, Schaake JC Jr (1983) Airborne snow water equivalent and soil moisture measurement using natural terrestrial gamma radiation. In: 1983 Technical Symposium East. International Society for Optics and Photonics, pp 208–213
Chang AT, Rango A (2000) Algorithm theoretical basis document (ATBD) for the AMSR-E snow water equivalent algorithm. NASA/GSFC, Nov 2000
Chang TC, Gloersen P, Schmugge T, Wilheit TT, Zwally HJ (1975) Microwave emission from snow and glacier ice. Ann Glaciol 16(74):23–39
Chang A, Foster J, Hall D, Rango A, Hartline B (1982) Snow water equivalent estimation by microwave radiometry. Cold Reg Sci Technol 5(3):259–267
Chang A, Foster J, Hall D (1987) Nimbus-7 SMMR derived global snow cover parameters. Ann Glaciol 9(9):39–44
Chang A, Foster J, Kelly R, Josberger E, Armstrong R, Mognard N (2005) Analysis of ground-measured and passive-microwave-derived snow depth variations in midwinter across the northern Great Plains. J Hydrometeorol 6(1):20–33
Clifford D (2010) Global estimates of snow water equivalent from passive microwave instruments: history, challenges and future developments. Int J Remote Sens 31(14):3707–3726
Colbeck S (1982) An overview of seasonal snow metamorphism. Rev Geophys 20(1):45–61
Colliander A, Chan S, Kim S-B, Das N, Yueh S, Cosh M, Bindlish R, Jackson T, Njoku E (2012) Long term analysis of PALS soil moisture campaign measurements for global soil moisture algorithm development. Remote Sens Environ 121:309–322. doi:10.1016/j.rse.2012.02.002
Cork H, Loijens H (1980) The effect of snow drifting on gamma snow survey results. J Hydrol 48 (1–2):41–51
Davenport I, Sandells M, Gurney R (2012) The effects of variation in snow properties on passive microwave snow mass estimation. Remote Sens Environ 118:168–175

Dietz AJ, Kuenzer C, Gessner U, Dech S (2012) Remote sensing of snow—a review of available methods. Int J Remote Sens 33(13):4094–4134
Dong J, Walker JP, Houser PR (2005) Factors affecting remotely sensed snow water equivalent uncertainty. Remote Sens Environ 97(1):68–82
Foster J, Rango A, Hall D, Chang A, Allison L, Diesen B (1980) Snowpack monitoring in North America and Eurasia using passive microwave satellite data. Remote Sens Environ 10 (4):285–298
Foster JL, Barton J, Chang AT, Hall DK (2001) Snow crystal and land cover effects on the scattering of passive microwave radiation for algorithm development. In: Europto Remote Sensing 2001. International Society for Optics and Photonics, pp 149–155
Frei A, Tedesco M, Lee S, Foster J, Hall DK, Kelly R, Robinson DA (2012) A review of global satellite-derived snow products. Adv Space Res 50(8):1007–1029
Gan TY, Kalinga O, Singh P (2009) Comparison of snow water equivalent retrieved from SSM/I passive microwave data using artificial neural network, projection pursuit and nonlinear regressions. Remote Sens Environ 113(5):919–927
Giddings JC, LaChapelle E (1962) The formation rate of depth hoar. J Geophys Res 67 (6):2377–2383
Grasty R (1982) Direct snow-water equivalent measurement by air-borne gamma-ray spectrometry. J Hydrol 55(1–4):213–235
Grenfell T, Putkonen J (2008) A method for the detection of the severe rain-on-snow event on Banks Island, October 2003, using passive microwave remote sensing. Water Resour Res 44 (3), W03425
Hall DK, Kelly REJ, Foster JL, Chang ATC (2006) Estimation of snow extent and snow properties. In: Encyclopedia of hydrological sciences. Wiley, New York
Hallikainen MT (1989) Microwave radiometry of snow. Adv Space Res 9(1):267–275
Hallikainen MT, Ulaby FT, Abdelrazik M (1986) Dielectric properties of snow in the 3 to 37 GHz range. IEEE Trans Antennas Propag 34(11):1329–1340
Hancock S, Baxter R, Evans J, Huntley B (2013) Evaluating global snow water equivalent products for testing land surface models. Remote Sens Environ 128:107–117
Harr RD (1981) Some characteristics and consequences of snowmelt during rainfall in western Oregon. J Hydrol 53(3):277–304. doi:10.1016/0022-1694(81)90006-8
Hirsch R, Ryberg K (2012) Has the magnitude of floods across the USA changed with global CO2 levels? Hydrol Sci J 57(1):1–9
Jackson TJ III (1993) Measuring surface soil moisture using passive microwave remote sensing. Hydrol Process 7(2):139–152
Jackson TJ, Cosh MH, Bindlish R, Starks PJ, Bosch DD, Seyfried M, Goodrich DC, Moran MS, Du J (2010) Validation of advanced microwave scanning radiometer soil moisture products. IEEE Trans Geosci Remote Sens 48(12):4256–4272
Jackson TJ, Bindlish R, Cosh MH, Zhao T, Starks PJ, Bosch DD, Seyfried M, Moran MS, Goodrich DC, Kerr YH, Leroux D (2012) Validation of soil moisture and ocean salinity (SMOS) soil moisture over watershed networks in the U.S. IEEE Trans Geosci Remote Sens 50(5):1530–1543. doi:10.1109/TGRS.2011.2168533
Jacobs JM, Myers DA, Whitfield BM (2003) Improved rainfall/runoff estimates using remotely sensed soil moisture. J Am Water Resour Assoc 39(2):313–324. doi:10.1111/j.1752-1688.2003.tb04386.x
Jones WK, Carroll TR (1983) Error analysis of airborne gamma radiation soil moisture measurements. Agric Meteorol 28(1):19–30
Jones LA, Kimball JS, Podest E, McDonald KC, Chan SK, Njoku EG (2009) A method for deriving land surface moisture, vegetation optical depth, and open water fraction from AMSR-E. In: Geoscience and Remote Sensing Symposium, 2009 I.E. International, IGARSS 2009. IEEE, pp III-916–III-919
Josberger EG, Mognard NM (2002) A passive microwave snow depth algorithm with a proxy for snow metamorphism. Hydrol Process 16(8):1557–1568

Josberger E, Mognard N, Lind B, Matthews R, Carroll T (1998) Snowpack water-equivalent estimates from satellite and aircraft remote-sensing measurements of the Red River Basin, north-central USA. Ann Glaciol 26:119–124

Kang DH, Barros AP, Dery SJ (2014) Evaluating passive microwave radiometry for the dynamical transition from dry to wet snowpacks. IEEE Trans Geosci Remote Sens 52(1):3–15

Kelly R (2009) The AMSR-E snow depth algorithm: description and initial results. J Remote Sens Soc Jpn 29(1):307–317

Kelly RE, Chang AT, Tsang L, Foster JL (2003) A prototype AMSR-E global snow area and snow depth algorithm. IEEE Trans Geosci Remote Sens 41(2):230–242

Kerr YH, Waldteufel P, Richaume P, Wigneron JP, Ferrazzoli P, Mahmoodi A, Al Bitar A, Cabot F, Gruhier C, Juglea SE (2012) The SMOS soil moisture retrieval algorithm. IEEE Trans Geosci Remote Sens 50(5):1384–1403

Knox JC (2000) Sensitivity of modern and Holocene floods to climate change. Quat Sci Rev 19 (1–5):439–457. doi:10.1016/S0277-3791(99)00074-8

Koster RD, Guo Z, Yang R, Dirmeyer PA, Mitchell K, Puma MJ (2009) On the nature of soil moisture in land surface models. J Climate 22(16):4322–4335

Kreller M, Plumb EW, Holloway E, Li S (2016) River ice and flood detection products derived from Suomi NPP VIIRS satellite data to support 2015 hydrologic forecast operations in Alaska. In: Paper presented at the American Meteorological Society Annual Meeting, New Orleans, LA, 10–14 Jan 2016

Leroux DJ, Das NN, Entekhabi D, Colliander A, Njoku E, Jackson TJ, Yueh S (2016) Active-passive soil moisture retrievals during the SMAP validation experiment 2012. IEEE Geosci Remote Sens Lett 13(4):475–479. doi:10.1109/LGRS.2015.2491643

Liu YY, Parinussa RM, Dorigo WA, De Jeu RAM, Wagner W, van Dijk AIJM, McCabe MF, Evans JP (2011) Developing an improved soil moisture dataset by blending passive and active microwave satellite-based retrievals. Hydrol Earth Syst Sci 15(2):425–436. doi:10.5194/hess-15-425-2011

Luojus K, Pulliainen J, Takala M, Lemmetyinen J, Kangwa M, Smolander T, Derksen C (2013) Global snow monitoring for climate research (GlobSnow) algorithm theoretical basis document—SWE-algorithm. p 35

Mätzler C (1987) Applications of the interaction of microwaves with the natural snow cover. Remote Sens Rev 2(2):259–387

Miller JE, Frink DL (1984) Changes in flood response of the Red River of the North basin, North Dakota-Minnesota. United States Government Printing Office, Washington, DC

Mognard M, Josberger EG (2002) Northern great plains 1996/97 seasonal evolution of snowpack parameters from satellite passive-microwave measurements. Ann Glaciol 34(1):15–23

Mote TL, Grundstein AJ, Leathers DJ, Robinson DA (2003) A comparison of modeled, remotely sensed, and measured snow water equivalent in the northern Great Plains. Water Resour Res 39 (8):1209

NALCMS (2015) 2010 North American Land Cover at 250 m spatial resolution. In: North American Land Change Monitoring System. Natural Resources Canada/Canadian Center for Remote Sensing (NRCan/CCRS), U.S.G.S.U., Instituto Nacional de Estadística y Geografía (INEGI), Comisión Nacional para el Conocimiento y Uso de la Biodiversidad (CONABIO), and Comisión Nacional Forestal (CONAFOR)

Njoku EG, Entekhabi D (1996) Passive microwave remote sensing of soil moisture. J Hydrol 184 (1–2):101–129. doi:10.1016/0022-1694(95)02970-2

Njoku EG, Jackson TJ, Lakshmi V, Chan TK, Nghiem SV (2003) Soil moisture retrieval from AMSR-E. IEEE Trans Geosci Remote Sens 41(2):215–229

NWS-HIC (2011) United States Flood Loss Report—Water Year 2011. In: NWS Annual Flood Loss Summary Reports to U.S. Army Corps of Engineers. NOAA National Weather Service-Hydrologic Information Center

Owe M, De Jeu R, Walker J (2001) A methodology for surface soil moisture and vegetation optical depth retrieval using the microwave polarization difference index. IEEE Trans Geosci Remote Sens 39(8):1643–1654
Peck EL, Bissell V, Jones E, Burge D (1971) Evaluation of snow water equivalent by airborne measurement of passive terrestrial gamma radiation. Water Resour Res 7(5):1151–1159
Peck EL, Carroll TR, VanDemark SC (1980) Operational aerial snow surveying in the United States/Etude de neige aérienne effectuée aux Etats Unis. Hydrol Sci J 25(1):51–62
Pulliainen J (2006) Mapping of snow water equivalent and snow depth in boreal and sub-arctic zones by assimilating space-borne microwave radiometer data and ground-based observations. Remote Sens Environ 101(2):257–269. doi:10.1016/j.rse.2006.01.002
Quiring SM, Ford TW, Wang JK, Khong A, Harris E, Lindgren T, Goldberg DW, Li Z (in press) The North American Soil Moisture Database: Development and Applications. Bulletin of the American Meteorological Society 97(8). doi:10.1175/BAMS-D-13-00263.1
Rahman MM, Lin Z, Jia X, Steele DD, DeSutter TM (2014) Impact of subsurface drainage on streamflows in the Red River of the North basin. J Hydrol 511:474–483
Ramage JM, Isacks BL (2002) Determination of melt-onset and refreeze timing on southeast Alaskan icefields using SSM/I diurnal amplitude variations. Ann Glaciol 34(1):391–398
Rango A II (1993) Snow hydrology processes and remote sensing. Hydrol Process 7(2):121–138
Rannie W (2016) The 1997 flood event in the Red River basin: causes, assessment and damages. Can Water Resour J 41(1–2):45–55. doi:10.1080/07011784.2015.1004198
Rasmussen PF (2016) Assessing the impact of climate change on the frequency of floods in the Red River Basin. Can Water Resour J 41(1–2):331–342. doi:10.1080/07011784.2015.1025101
Reichle RH (2008) Data assimilation methods in the Earth sciences. Adv Water Resour 31 (11):1411–1418
Reichle RH, Koster RD (2004) Bias reduction in short records of satellite soil moisture. Geophys Res Lett 31(19):L19501
Rijal S, Zhang X, Jia X (2013) Estimating surface soil water content in the Red River Valley of the North using Landsat 5 TM data. Soil Sci Soc Am J 77(4):1133–1143
Rogers P, Kaiser J, Kellenbenz D, Ewens M (2013) A comparative hydrometeorological analysis of the 2009, 2010, and 2011 Red River of the North Basin Spring floods. National Weather Service, Central Region Technical Attachment Number 13-03
Santi E, Pettinato S, Paloscia S, Pampaloni P, Macelloni G, Brogioni M (2012) An algorithm for generating soil moisture and snow depth maps from microwave spaceborne radiometers: HydroAlgo. Hydrol Earth Syst Sci 16(10):3659–3676
Schaefer GL, Cosh MH, Jackson TJ (2007) The USDA natural resources conservation service soil climate analysis network (SCAN). J Atmos Oceanic Tech 24(12):2073–2077
Schwert DP (2003) A geologist's perspective on the Red River of the North: history, geography, and planning/management issues. In: Proceedings of 1st International Water Conference, Red River Basin Institute, Moorhead, MN
Shanley JB, Chalmers A (1999) The effect of frozen soil on snowmelt runoff at Sleepers River, Vermont. Hydrol Process 13(12–13):1843–1857. doi:10.1002/(SICI)1099-1085(199909)13:12/13<1843::AID-HYP879>3.0.CO;2-G
Singh PR, Gan TY (2000) Retrieval of snow water equivalent using passive microwave brightness temperature data. Remote Sens Environ 74(2):275–286
Stadnyk T, Dow K, Wazney L, Blais E-L (2016) The 2011 flood event in the Red River Basin: causes, assessment and damages. Can Water Resour J 41(1–2):65–73. doi:10.1080/07011784.2015.1008048
Sturm M, Holmgren J, Liston GE (1995) A seasonal snow cover classification system for local to global applications. J Climate 8(5):1261–1283
Sturm M, Taras B, Liston GE, Derksen C, Jonas T, Lea J (2010) Estimating snow water equivalent using snow depth data and climate classes. J Hydrometeorol 11(6):1380–1394
Takala M, Luojus K, Pulliainen J, Derksen C, Lemmetyinen J, Kärnä J-P, Koskinen J, Bojkov B (2011) Estimating northern hemisphere snow water equivalent for climate research through

assimilation of space-borne radiometer data and ground-based measurements. Remote Sens Environ 115(12):3517–3529
Tedesco M, Narvekar PS (2010) Assessment of the NASA AMSR-E SWE Product. IEEE J Sel Top Appl Earth Observ Remote Sens 3(1):141–159
Tedesco M, Pulliainen J, Takala M, Hallikainen M, Pampaloni P (2004) Artificial neural network-based techniques for the retrieval of SWE and snow depth from SSM/I data. Remote Sens Environ 90(1):76–85
Todhunter PE (2001) A hydroclimatological analysis of the Red River of the North Snowmelt Flood Catastrophe of 1997. J Am Water Resour Assoc 37:1263–1278
Ulaby FT, Stiles WH (1980) The active and passive microwave response to snow parameters: 2. Water equivalent of dry snow. J Geophys Res Oceans 85(C2):1045–1049. doi:10.1029/JC085iC02p01045
Ulaby FT, Moore RK, Fung AK (1981) Microwave remote sensing: active and passive. Microwave remote sensing, fundamentals and radiometry, vol 1. Addison-Wesley, Reading, MA
Vuyovich C, Jacobs JM (2011) Snowpack and runoff generation using AMSR-E passive microwave observations in the Upper Helmand Watershed, Afghanistan. Remote Sens Environ 115 (12):3313–3321
Vuyovich CM, Jacobs JM, Daly SF (2014) Comparison of passive microwave and modeled estimates of total watershed SWE in the continental United States. Water Resour Res 50 (11):9088–9102
Wagner W, Hahn S, Kidd R, Melzer T, Bartalis Z, Hasenauer S, Figa-Saldaña J, de Rosnay P, Jann A, Schneider S (2013) The ASCAT soil moisture product: a review of its specifications, validation results, and emerging applications. Meteorol Z 22(1):5–33
Walker A, Goodison B (1993) Discrimination of a wet snowcover using passive microwave satellite data. Ann Glaciol 17:307–311
Wang JR, Tedesco M (2007) Identification of atmospheric influences on the estimation of snow water equivalent from AMSR-E measurements. Remote Sens Environ 111(2–3):398–408. doi:10.1016/j.rse.2006.10.024
Wang Y, Colby J, Mulcahy K (2002) An efficient method for mapping flood extent in a coastal floodplain using Landsat TM and DEM data. Int J Remote Sens 23(18):3681–3696
Wazney L, Clark SP (2016) The 2009 flood event in the Red River Basin: causes, assessment and damages. Can Water Resour J 41(1–2):56–64. doi:10.1080/07011784.2015.1009949
Wilson BA, Rashid H (2005) Monitoring the 1997 flood in the Red River Valley using hydrologic regimes and RADARSAT imagery. Can Geogr 49(1):100–109
Winther JG, Hall DK (1999) Satellite-derived snow coverage related to hydropower production in Norway: present and future. Int J Remote Sens 20(15–16):2991–3008. doi:10.1080/014311699211570
WMO (2009) WMO-No. 168. Guide to hydrometeorological practices. World Meteorological Organization, Geneva, Switzerland
Zhan X, Sohlberg R, Townshend J, DiMiceli C, Carroll M, Eastman J, Hansen M, DeFries R (2002) Detection of land cover changes using MODIS 250 m data. Remote Sens Environ 83 (1):336–350

Chapter 3
The NASA Global Flood Mapping System

F. Policelli, Dan Slayback, Bob Brakenridge, Joe Nigro, Alfred Hubbard, Ben Zaitchik, Mark Carroll, and Hahn Jung

3.1 Introduction

Flooding is the most frequently occurring natural disaster, affecting more people than any other weather-related disaster. According to the Center for Research on the Epidemiology of Disasters (CRED), there were an average of 171 flood disasters per year during 2005–2014, up nearly 35 % from the previous decade, and

The original version of this chapter was revised. An erratum to this chapter can be found at DOI 10.1007/978-3-319-43744-6_12

F. Policelli, B.S., M.S. (✉)
Hydrological Sciences Laboratory, NASA GSFC, Code 617, 8800 Greenbelt Road, Greenbelt, MD 20771, USA

Department of Earth and Planetary Sciences, Johns Hopkins University, 3400 N. Charles Street, Baltimore, MD 21218, USA
e-mail: fritz.s.policelli@nasa.gov

D. Slayback, Ph.D. • J. Nigro, M.A. • A. Hubbard • M. Carroll, B.S., M.P.S., Ph.D.
SSAI and Biospheric Sciences Laboratory, NASA GSFC, Code 618, 8800 Greenbelt Road, Greenbelt, MD 20771, USA
e-mail: dan.slayback@nasa.gov; joseph.d.nigro@nasa.gov; alfred.b.hubbard@nasa.gov; mark.carroll@nasa.gov

B. Brakenridge, Ph.D.
CDSMS/INSTAAR, University of Colorado, 4001 Discovery Drive, Boulder, CO 80309, USA
e-mail: Robert.Brakenridge@Colorado.edu

B. Zaitchik, Ph.D.
Department of Earth and Planetary Sciences, Johns Hopkins University, 3400 N. Charles Street, Baltimore, MD 21218, USA
e-mail: zaitchik@jhu.edu

H. Jung, Ph.D.
Hydrological Sciences Laboratory, NASA GSFC, Code 617, 8800 Greenbelt Road, Greenbelt, MD 20771, USA

SSAI, 10210 Greenbelt Road, Lanham, MD 20706, USA
e-mail: hahnchul.jung@nasa.gov

V. Lakshmi (ed.), *Remote Sensing of Hydrological Extremes*, Springer Remote Sensing/Photogrammetry, DOI 10.1007/978-3-319-43744-6_3

2.3 billion people were affected by these events during 1995–2015 (CRED and UNISDR 2015). CRED defines a disaster as a "situation or event, which overwhelms local capacity, necessitating a request to national or international level for external assistance ... An unforeseen and often sudden event that causes great damage, destruction and human suffering." Flooding affects populations quite frequently on all of the Earth's continents except Antarctica, but it remains stubbornly difficult to model and forecast with high precision. To respond to these events, disaster managers need reliable, low latency, and frequently updated information on the location and extent of inundation. Because of its ability to provide a synoptic perspective on a large range of scales and resolutions, satellite-based remote sensing offers the most practical solution to providing this information for many flooding events. To address this need and to provide useful products for the Earth science research community, we have developed the automated NASA Global Flood Mapping system. Elements of this system have been mentioned in previous publications (Revilla-Romero 2015; Brakenridge et al. 2012). Here we provide background on the capabilities and limitations of satellite remote sensing in which our system must operate and a full description of the system. We also discuss the remaining challenges for fully implementing the system, our mitigation strategies for these challenges, recent product uses, product evaluation results, and future prospects.

3.2 Remote Sensing of Water

A number of factors influence the detection and mapping of surface water using remote sensing techniques, including the distribution of solar and Earth irradiance across the electromagnetic (EM) spectrum, the existence of wavelength bands or "atmospheric windows" in the EM spectrum through which solar or surface-emitted EM radiation can pass, differential reflection or emission of water in these bands relative to neighboring land cover types, and differential scattering effects of atmospheric constituents. Solar irradiance peaks in the visible portion of the EM spectrum. There is a relatively clear atmospheric window in these wavelengths, and water reaches its greatest reflectivity at about 500 nm; this is what gives water its intrinsic blue color. However, soils and vegetation typically reflect solar radiation more strongly than water, and consequently it is the dark areas within spectral reflectance images that indicate water (Fig. 3.1). In blue and green wavelengths (Fig. 3.1a and b), the difference between the reflectance of water and its surroundings is somewhat less than in longer wavelengths and photons are preferentially scattered by the atmosphere, making water difficult to distinguish from its environment. At progressively longer wavelengths, the number of solar photons drops off; however, the reflectance of vegetation and soils is generally much higher than water in these wavelengths. As a consequence, water can be readily distinguished from its surroundings in red wavelengths (Fig. 3.1c) and through the atmospheric windows

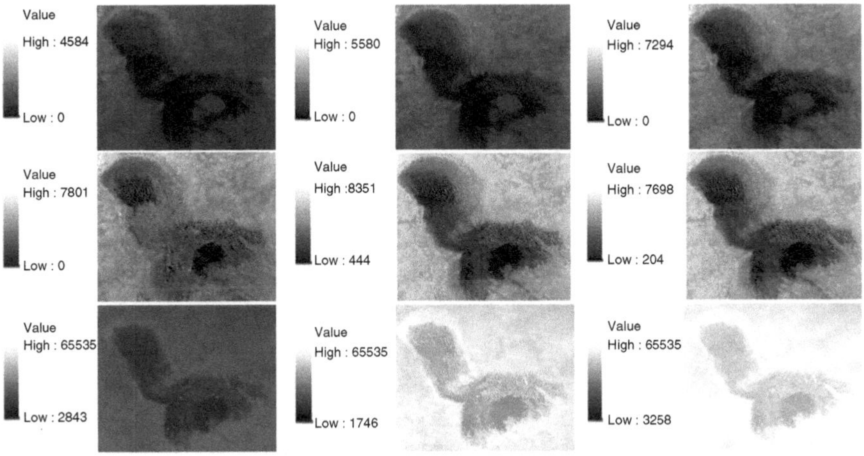

Fig. 3.1 MODIS data for Lake Chad, Central Africa (reflectance × 10,000), May 19, 2001. Row 1: (**a**) Band 3 (*blue*), (**b**) Band 4 (*green*), (**c**) Band 1 (*red*), Row 2: (**d**) Band 2 (NIR), (**e**) Band 6 (SWIR), (**f**) Band 7 (SWIR), Row 3: (**g**) Band 23 (MWIR), (**h**) Band 25, (MWIR), and (**i**) Band 31 (TIR)

in the near infrared (NIR, Fig. 3.1d) and short wavelength infrared (SWIR, Fig. 3.1e and f) portions of the EM spectrum.

Through the mid-wavelength infrared (MWIR) part of the spectrum (Fig. 3.1 g and h) there are progressively fewer solar photons to reflect; however, there are increasing numbers of emitted photons, and detection and mapping of water is still possible, though with some loss of detail. At the longer wavelengths of the thermal infrared (TIR) portion of the spectrum (Fig. 3.1i), there are significantly decreased numbers of solar photons to reflect; however, the number of emitted photons increases, again enabling detection and mapping of water with some loss of detail (Fig. 3.1j). At temperatures normally encountered on the Earth's surface, the wavelength of emitted energy is strongest in the thermal infrared region of the EM spectrum (8–15 μm); within this range, the clearest atmospheric windows occur around 11 and 12 μm. The Stefan–Boltzmann law states that the emissive power of a substance is the product of its emissivity and its temperature raised to the fourth power. The emissivity of water in the thermal infrared region is in the same range as a wide variety of soils and vegetation and is not itself a strong discriminator (http://www.icess.ucsb.edu/modis/EMIS/html/em.html). Yet water can often be distinguished from its surrounding environment using thermal remote sensing because of its differing temperature. This is due to the generally higher heat capacity of water relative to its surroundings, resulting in slower temperature changes in response to diurnal inputs. This can mean that during data acquisitions with thermal infrared sensors, water may appear darker (colder) than its environment during daytime acquisitions, and lighter (warmer) during nighttime acquisitions, depending on the precise nature of the temperature changes, the water body, and the surrounding environment being observed (see Fig. 3.2).

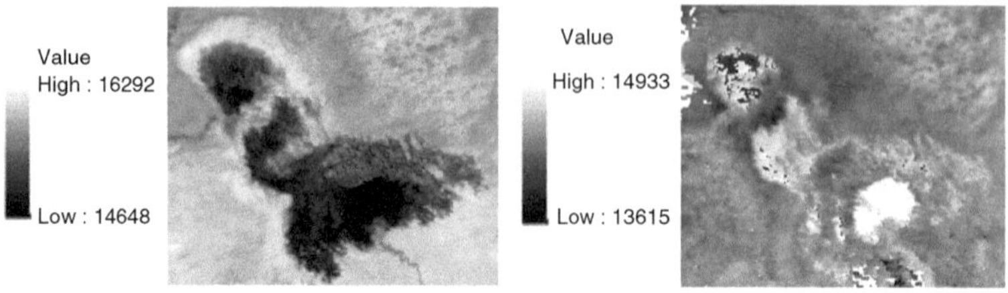

Fig. 3.2 8-day clear sky average land surface temperature maps of Lake Chad derived from MODIS Aqua thermal infrared data. Units are 50 × Degrees K. (**a**) daytime acquisition, (**b**) nighttime acquisition

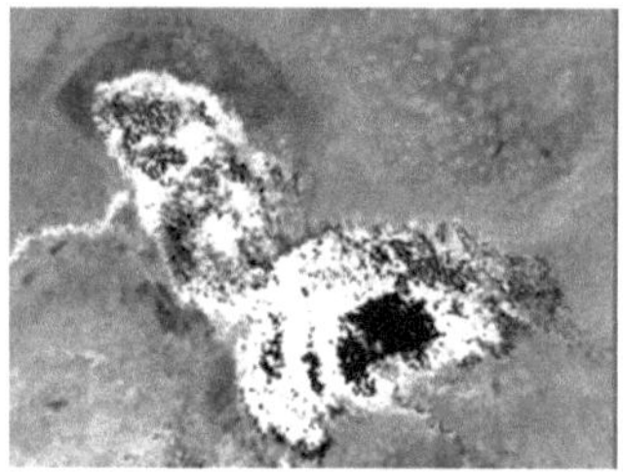

Fig. 3.3 Lake Chad region. Dark areas are identified as open water using the Dartmouth Flood Observatory's classification method

Efforts to use remote sensing methods to map surface water and flood extent must work within the opportunities for and constraints against observing water as described earlier, and further limitations imposed by the currently operating fleet of satellites (discussed more later). Coauthor G. R. Brakenridge at the "Dartmouth Flood Observatory" (DFO, http://floodobservatory.colorado.edu/) has found a ratio of the reflectance from the MODIS NIR and Red 250 m resolution bands, in the form of (NIR (858.5 nm) + A)/(Red (845 nm) + B) to be very useful for rapid water mapping purposes. This takes advantage of the increase in absorption by water in the NIR band relative to the red band and the decrease in absorption by other land surface types in the NIR band relative to the red band and provides a generally reliable distinction between open water and other land covers (see Fig. 3.3).

Currently operating optical imaging satellite missions and sensors useful for surface water and flood mapping include the NASA/ USGS Landsat 7 and 8 satellites, the MODerate resolution Imaging Spectro-radiometer (MODIS) instruments on the NASA Aqua and Terra satellites, the Advanced Land Imager (ALI) sensor aboard the NASA Earth Observing Mission 1 (EO-1), the ASTER instrument on the NASA Terra satellite, the VIIRS (Visible Infrared Imaging Radiometer) instrument on the NASA/NOAA Suomi National Polar-orbiting Partnership (NPP) satellite, the French Space Agency's SPOT (Satellite Pour l'Observation de la Terre) satellites, the European Space Agency (ESA)'s Sentinel-2, and high resolution commercial imaging systems such as Digital Globe's IKONOS, Quick Bird, GeoEYE, and WorldView satellites.

Each of these systems has capabilities that can contribute unique information on flooding to disaster managers, who often need frequently updated, finely detailed information on the location of floodwaters, which may cover very large areas. However, there is a technology related trade-off between (1) spatial resolution on the one hand and (2) temporal resolution and areal coverage on the other in the design of Earth sensing satellites and sensors. Satellite and sensor systems (such as the Aqua and Terra MODIS sensors) designed to provide data at relatively high temporal resolution (each MODIS sensor provides a day and a night acquisition for nearly the entire globe each day) and large areal coverage (the swath width of the MODIS sensor is 2330 km) generally have to forego high spatial resolution (the highest spatial resolution bands on MODIS are 250 m, and there are only two spectral bands at that resolution). In contrast, a higher resolution imaging system such as the Operational Land Imager on the Landsat 8 satellite provides a single panchromatic band at 15 m resolution and eight spectral bands at 30 m. However, the revisit time rises to 16 days at the equator and the swath width decreases to 185 km. What this means in practice for supporting disaster responders with data from a single sensor during flooding events is that data can be provided either at moderate to course resolution very frequently or at higher resolution much less frequently.

Of all the systems on orbit providing data that can potentially be accessed by our project, the MODIS sensors provide the highest resolution data available on a nearly global basis each day. This is the reason we decided to concentrate our initial efforts on providing flood maps based on NASA MODIS data, primarily using the two highest resolution (250 m) bands, Band 1 (red) and Band 2 (near infrared). We endeavor to provide products for most of the globe, updated every day with the least amount of latency we can achieve, and distribute these in easy-to-use formats.

3.3 The NASA Global Flood Mapping System: Description

The authors have collaborated to develop and operate an experimental, fully automated, near global, near real-time (NRT) surface water extent and flood mapping system. The system has been running at NASA's Goddard Space Flight Center since November 2011 and many significant improvements have been implemented since that time. The system ingests data from the MODIS sensors on the NASA Aqua and Terra satellites, which has been processed by the NASA LANCE-MODIS system (Ref. http://lance-modis.eosdis.nasa.gov) and further processes the data to produce and distribute daily maps of the Earth's surface water, including floods. Data are received by LANCE through satellite downlink, georeferenced, composited into 10° by 10° tiles and provided to our system within 3–4 h of data acquisition. The 10° by 10° tiles are a convenient size for further processing and consistent with the long-term record of flood maps produced manually by DFO. Key elements of the NASA Global Flood Mapping system include an initial screening for clouds and nonwater land cover types, compositing of the multiple available observations over a few days to minimize cloud-shadow false-positives, identification of water using the 250 m resolution red and NIR bands on MODIS, and masking of cloud and terrain shadows to produce surface water extent maps, and further masking for "normal" water to produce flood maps. We use a MODIS-based product (MOD44W, Carroll et al. 2009) developed by the University of Maryland as the "normal" water mask.

There are numerous remote sensing-based water detection algorithms described in the literature including the Carroll et al. 2009 classification tree method; the Pekel et al. 2014 Hue, Saturation, and Value (HSV) transformation approach; and the Gao 1996 Normalized Difference Water Index (NDWI) formulation. Coauthor Brakenridge (DFO) developed the water identification algorithm we have adopted using an empirical approach described later specifically for rapid flood mapping through an iterative process. The NASA Global Flood Mapping System applies this approach for land surfaces up to 70° latitude in the Northern hemisphere and up to 50° latitude in the Southern hemisphere (see Fig. 3.7a). The current algorithm performs well in this global scale region of coverage (which includes most of humanity) when compared with independent water classifications using Landsat and other data sources. We believe there is opportunity to improve the algorithm for specific regions and to identify "fringe" pixels at the water–land interface as water when the estimated water fraction exceeds 50 % (or other fixed percentage).

The initial cloud screening applies a threshold on SWIR reflectance. For a given pixel, if Band 7 (2105–2155 nm) reflectance < Cloud Threshold then the pixel is not likely a cloud and is considered for further processing. Brakenridge has found a useful Cloud Threshold reflectance to be approximately 0.07. The initial screening for nonwater land cover types applies a threshold on red band reflectance. For a given pixel, if Band 1 (620–670 nm)<Soil and Vegetation Threshold then the pixel is not likely soil or vegetation and is considered for further processing. Soil and vegetation typically have a higher reflectivity in red than water. Brakenridge has found a useful Soil and Vegetation Threshold reflectance to be approximately 0.2. Finally, if (Band 2 = NIR + A)/(Band 1 = Red + B)<Water/Land Threshold then the pixel is labeled as water. Brakenridge has found useful approximate values of A = 0.001, B = 0.1, and a Water/Land Threshold of 0.7.

To provide the most spatially complete cloud-free imagery possible, the data are composited temporally at a discrete number of time periods (1, 2, and 3 days). If a pixel (1) passes the initial screenings for clouds and nonwater land cover types, (2) is identified as water using the MODIS red and NIR bands as described earlier, and (3) does so for half or more of the data acquisitions for a given composite product (given there are generally two available observations per day: one or two water identifications for the 1-day product, two or more for the 2-day product, and three or more for the 3-day product), then the pixel is finally classified as water. A second-order 14-day composite product is also generated using the 3-day products from the most recent 14 days, to provide a recent historical view of flooding and surface water extent, which largely overcomes patchiness due to clouds in products of lesser time extent. Each product also includes an ancillary layer indicating persistent cloud cover, by compositing the MOD35 MODIS cloud

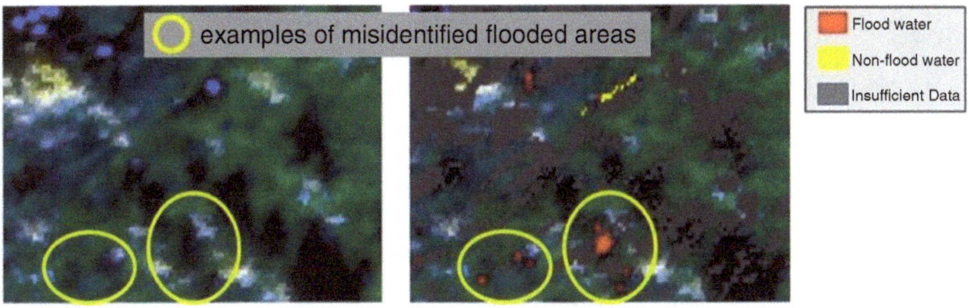

Fig. 3.4 Misidentification of cloud shadows as flood water. *Left*: clouds and cloud shadows in MODIS image circled in *yellow*. *Right*: cloud shadows misidentified as flood water in same scene

mask (Ackerman et al. 2015) over the appropriate compositing period (1, 2, or 3 days).

Shadows from clouds and terrain are difficult to distinguish from water in MODIS bands 1 and 2 (Fig. 3.4) and must be removed to avoid false positives.

Our masking for cloud shadows (Fig. 3.5) primarily relies on the compositing process described earlier, with the imperfect result that if cloud (or other) shadows occur in the same pixel multiple times (e.g., two for the 2-day product), that pixel will be tagged as water. This does occur in some regions and especially during seasons of low sun angle. Fortunately, cloud cover generally does change over time, so cloud-shadow false-positives are infrequent in the 2-day product, and almost entirely eliminated in the 3-day product. For the 1-day product, because the potential for false positives from cloud shadows is enhanced, we also geometrically project cloud shadow locations based on cloud position and height as given in the MOD35 (cloud mask) and MOD06 (cloud properties) products, and remove provisionally identified water found under the projected clouds. We do not apply this to the multiday products because the result is often not optimal, due to assumptions we must make, limitations in the cloud products, and the difference in resolution (the cloud products are at 1 km compared with the flood maps at 250 m). However, in the future, we expect to improve these algorithms so that they can be reliably applied to the multiday products (while minimizing the removal of real water), and used to generate cleaner 1-day products. Use of Synthetic Aperture RADAR (SAR) data to penetrate thicker cloud cover may also be used in the future to supplement the MODIS data; this type of data has recently become readily available from the European Space Agency's Sentinel-1 platform and could significantly enhance our current system.

We have implemented an initial version of terrain shadow masking (Fig. 3.6) which uses digital elevation data from the NASA/METI (Japan's Ministry of Economy, Trade, and Industry) GDEM2 product and solar viewing geometry information provided with the MODIS data to project and mask areas of expected terrain shadowing. Due to assumptions required for near real-time production purposes, and possible inaccuracies in the DEM, the result is not perfect, but typically reduces terrain-shadow false water detections by 70–90 %. This

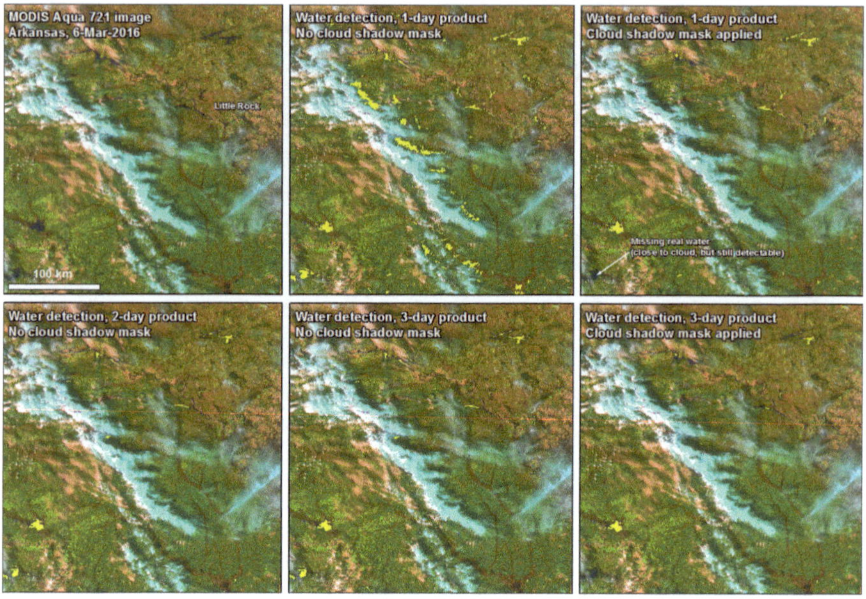

Fig. 3.5 Cloud shadow masking. System-identified water is shown in yellow. First row: (**a**) MODIS image of Arkansas, USA in the region of Little Rock with significant cloud cover. (**b**) 1-day surface water extent product, without projected cloud mask. (**c**) 1-day surface water extent product, with cloud mask applied. Note significant removal of misidentified cloud shadow at the expense of losing correctly identified water in lower left. Second row: (**d**) and (**e**) 2-day and 3-day water extent products without application of cloud mask. Note significant reduction of misidentified cloud shadow while not losing identification of real water in lower left. (**f**) 3-day surface water extent product with cloud mask applied (not a standard system product). Note loss of water identification in lower left

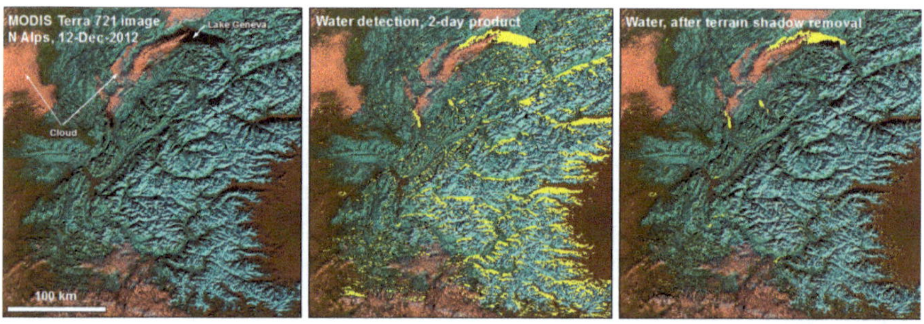

Fig. 3.6 Terrain shadow masking. System-identified water is shown in *yellow*. (**a**) MODIS image of the Alps in the area of Lake Geneva. (**b**) 2-day water extent product. Note proliferation of terrain shadows misidentified as water. (**c**) 2-day water extent product with terrain shadow mask applied. Note significant decrease of misidentified water while retaining correct identification of Lake Geneva

significantly reduces the 'noise' that will otherwise litter mountainous areas in winter, when shadows can be substantial at the nominal overpass times of 10:30 a.m. and 1:30 p.m.

The primary product from the NASA Global Flood Mapping System is the MODIS Water Product (MWP) provided in raster (geotiff) format. This product tags pixels with one of four values: (0) insufficient data to determine surface condition, (1) no water detected, (2) normal water detected (matching water layer of the MOD44W product), and (3) flood water detected (not matching MOD44W layer). From this product, we also generate rasters and vector layers for: MODIS Surface Water (MSW), showing all water detected, MODIS Flood Water (MFW), showing only flood water, and the MODIS Flood Map (MFM) product, which is simply an annotated $10 \times 10°$ graphic product (png format). These products are made available through a web interface at http://oas.gsfc.nasa.gov/floodmap (Fig. 3.7a). The web interface allows users to choose the $10 \times 10°$ area of interest for any dates available in our archive (Fig. 3.7b). Users can also access products with the different compositing options (1, 2, 3, or 14 days), view a slide show of maps from the past 10 days, review product-related information, and correspond with the project team. Due to current limitations on data processing and storage, the option to view 1-day composites is only enabled (manually) during major flooding events and other times of interest. The data is also automatically delivered to the Dartmouth Flood Observatory, where additional value-added products are created and distributed, including automated 14-day (forward running) and annual surface water products in .shp (GIS vector) formats.

3.4 Remaining Challenges and Mitigation Strategies

Clouds are a critical limiting factor for any type of land surface imaging using optical instruments. The MOD35 cloud mask used by our project is imperfect and thus is only provided as an ancillary dataset for the multiday composite products and used by necessity for the single day product. It is not uncommon for the algorithm to detect water in pixels that are marked as cloud in the cloud mask product. Typically, this occurs when the cloud is thin, and thus we are still able to detect dark water under it in the red and near-infrared bands (this would be more difficult at shorter wavelengths). If the cloud mask suggests that more than half of the available observations for a given composite period are cloudy (e.g., three or four in a 2-day product), the MWP is coded with an "insufficient data" value. Even so, if the algorithm does detect water in 2 (or more) observations for such pixels, then the "insufficient data" value is overwritten with values indicating detected water (or flood). To summarize, we are never completely certain when we can or cannot see the surface, but the pattern of "insufficient data" values typically provides useful spatial context that makes clear if a region was cloudy or not. In the future, we hope to improve the use of the MODIS cloud mask and cloud properties products to provide more robust flood mapping products, for example, by using the

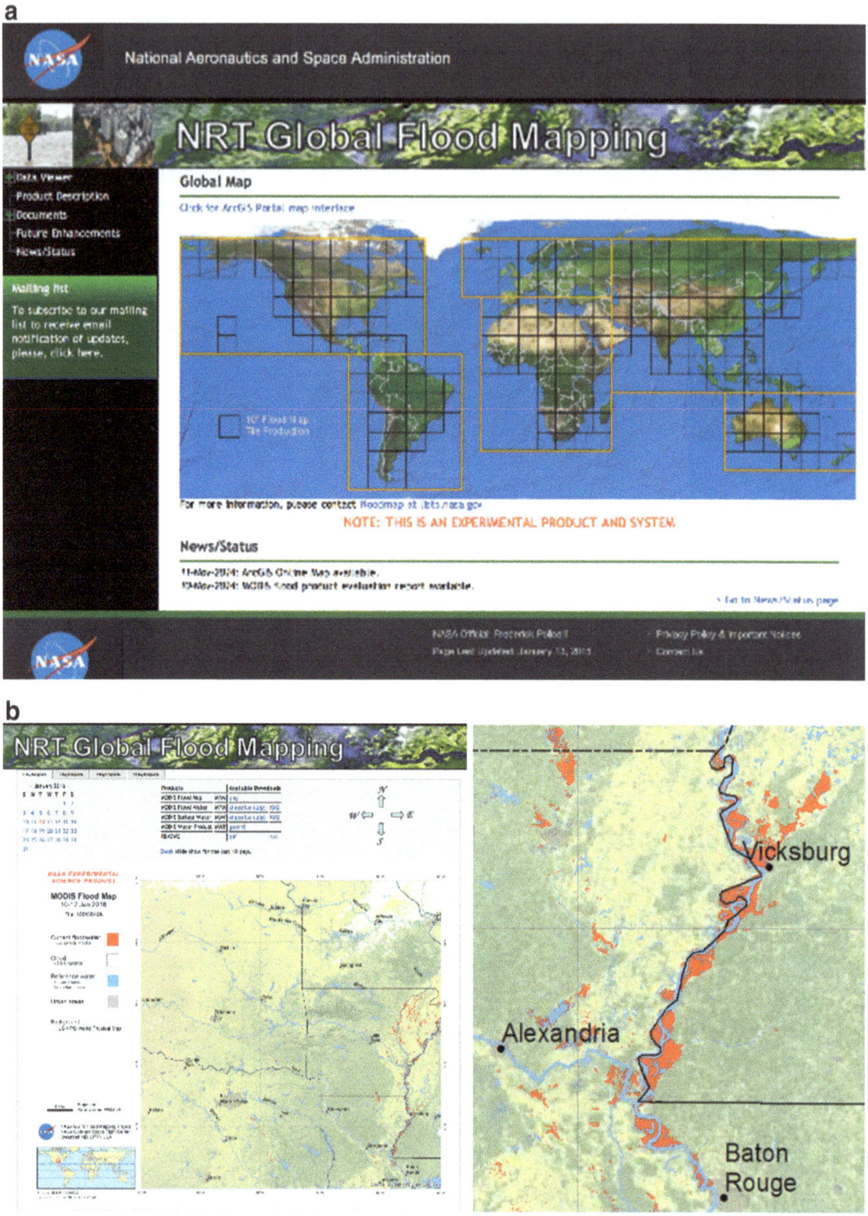

Fig. 3.7 (**a**) Front page of the web interface for the NASA Global Flood Mapping System, (**b**) product viewing example from the web interface for the NASA Global Flood Mapping System; January 2016 flooding of the Mississippi river, (**c**) limited zoom capability of current interface

cloud optical depth information from the latter to apply customized water detection algorithms for surfaces below thin clouds, and by making use of the probability of cloud detection information from the former for use in the cloud shadow masking algorithm.

We expect our approach to terrain shadow masking described before to be enhanced by further masking of areas that are substantially elevated above the local drainage network, and therefore unlikely to flood. This will be made possible by processing the HydroSHEDS DEM (http://hydrosheds.cr.usgs.gov, a hydrologically corrected version of the NASA/DoD (Department of Defense) Shuttle Radar Topography Mission (SRTM) land surface elevation dataset) with the Height Above Nearest Drainage Algorithm (HAND; Nobre et al. 2011). This algorithm calculates a drainage network based on the drainage direction of each pixel, and then computes the elevation difference between each pixel and the point in the drainage network to which that pixel drains. The result is an elevation surface with values relative to the local drainage network. An elevation threshold can then be applied to this HAND DEM to create a mask that covers areas that are substantially elevated above the local drainage features, and therefore not likely to flood. Two known problems exist with this method. First, while HAND effectively identifies the features that create terrain shadows, if the actual shadows are cast onto pixels beneath the elevation threshold, the HAND-based mask will not remove the false positives. In other words, confounding terrain shadows may still fall on areas that are likely to flood. This is common in narrow riverbeds in mountainous regions and with shadows cast by tall features. Using the HAND mask in conjunction with the existing terrain shadow mask partially addresses this concern. The second issue is that small water bodies that are elevated above the main drainage network may be improperly masked. This was found to be an issue in the glacial regions of the Himalayas. Development and testing of this algorithm has been completed and it will be included in the next version of the system. Dark volcanic soils have also been found to be misidentified as water (Fig. 3.8) in the current system and will need to be masked.

Detection of water under dense vegetation cover is also a very challenging problem and a significant error of omission for the NASA Global Flood Mapping System and other similar approaches to water extent and flood mapping. The project team is investigating several approaches, including use of thermal infrared and Synthetic Aperture RADAR (SAR) data, to introduce mapping of these waters into the current system. As discussed earlier, under certain conditions, thermal infrared data can detect water because of temperature differences between the water and the surrounding land (see Fig. 3.2). Specifically, some researchers have found utility from this method for mapping water, including under vegetation canopies, in semiarid and arid regions where the water temperature changes much more slowly due to the diurnal heating cycle than adjacent dry soil (Leblanc et al. 2011). SAR backscatter coefficient data is extremely useful to distinguish flooded vegetation, nonflooded areas, and open water (Jung and Alsdorf 2010). Certain types of flooded vegetation in SAR amplitude images yield brighter returns than nonflooded areas because the radar pulse is returned to the antenna when it reflects from water

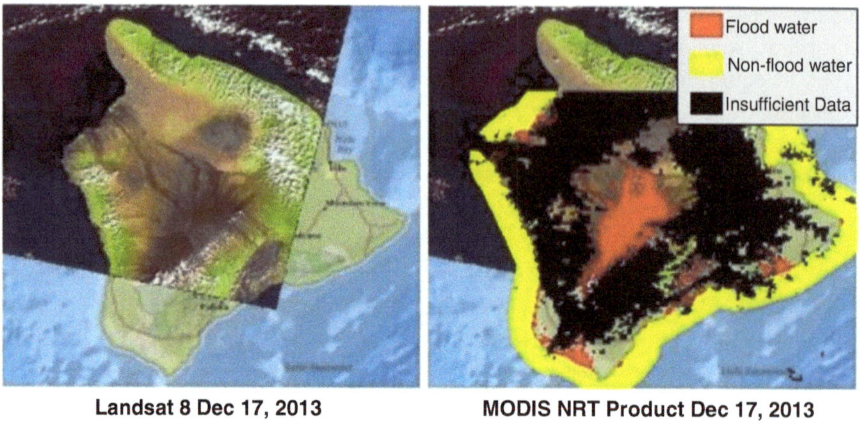

Fig. 3.8 Misidentification of volcanic soils as water. *Left*: Landsat 8 image of Hawaii showing (*dark*) volcanic soils. *Right*: NASA MODIS flood map of Hawaii showing locations of misidentified floodwater (*red*)

surfaces and scatters from inundated vegetation (i.e., the 'double bounce effect'). In contrast, open water shows little backscattering return (i.e., a specular reflection). Furthermore, SAR data can provide information on surface water elevation changes through analysis of variations of interferometric coherence (Jung and Alsdorf 2010; Jung et al. 2010).

Lakes and rivers with a high sediment load, which may appear chalky blue or muddy brown in visible imagery, are often not detected as water by the current MODIS algorithm. This is a fundamentally difficult issue because such water bodies can appear spectrally quite similar to certain types of uninundated land surfaces in the highest resolution MODIS bands. Use of both SAR and thermal IR bands may help address this issue.

Finally, the current MOD44W reference water is not optimal for our purposes because it is seasonally static and in places out of date (some indicated lakes have changed shape or no longer exist, while others have been formed; some rivers have changed course), and thus does not reflect normal seasonal lake and river water height variations. Efforts are underway by the authors to develop an ephemeral water map as an optional mask for our product users.

3.5 Product Use

The target audiences for the NRT flood products are (1) organizations involved in disaster management and humanitarian relief and (2) the scientific research community. Recent users in the former category include: UN OCHA, The World Food Program, UNOSAT, the Global Disaster Alert and Coordination System, U.S. FEMA, Swiss Re, GeoSUR in Latin America, and the UK-based NGO

MapAction. In the latter category are a handful of researchers working to use the data to improve the results of hydrological models or for understanding regional flood history. Though they are few in number, their work may prove to be extremely valuable both to our understanding of the Earth's surface water and to our ability to produce accurate near-term forecasts of hydrological conditions, including flooding.

Because our website is open to the public and we do not track users, it is not possible to know who all of the users are; however, usage statistics were collected for the website to provide insight into the size of the user community. Not including traffic from automated programs, the website received an average of 588 unique visitors each month from June 2015 through February 2016, the period of data collection. Only 48.5 % of visits to the website last longer than 30 s, which may mean that roughly half of the unique visitors do not stay long enough to significantly interact with the website. Therefore, the number of unique visitors per month that visit the website and view some of the available information is probably in the hundreds but is less than 500. The average bandwidth per month is 55.07 GB, which means an average of 93.7 MB per unique visitor. Figure 3.9 shows the average number of pages accessed from each country during the period of data collection.

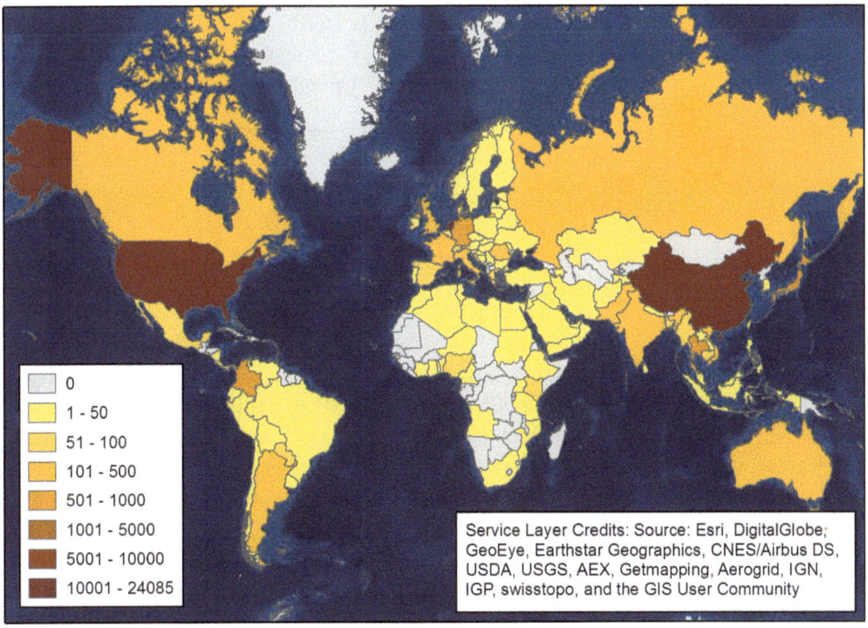

Fig. 3.9 Average number of pages accessed from each country from June 2015 through February 2016. The most prolific users are the US (24,085 pages/month), China (13,528 pages/month), and the Netherlands (17,717 pages/month)

3.6 Product Evaluation

Except through the use of other remote sensing products, or of severely limited numbers (in most cases nonexistent) of ground observations, there are few options for evaluating the results of satellite-based water and flood mapping. Beatriz Revilla-Romero et al. (2015) compared the NRT flood mapping products with the European Union Joint Research Center's Global Flood Detection System (GFDS) flood maps, which are based on the use of satellite-based passive microwave systems. The authors found that for the ten major flood events studied, the two systems provided highly complementary results. While the microwave system consistently detected more water, particularly during heavy cloud cover, the higher resolution optical data from the MODIS system provided considerably more detail valuable to disaster managers.

A second evaluation, conducted by the flood mapping team members at NASA GFSC and DFO, was largely restricted to visual inspection and qualitative assessments of the automated mapping results compared to the MODIS products and augmented by other sources. This evaluation focused on ~50 flood sites and ~50 nonflood sites containing "permanent water." The selection of these sites was based on several criteria to better understand how different factors influence the results. These criteria included: (1) varying latitudes, (2) different times of the year, (3) areas of high and low percentage cloud cover, (4) areas in which the flood mapping team had active projects (and thus complementary datasets useful for evaluation), and (5) flooded areas that coincide with various land cover types, represented by the NASA MODIS 500 m IGBP Land Cover Type product. The evaluation analyst used a scale of 1–5 to rate the quality of the products as follows:

1—poor = water not detected or water erroneously detected.
2—fair = less than half of the water is detected.
3—good = about half of the water is detected.
4—excellent = most of the water is detected.
5—almost perfect = nearly all of the water is detected.
TMC—Too Many Clouds = no flood or permanent water detected.

The results for both flood and nonflood sites are listed in Tables 3.1 and 3.2, respectively. The full evaluation report is online and can be accessed at (http://oas.gsfc.nasa.gov/floodmap/documents/NASAGlobalNRTEvaluationSummary_v4.pdf). Overall, the flood event and permanent water detection capabilities of the NASA Global MODIS Flood Mapping products worked very well (see example in Fig. 3.10).

Performance patterns did not seem to be influenced by latitude or time of the year, although IGBP land cover type may be a factor. The evaluation did not reveal any unknown deficiency in the water detection algorithm, as most of the errors of omission seem to be results of cloud cover and inundated vegetation, while areas of extreme terrain, volcanic material, and cloud shadows lead to errors of commission. These causes, along with the examples provided in the report, are listed in Table 3.3.

Table 3.1 Evaluation of products in flood conditions

Rating	Count	%
5—almost perfect	11	21
4—excellent	10	19
3—good	2	4
2—fair	1	2
1—poor	11	21
TMC—too many clouds	17	33
Outside product coverage area	1	Eliminated from equation
Totals	53	100

Table 3.2 Evaluation of products in permanent water conditions

Rating	Count	%
5—almost perfect	16	29
4—excellent	9	16
3—good	7	13
2—fair	2	4
1—poor	4	7
TMC—too many clouds	17	31
Outside product coverage area	1	Eliminated from equation
Totals	56	100

A main conclusion of the product evaluation was that some opportunities for improvement lie in the possibility to adjust the band math portion of the algorithm for different terrain and/or land cover types. Investigating change detection approaches may also prove fruitful.

3.7 Future Prospects

In the near term, we expect the current product suite and utility of the NASA Global Flood Mapping System to be enhanced through a number of the strategies discussed earlier, including the HAND masking of false positives caused by terrain shadows on areas of high slope, by transitioning to use of swath data to provide improved shadow masking and decreased product latency, and by the implementation of a new web Portal to enable inclusion of additional flood mapping products (including those derived from Landsat data) and supplemental datasets useful to disaster managers (including roads, schools, hospitals, etc.). Our goals in this regard include supporting improved situational awareness for disaster managers, improved online use of the data, and facilitating development of more customized user interfaces. Other ongoing efforts include the development of an ephemeral water mask as an optional aid to distinguishing flood from normal water fluctuations, and

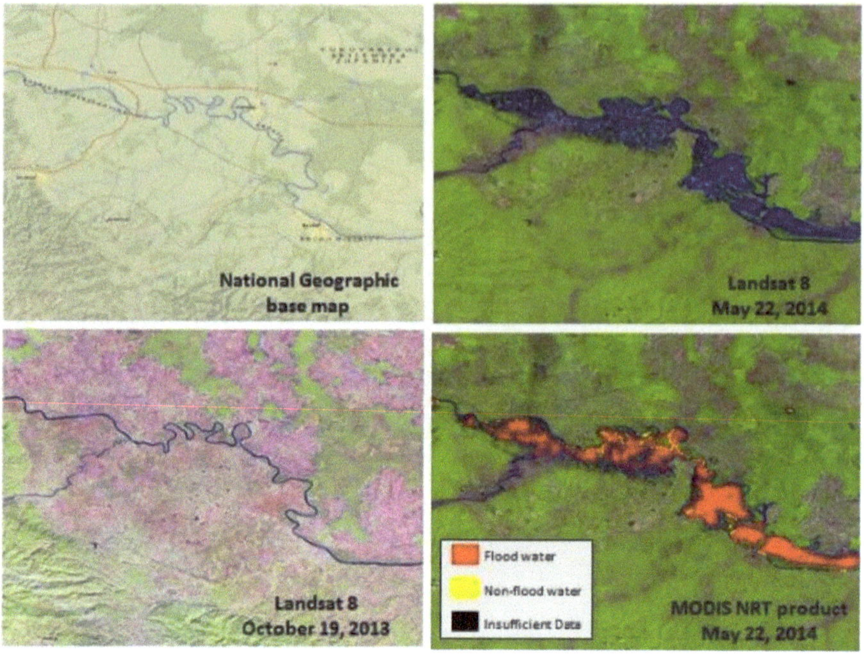

Fig. 3.10 Example of successful mapping of flood waters in Bosnia and Herzegovina by the NASA Global Flood Mapping System. Before (*lower left*) and after (*upper right*) images of flooding are shown in the 30 m resolution Landsat images. Areas identified as flooding are shown in red in the 250 m resolution MODIS flood maps produced by the system

Table 3.3 Causes of erroneous flood commission/omission

Causes of commission	Causes of omission
Terrain variation (Oman, Nepal)—terrain shadow leading to false flood water positives	Cloud cover (Peru, Cambodia)—obscuring spectral response of flood water
Volcanic material (Hawaii, Syria, Arizona)—spectral characteristics of volcanic materials leading to false flood positives	Inundated vegetation (Namibia)—obscures surface water
Cloud shadow (Ireland, Australia)—shadow leading to false flood positives	Short-lived floods, small floods, and sediment-rich areas of water (not encountered in the evaluation locations)

transitioning the current system to an operational organization. Important system enhancements to the NASA Global Flood Mapping System which we hope to be able to implement in the intermediate term include improved cloud shadow masking, masking of volcanic soils, updating of the current normal water mask, and processing of water extent and flood maps from the entire MODIS data archive.

Additionally, when the NASA MODIS sensors stop providing data (they are both already operating well beyond their design lives), products derived from VIIRS sensor data provided by the NASA Suomi/ NPP mission and future

NOAA NPOES missions will be necessary for the kind of frequent, synoptic, large-scale flood mapping our project conducts. Further in the future we expect useful new data sources to become available to our project, possibly including additional RADAR sources, near real-time hi-resolution commercial data, and data from planned missions such as the NASA INSAR mission, the ESA Sentinel -1B and -2B missions, and others. We also expect to improve the reliability of the MODIS system's identification of floodwater to enable automated tasking and processing of other satellites and datasets for areas identified by the system. Additionally, we expect to develop appropriate methods of estimating uncertainty, an unmet need that will be particularly valuable for our research users.

Finally, we believe that many aspects of the system can be improved by closer partnerships with end-users and other stakeholders. For instance, end-users could guide us toward producing more useful data products and formats, and help us identify and mitigate local factors such as DEM inaccuracies to support improved terrain shadow masking. It is also very likely that the products could be significantly improved by working with local officials to improve the definition of the normal water extent.

References

Ackerman S et al (2015) MODAPS Services—About, doi:10.5067/MODIS/MOD35_L2.006

Brakenridge G, Syvitski J, Overeem I, Higgins S, Kettner A, Stewart-Moore J, Westerhoff R (2012) Global mapping of storm surges and the assessment of coastal vulnerability. Nat Hazards 66:1295–1312

Carroll M, Townshend J, DiMiceli C, Noojipady P, Sohlberg R (2009) A new global raster water mask at 250 m resolution. Int J Digital Earth 2:291–308

CRED and UNISDR (2015). http://cred.be/sites/default/files/HCWRD_2015.pdf

Gao B (1996) NDWI—a normalized difference water index for remote sensing of vegetation liquid water from space. Remote Sens Environ 58:257–266

Jung HC, Alsdorf D (2010) Repeat-pass multi-temporal interferometric SAR coherence variations with Amazon floodplain and lake habitats. Int J Remote Sens 31:881–901

Jung H, Hamski J, Durand M, Alsdorf D, Hossain F, Lee H, Hossain A, Hasan K, Khan A, Hoque A (2010) Characterization of complex fluvial systems using remote sensing of spatial and temporal water level variations in the Amazon, Congo, and Brahmaputra rivers. Earth Surf Process Landf 35:294–304

Leblanc M, Lemoalle J, Bader J, Tweed S, Mofor L (2011) Thermal remote sensing of water under flooded vegetation: new observations of inundation patterns for the 'Small' Lake Chad. J Hydrol 404:87–98

Nobre A, Cuartas L, Hodnett M, Rennó C, Rodrigues G, Silveira A, Waterloo M, Saleska S (2011) Height above the nearest drainage—a hydrologically relevant new terrain model. J Hydrol 404:13–29

Pekel J, Vancutsem C, Bastin L, Clerici M, Vanbogaert E, Bartholomé E, Defourny P (2014) A near real-time water surface detection method based on HSV transformation of MODIS multi-spectral time series data. Remote Sens Environ 140:704–716

Revilla-Romero B, Hirpa F, Pozo J, Salamon P, Brakenridge R, Pappenberger F, De Groeve T (2015) On the use of global flood forecasts and satellite-derived inundation maps for flood monitoring in data-sparse regions. Remote Sens 7:15702–15728

Chapter 4
Congo Floodplain Hydraulics using PALSAR InSAR and Envisat Altimetry Data

Ting Yuan, Hyongki Lee, and Hahn Chul Jung

4.1 Introduction

Tropical wetlands are one of the most important sources of global carbon and methane emissions and play a significant role in regulating water balance and maintaining biodiversity (Barbier 1994; Neue et al. 1997; Hayashi et al. 1998). The water flow through the tropical wetlands, which governs biogeochemical process, sediment delivery, and nutrient exchange, is probably the most important controlling factor (Mitsch and Gosselink 2007). However, the complexity of floodwater flows has not been well captured because floodwater moves laterally across wetlands and its movement is not bounded like that of typical channel flow. Water flow across wetlands is more complex than implied by one-dimensional, point-based measurements.

As an active microwave sensor, Synthetic Aperture Radar (SAR) has all-day and all-weather imaging capability. In densely vegetated wetlands, the ability of radar pulse to penetrate canopy depends on its electromagnetic wavelength. It is well known that a SAR system with longer wavelength has better canopy penetration capability than shorter wavelength. Therefore, the L-band SAR instrument with a wavelength of 23.6 cm onboard JERS-1 and ALOS-1 has been successfully used to map floodplain inundation extents in river basins with dense vegetation canopy,

T. Yuan (✉) • H. Lee
Department of Civil and Environmental Engineering, University of Houston, 5000 Gulf Freeway, Building 4, Room 216, Houston, TX 77204-5059, USA

National Center for Airborne Laser Mapping, University of Houston, Houston, TX, USA
e-mail: tyuan@uh.edu

H.C. Jung
Hydrological Sciences Laboratory, NASA Goddard Space Flight Center, Greenbelt, MD 20771, USA

Science Systems and Applications, Inc., Lanham, MD 20706, USA

V. Lakshmi (ed.), *Remote Sensing of Hydrological Extremes*, Springer Remote Sensing/Photogrammetry, DOI 10.1007/978-3-319-43744-6_4

such as the Congo and Amazon basins (Hess et al. 1999, 2003; Rosenqvist and Birkett 2002; Lee et al. 2014). Apart from inundation areas in the wetlands, L-band SAR amplitude images have also been used to estimate water level changes or water depth in the wetlands based on the fact that the double-bounce backscattering is enhanced when water level increases beneath the vegetation canopy (Kim et al. 2014; Lee et al. 2015; Yuan et al. 2015). However, the applicability of SAR amplitude images in estimating water level change is limited in dense forest area because the hydrological factor becomes less dominant contributing factors to SAR backscattering variations (Kim et al. 2014; Yuan et al. 2015).

Interferometric SAR (InSAR) technique has been successfully used to map relative $\partial h/\partial t$ in the wetlands with high spatial resolution (~30 m) and centimetric accuracy (Alsdorf et al. 2001; Lu and Kwoun 2008; Wdowinski et al. 2008). However, the interferograms can only provide a spatially relative gradient of water level changes. Therefore, a vertical reference is needed to convert the relative changes into absolute changes (Kim et al. 2009). Since most of the floodplains are not monitored by in situ gauges, satellite altimetry-derived water level change has been used as the vertical reference instead (Kim et al. 2009). Therefore, by integrating of satellite radar altimetry and InSAR data, mapping spatially detailed absolute $\partial h/\partial t$ in the wetlands becomes possible.

The Congo wetlands play an important role in global water and carbon cycles, and regulate global and regional climate and environments (Barbier 1994; Neue et al. 1997; Hayashi et al. 1998). The Congo River meandering over 4374 km is the world's second largest river in terms of discharge (40,200 m^3/s) and drainage area (3.7 million km^2) (Lehner et al. 2008). Extensive wetlands and swamps, estimated to be 190,000 km^2, expand along the banks of the river and its tributaries (Hughes 1992; Keddy et al. 2009). Despite its enormous size and its significance to global and regional climate and environment, the Congo Basin is the least known wetland system (Keddy et al. 2009; Alsdorf et al. 2016). This is mainly due to the historic instability and conflicts that have plagued in this region. There used to be more than 400 stream gauges in the Congo Basin until 1960 while there are currently only about ten operating gauges (Alsdorf et al. 2016). In addition to the fact that most of the in situ gauges are installed in the rivers, not in the wetlands, the two-dimensional flow in the wetlands cannot be measured with the gauges (Alsdorf et al. 2007).

Only a couple of studies have used InSAR and satellite altimetry data to explore the hydraulics of the Congo's floodplains. Jung et al. (2010) used two interferograms generated with L-band JERS-1 images obtained in 1996 over the interfluvial wetlands near the confluence of the mainstem and its major tributaries including the Ubangi and Sangha, and reported that the flow patterns are not well defined and have diffuse patterns. Lee et al. (2015) generated two PALSAR interferograms over the similar central regions and revealed dense fringe patterns showing increasing $\partial h/\partial t$ toward the mainstem. Since multitemporal water level changes and hydraulic processes have not been extensively investigated, we aim at generating multiple absolute $\partial h/\partial t$ maps over the middle reach of the Congo near the city of Lisala using PALSAR images, and quantifying and characterizing the flow dynamics in the floodplains.

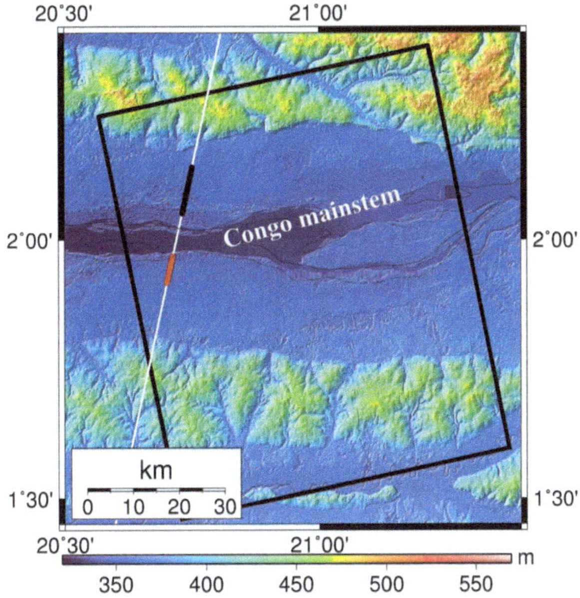

Fig. 4.1 Map of the study area with Envisat altimeter pass 300 (*white line*). The *red* and *black* dots represent the intersections of Envisat altimeter pass and the wetlands in the southern and northern floodplains, respectively. ALOS PALSAR image coverage is shown with *black* box. Background is topography from C-band SRTM DEM

4.2 Study Area

The study area is located along the middle reach of the Congo mainstem, as shown in Fig. 4.1, near the city of Lisala, the capital of Mongala District in the Democratic Republic of the Congo. Large islands can be seen in the river, which is typical in the Congo River. The floodplains next to the mainstem and the islands are covered with seasonally flooded forest (Hansen et al. 2008). The river surface slope is relatively high (6~8 cm/km) compared to other reaches of the Congo River (O'Loughlin et al. 2013).

4.3 Datasets

4.3.1 Envisat Altimetry

We used Envisat RA2 Geophysical Data Record (GDR) of pass 300 from Oct 2002 to Sep 2010 in this study. The revisit period of Envisat is 35 days. The Envisat GDR contains 18-Hz retracked range observations using OCEAN, ICE-1, ICE-2, and SEA-ICE retracking algorithms. The along-track sampling of 18-Hz Envisat altimetry measurements is about 350 m. Among these retracked measurements, we adopted ICE-1 retracked measurements which have been shown to provide better performance over inland water bodies (Frappart et al. 2006; Yuan et al. 2015). The instrument corrections, media corrections including troposphere and ionosphere corrections, and geophysical corrections including solid Earth and pole tides were also applied. For cycles 44–70, we applied 5.6 m level Ultra Stable Oscillator (USO) anomalies using European Space Agency's (ESA) algorithm in the form of a table (ESA 2007).

4.3.2 ALOS PALSAR

Sixteen fine-beam mode ALOS PALSAR images were used in this study, as summarized in Table 4.1. Seven of them were obtained with Fine Beam Single (FBS) mode (HH polarization) while the other nine were obtained with Fine Beam Dual (FBD) mode (both HH and HV polarizations). We adopted the HH polarization mode SAR images because it is known to be sensitive to water level change beneath the vegetation (Pope et al. 1997; Rosenqvist et al. 2007) and maintain better coherence for interferometric processing (Kim et al. 2009). It should be noted that the FBS and FBD images are generated with different bandwidths of 28 and 14 MHz, respectively (Shimada et al. 2009). Therefore, the spatial resolution in range direction of FBS image is two times higher than that of the FBD image. To obtain consistent resolution, the FBD images were oversampled by a factor of 2 in the range direction.

Table 4.1 List of PALSAR scenes used in this study

Scene ID	Operation mode	Date	Path	Frame	Polarization mode
ALPSRP074660020	FBD	2007/6/19	640	20	HH/HV
ALPSRP121630020	FBS	2007/8/4	640	20	HH
ALPSRP269250020	FBD	2007/9/19	640	20	HH/HV
ALPSRP262540020	FBS	2007/12/20	640	20	HH
ALPSRP101500020	FBS	2008/3/21	640	20	HH
ALPSRP222280020	FBD	2008/5/6	640	20	HH/HV
ALPSRP155180020	FBD	2008/6/21	640	20	HH/HV
ALPSRP114920020	FBS	2008/12/22	640	20	HH
ALPSRP081370020	FBD	2009/6/24	640	20	HH/HV
ALPSRP215570020	FBD	2009/9/24	640	20	HH/HV
ALPSRP249120020	FBS	2010/2/9	640	20	HH
ALPSRP182020020	FBD	2010/5/12	640	20	HH/HV
ALPSRP128340020	FBD	2010/6/27	640	20	HH/HV
ALPSRP088080020	FBD	2010/9/27	640	20	HH/HV
ALPSRP228990020	FBS	2010/12/28	640	20	HH
ALPSRP235700020	FBS	2011/2/12	640	20	HH

4.4 Results and Discussions

4.4.1 Inundated Areas from PALSAR Backscattering Coefficients

The energy backscattered to the SAR antenna is sensitive to surface roughness, dielectric constant, and terrain slope (Bayer et al. 1991; Mattia et al. 1997; Freeman and Durden 1998). Accordingly, SAR intensity images have been utilized to map inundated areas based on different radar backscattering mechanisms depending on different vegetation types. Over open water surface, the radar pulse follows a specular travel path, which results in little energy backscattered to the SAR antenna. In cases of dry vegetation, the backscattering is dominated by volume backscattering from the vegetation canopy. On the other hand, over inundated but not submerged vegetation, the water surface and vegetation trunk allow the radar pulse following a double bounce travel path to the antenna. This double-bounce backscattering from the inundated vegetation results in brighter SAR intensity, compared to nonflooded vegetation. This phenomenon has been reported by a number of studies, such as (Hess et al. 1995; Pope et al. 1997; Lu and Kwoun 2008). Based on the fact that variation of hydrologic condition influences the signal amplitude backscattered to the SAR antenna, SAR images have been used to delineate flooded and nonflooded areas in the wetlands (Hess et al. 1995, 2003; Rosenqvist and Birkett 2002; Lee et al. 2014).

To identify inundated areas, backscattering coefficient (σ_0) from PALSAR images was used in this study. After we generated single-look complex (SLC) images from SAR raw data, all of the SLC images were first coregistered to a common SLC image obtained on September 19 2007. Second, all of the SAR images were multilooked by a factor of 4×9 to reduce the speckle noise. Next, the multilooked SAR intensity images were geocoded using the 1-arcsec Shuttle Radar Topography Mission (SRTM) Digital Elevation Model (DEM). Finally, the backscattering coefficients were calculated using Equation (4.1) with the geocoded SAR images, following the method in (Yuan et al. 2015). A median filter of 5×5 was applied to further reduce the speckle noise.

$$\sigma_0 = 10\log_{10}\left(DN^2\right) + CF\ (\text{dB}) \tag{4.1}$$

where *DN* is the digital number of the geocoded SAR intensity image, and *CF* is the radiometric calibration factor. Value of *CF* is −51.8 for FBD HH data and −51.9 for FBS HH data (Werner et al. 2000).

Figure 4.2 shows the spatial and temporal variations of the calculated σ_0. The Congo mainstem shows constantly low σ_0 in all of the σ_0 images because of specular backscattering from open water surface. Significant temporal variations of σ_0 can be observed in the floodplains located along the mainstem. For example, SAR image obtained on December 20 2007 shows the highest σ_0 while SAR image obtained on February 11 2011 shows the lowest σ_0. Threshold classifier is a simple

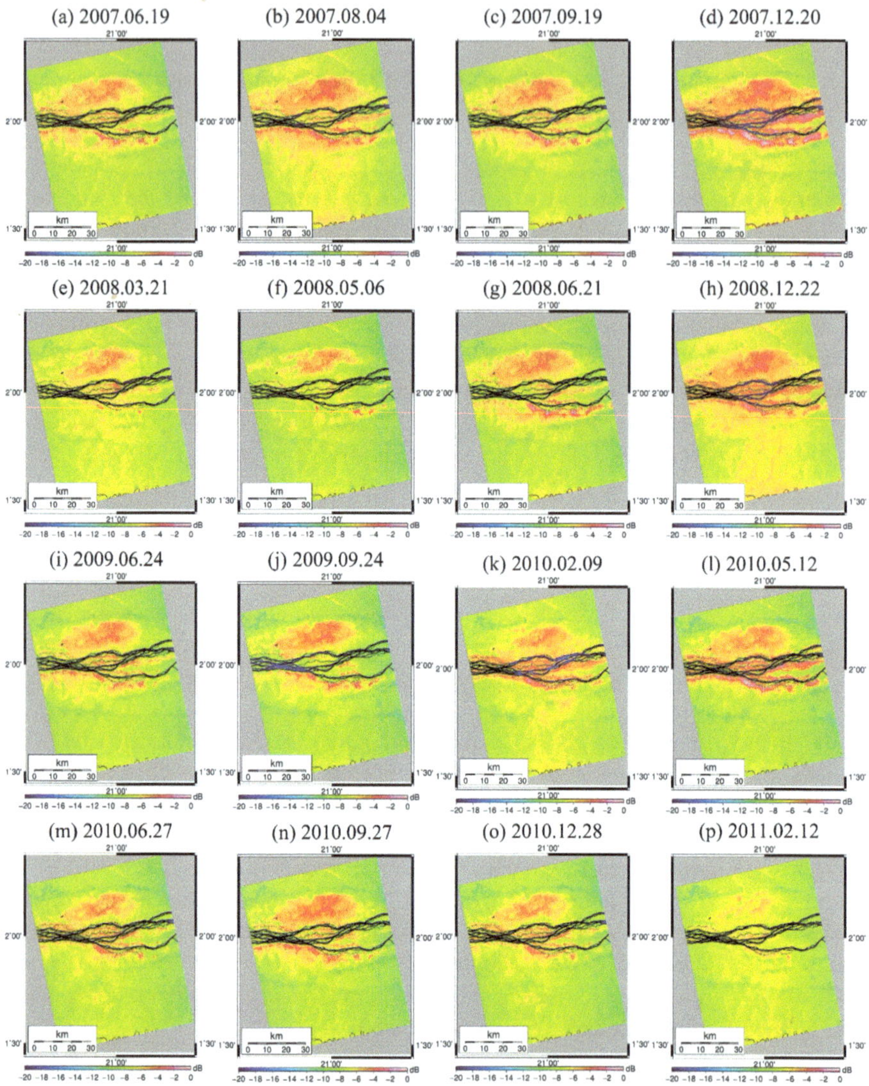

Fig. 4.2 Maps PALSAR backscattering coefficients showing seasonal variations over the floodplains

yet effective method to delineate inundation extents and vegetation types, which has been implemented in a number of studies, e.g. (Hess et al. 2003; Rosenqvist 2008; Lee et al. 2014). Here, we used a threshold of −6 dB to classify inundated and noninundated forest areas (Lee et al. 2014). The threshold of −6 dB has been determined based on the feasibility of generating water level change time series using Envisat altimetry, and used in the central Congo floodplain delineating the flooded and nonflooded forests with PALSAR ScanSAR images (Lee et al. 2014).

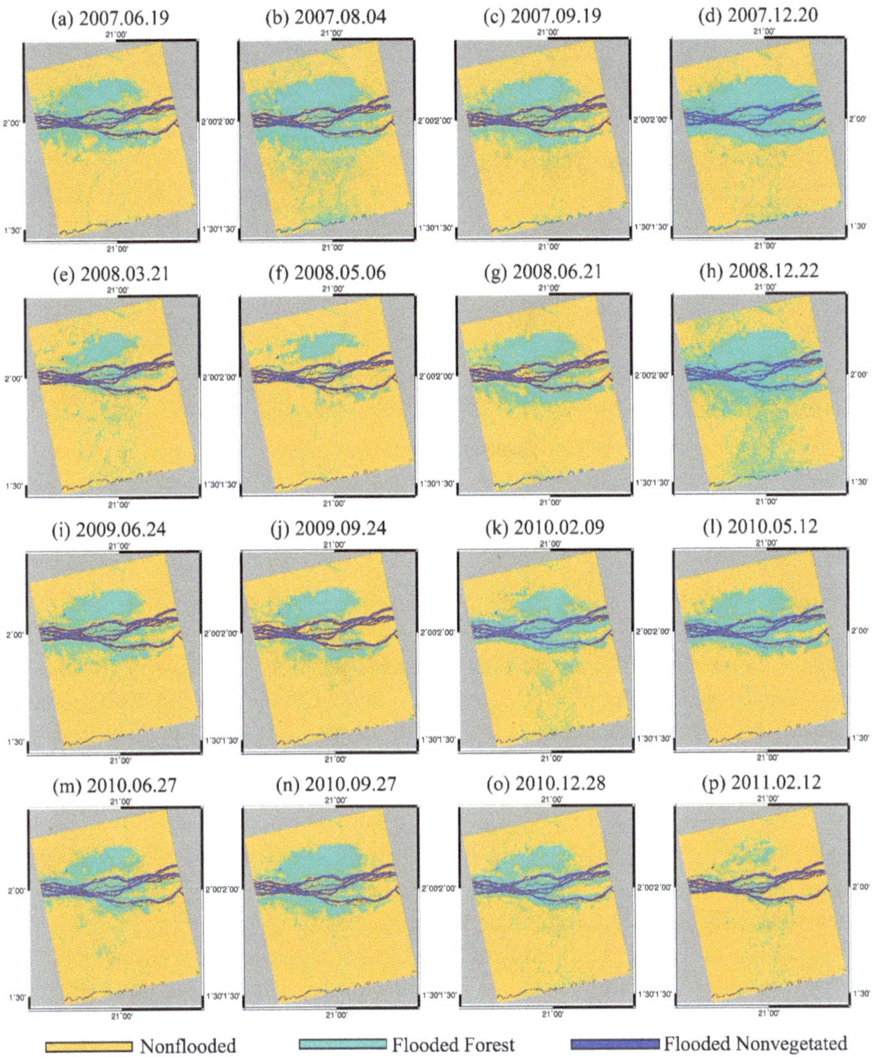

Fig. 4.3 Classification maps of flooded nonvegetated (*blue*), flooded forests (*cyan*), and nonflooded areas (*brown*)

Then, the SAR pixels with σ_0 lower than -14 dB were classified as the nonvegetated flooded areas (open water surface) as in (Hess et al. 2003). The estimated inundation maps are shown in Fig. 4.3. The temporal variation of total inundated areas is shown in Fig. 4.4. The maps of inundated areas will later be used to determine if the interferometric phases were dominated by water level change or atmospheric artifacts.

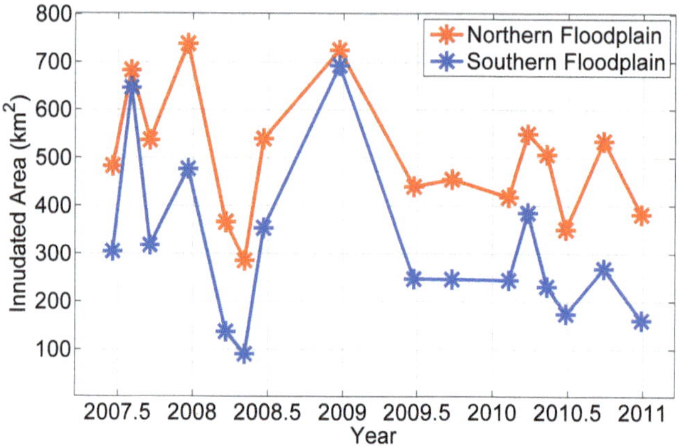

Fig. 4.4 Temporal variation of total inundated areas

4.4.2 InSAR Processing and Coherence Analysis

Two-pass repeat interferometry method has been applied to process the SAR images. The interferometric processing uses two SAR images, obtained from different times with slightly different imaging position. The two SAR images were first coregistered based on the correlation between the SAR intensity images. An interferogram was then generated by multiplying the first SAR image (master image) with the complex conjugate of the second coregistered SAR image (slave image). Flat earth phase was removed with the baseline and its variation throughout the interferogram using the satellite orbit information. The topographic phase in the interferogram was simulated using 1 arc-sec resolution SRTM DEM data and then was subtracted from the interferogram. After removing the topographic component from the interferogram, a differential interferogram was generated which represents the relative $\partial h/\partial t$ in the floodplains. The residual phase trend throughout the differential interferogram caused by inaccurate orbit information was estimated using a second-order polynomial and then removed from the differential interferogram.

After all the processes afore mentioned, 15 wrapped differential interferograms were generated as shown in Fig. 4.5. The interferograms were generated using the Small Temporal Baseline Subset (STBAS) strategy (Hong et al. 2010). The STBAS method generates interferograms with small temporal baseline criterion in order to obtain better coherence in wetland environment (Kim et al. 2013). Following the STBAS method, we generated the interferograms using every consecutive SAR image pairs except one interferogram shown in Fig. 4.5d generated using SAR images obtained on December 20 2007 and June 21 2008. The SAR image obtained on June 21 2008 was used as the slave image instead of the SAR images obtained on March 21 2008 or May 6 2008 which would have led to the interferogram with shorter temporal baseline. Because the SAR images obtained on March 21 2008 or

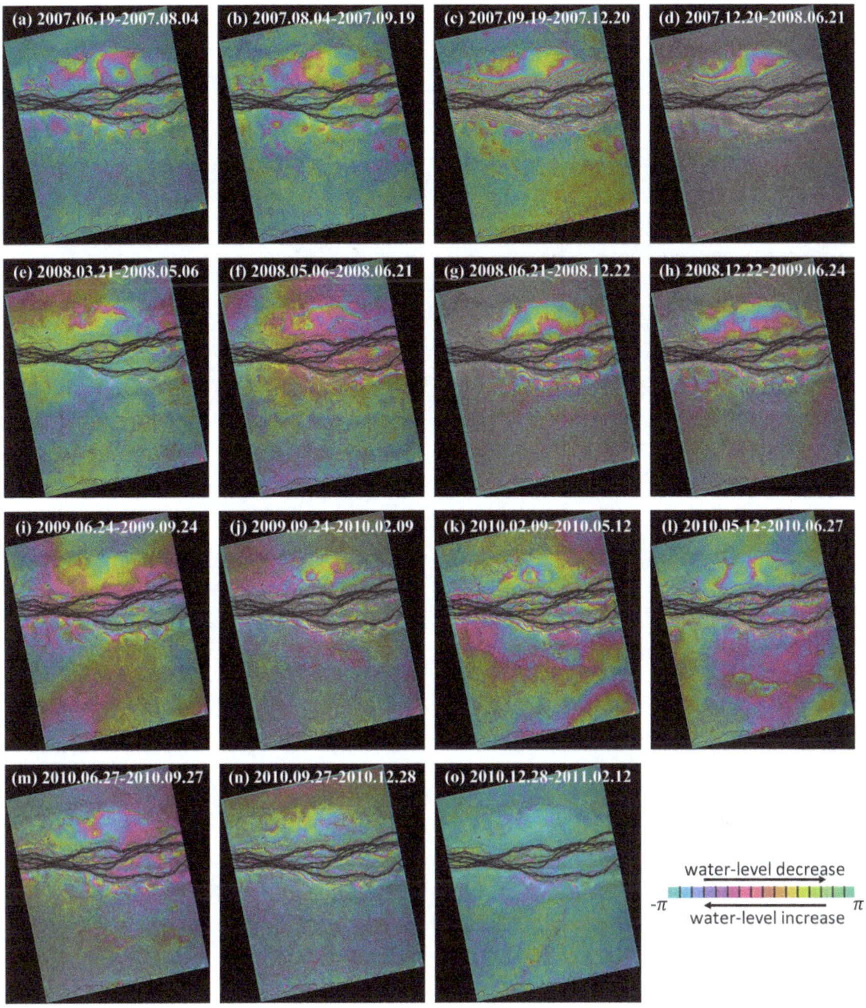

Fig. 4.5 Wrapped PALSAR differential interferograms

May 6 2008 represent the dry season with least inundated extents (see Fig. 4.3), the interferograms generated with those SAR images as the slave images lost coherence. Thus, we have generated the interferogram with the SAR image obtained on June 21 2008 as the slave image instead. The temporal and perpendicular baselines of the interferograms and the coherences over the floodplains are summarized in Table 4.2 (the alphabet in the first column is used to refer each interferogram hereafter). The coherence values were calculated only over the floodplains, located in the south and north of the mainstem. The extents of the southern and northern floodplains were delineated using the SAR image obtained on December 20 2007, which has the highest σ_0. The coherences listed in Table 4.2 are overall higher than

Table 4.2 List of ALOS PALSAR interferometric pairs

Pair	Master date	Slave date	Temporal baseline (days)	Perpendicular baseline (m)	Mean coherence in southern floodplain	Mean coherence in northern floodplain
(a)	20070619	20070804	46	273.20	0.34	0.33
(b)	20070804	20070919	46	−449.90	0.29	0.25
(c)	20070919	20071220	92	−36.37	0.29	0.29
(d)	20071220	20080621	92	−248.69	0.20	0.19
(e)	20080321	20080506	46	579.67	0.23	0.28
(f)	20080506	20080621	46	456.64	0.26	0.31
(g)	20080621	20081222	184	−828.84	0.22	0.23
(h)	20081222	20090624	184	397.32	0.19	0.20
(i)	20090624	20090924	92	−270.75	0.28	0.38
(j)	20090924	20100209	138	275.42	0.23	0.23
(k)	20100209	20100512	92	283.69	0.20	0.20
(l)	20100512	20100627	46	114.33	0.28	0.32
(m)	20100627	20100927	92	388.08	0.25	0.30
(n)	20100927	20101228	92	−544.35	0.24	0.24
(o)	20101228	20110212	46	364.13	0.25	0.26

0.2 (except for the interferograms (d) and (h)), indicating that the interferometric phases are overall well maintained using STBAS.

The fringe patterns of the interferograms in the wetlands can be characterized by spatial patterns of water level changes and atmospheric artifacts. The interferometric phases due to water level changes and atmospheric artifacts have similar wavelength (Hong et al. 2010). Thus, the approach used in Mora et al. (2002) to remove the atmosphere artifacts by applying the high-pass filter in time domain is not possible in wetlands environment. However, we can use the σ_0 images to determine if the fringes in the interferogram are likely to be caused by water level change or atmospheric delay (Lu and Kwoun 2008). In addition, it is expected that the fringes due to water level change should exhibit seasonal dynamics which is different from the atmospheric artifacts, characterized by high spatial correlation but low temporal correlation (Mora et al. 2002). Based on these two criteria, three fringe patterns caused by water level changes were identified.

The first pattern is the fringes parallel to the Congo mainstem flow direction observed in the southern floodplain, as shown in Fig. 4.5c, d, g, h, j–l, and in the northern floodplain, as shown in Fig. 4.5c, d, j, and k. These fringes are likely due to water level changes, not atmosphere delays. First of all, the atmosphere artifacts cannot result in such repeated pattern along the river. In addition, the location of the fringes corresponds to the location with significant σ_0 variations, implying variations in hydrologic condition. Furthermore, this parallel fringe patterns are commonly observed in a number of interferograms along the mainstem, which further indicates that it is not due to the atmosphere effect. The second fringe pattern that could be due to water level change is the broad flooding pattern perpendicular to the Congo mainstem in the northern floodplain which is observed in interferograms shown in Fig. 4.5a, b, and l. Similar to the first fringe pattern mentioned earlier, the atmosphere effect cannot result in this repeated fringe pattern. The third phase that could be due to water level change is the fringes observed along the boundary between the northern floodplain and the upland, as can be seen from Fig. 4.5f and m. The broad radial phase patterns in southwestern edge of Fig. 4.5i, and southeastern edge of Fig. 4.5k, and the broad red phase pattern in Fig. 4.5l, are likely caused by atmospheric delay, not water level changes because of negligible variation in σ_0 as shown in Fig. 4.2.

4.4.3 InSAR-Derived $\partial h/\partial t$

Measurements of $\partial h/\partial t$ from InSAR were obtained by unwrapping the differential interferograms in Fig. 4.5. Six interferograms (a, c, d, f, l, m) were selected for unwrapping as examples for each fringe pattern identified in Sect. 4.4.2. The interferograms (c) and (d) were selected for the first phase pattern (dense fringes parallel to the mainstem), (a) and (l) for the second phase pattern (broad fringes perpendicular to the mainstem), and (f) and (m) for the third fringes pattern observed along the boundary of northern floodplain.

Because InSAR-derived $\partial h/\partial t$ only represents spatially relative water level changes in the wetlands (Alsdorf et al. 2007; Kim et al. 2009; Lu et al. 2009), a vertical reference (or offset) is necessary to convert the relative water level changes from InSAR to absolute water level changes. Since there is no in situ gauge in the study area, the offset estimation has been performed using two different methods. The first method is to use the water level change observed from Envisat altimetry to estimate the offset for the interferograms in Fig. 4.5c and d. Profile comparisons of Envisat altimetry and InSAR measurements are shown in Fig. 4.6 in order to examine the feasibility of using altimetry-derived water level changes to estimate the offsets. The Envisat altimetry measurements have been interpolated to the SAR acquisition times. However, it should be noted that satellite altimetry measurements in dry season can be erroneous such as the black dots in Fig. 4.6d, which have been excluded from the offset estimation. The comparison shows a good agreement between altimetry and interferograms (c, d) from high water level season. The R-squared values are high varying from 0.8 to 0.93 and the root-mean square differences are from 7 to 11 cm. After unwrapping these interferograms and applying the offsets estimated from Fig. 4.6, absolute $\partial h/\partial t$ maps were generated as shown in Fig. 4.7a and b. The absolute water level change maps in Fig. 4.7a and b suggest that the proximal (channel marginal) floodplain close to the mainstem has the largest water level change during high water season. The absolute $\partial h/\partial t$ range up to from 1.2 to 1.4 m over both the southern and northern floodplains.

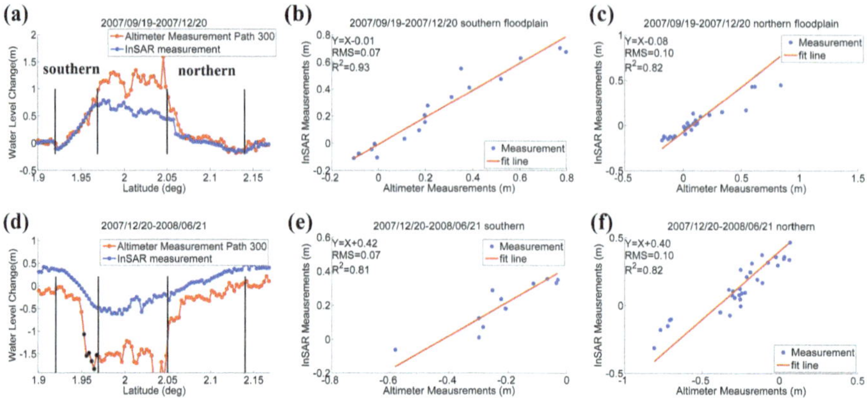

Fig. 4.6 (**a**) and (**d**) are the profiles of Envisat altimetry-derived and InSAR-derived water level changes extracted along the Envisat altimetry ground track from interferograms shown in Fig. 4.5c and d. Measurements within the southern floodplain and northern floodplain are marked as *red* dots and *black* dots in Fig. 4.1. (**b**) and (**e**) show the statistical comparison between them over the southern floodplain, and (**c**) and (**f**) illustrate the comparison over the northern floodplain

For the interferograms (a, f, l, m), since they were not obtained in high water season, the Envisat altimetry measurements become too noisy to be used to estimate the offsets. However, by visually examining the two SAR images (obtained on June 19 2007 and August 4 2007) used to generate the interferogram (a), the SAR image obtained on August 4 2007 has higher σ_0 than the SAR image obtained on June 19 2007. The higher σ_0 is expected to be due to higher water level. Thus, the absolute $\partial h/\partial t$ is also expected to show positive (or increasing) water level change. With this positive water level change direction, we generated the absolute $\partial h/\partial t$ map in Fig. 4.7c by assigning 0 to the pixel which has the lowest $\partial h/\partial t$. In this case, as it cannot be assumed that the floodplains have been completely drained on June 19 2007 based on its σ_0 image, we intend to use this map only to examine whether $\partial h/\partial t$ has increased or decreased, and the directions of water flow. Similarly, we assigned 0 to the pixel with the lowest $\partial h/\partial t$ for the interferograms (f, m), and obtained the absolute $\partial h/\partial t$ maps shown in Fig. 4.7e and f. For interferogram (l), the SAR image obtained on May 12 2010 has higher σ_0 than the SAR image obtained on June 27 2010. With decreasing σ_0, interferogram (l) is expected to show water level decreasing. Accordingly, we assigned 0 to the pixel with the highest $\partial h/\partial t$ for interferogram (l) to obtain the absolute $\partial h/\partial t$ map shown in Fig. 4.7d. Although Fig. 4.7c–f are not the true absolute $\partial h/\partial t$ maps, they still reflect the relative $\partial h/\partial t$ within the floodplain. By visually observing the values in Fig. 4.7c–f, the relative $\partial h/\partial t$ in the floodplain are at decimeter level.

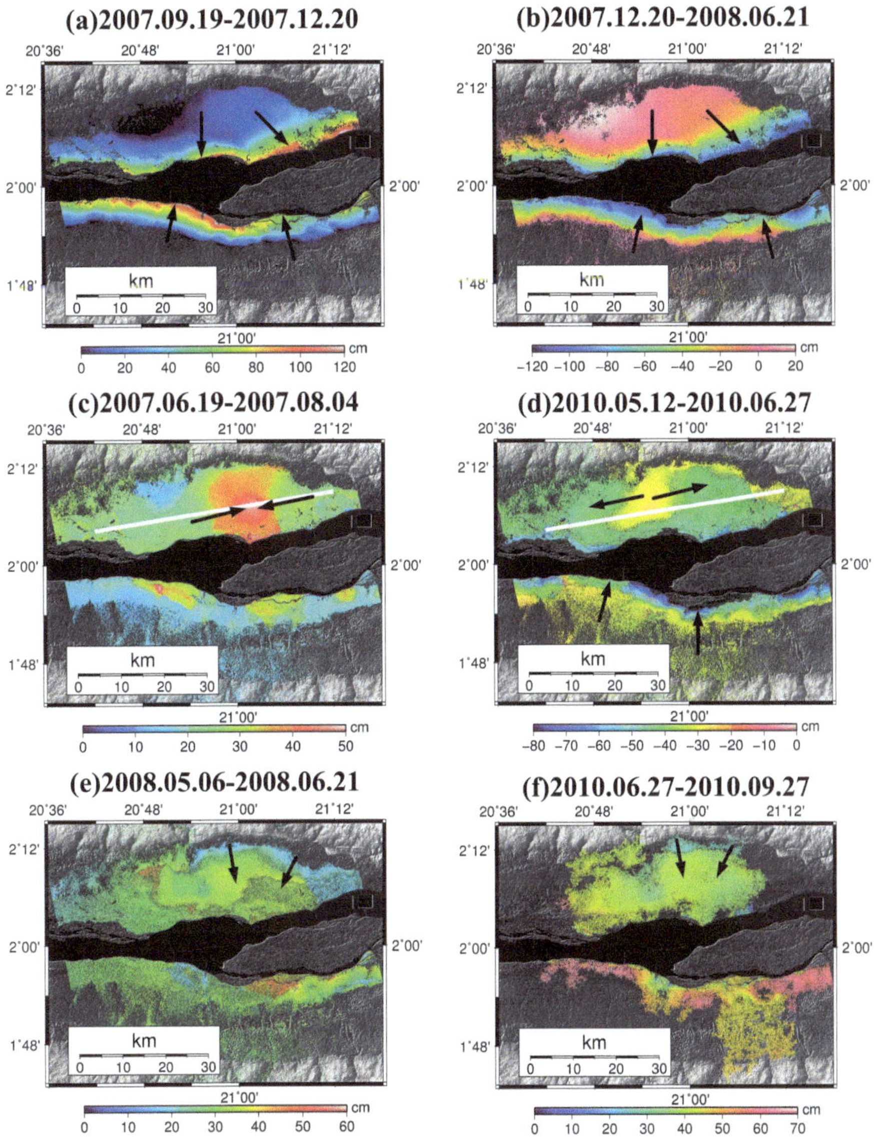

Fig. 4.7 Absolute water level change map over the interfluvial wetlands from integration of Envisat altimetry and PALSAR InSAR (**a**) 2007.09.19-2007.12.20, (**b**) 2007.12.20-2008.06.21. (**c**)-(**f**) show the absolute $\partial h/\partial t$ maps by assigning 0 to pixel with the highest or lowest water level changes. *Black arrows* show the water flow direction based on mass continuity

4.4.4 Hydraulic Processes from Absolute $\partial h/\partial t$

Based on the six absolute $\partial h/\partial t$ maps shown in Fig. 4.7, the water flow hydraulics in the Congo floodplain can be investigated based on mass continuity as in Alsdorf et al. (2007) and Jung et al. (2010). It is based on the fact that the location with greater $\partial h/\partial t$ receives more water than the location with smaller $\partial h/\partial t$. Thus, the water should flow from smaller $\partial h/\partial t$ to greater $\partial h/\partial t$.

During high water season as shown in Fig. 4.7a and b, both in the northern and southern floodplains, there is an overall increasing $\partial h/\partial t$ trend from upland to the mainstem. In other words, the proximal floodplain closer to the mainstem has greater $\partial h/\partial t$ than the distal (valley marginal) floodplain closer to the upland. Back flow arrows in Fig. 4.7a and b are based on the continuity with directions pointing toward areas of greater water accumulation during water increasing times in Fig. 4.7a and pointing toward evacuation during water decreasing times in Fig. 4.7b.

During low water season, we observed broad and diffuse pattern of $\partial h/\partial t$ as discussed in Sect. 4.4.3. Hence, no obvious hydraulic processes based on subtle $\partial h/\partial t$ can be observed. Figure 4.7c and d show broad water level change pattern parallel to the Congo mainstem in the northern floodplain. Figure 4.7e and f show water level change at the boundary in the northern floodplain, showing water flow from upland toward the floodplain.

As in Fig. 4.8, we also examined the relationship between our $\partial h/\partial t$ measurements and the 1 arc-sec resolution SRTM topography for the broad and diffuse $\partial h/\partial t$ patterns observed in Fig. 4.7c and d. The $\partial h/\partial t$ map, shown in Fig. 4.7c, during the water filling period shows a flow convergence toward the red circular region. The $\partial h/\partial t$ and SRTM profile comparison in Fig. 4.8a shows that the greater $\partial h/\partial t$ values are within the topographic depression area. On the contrary, the $\partial h/\partial t$ map in Fig. 4.7d, during the water draining period shows a flow divergence from the identical depression area based on Fig. 4.8b.

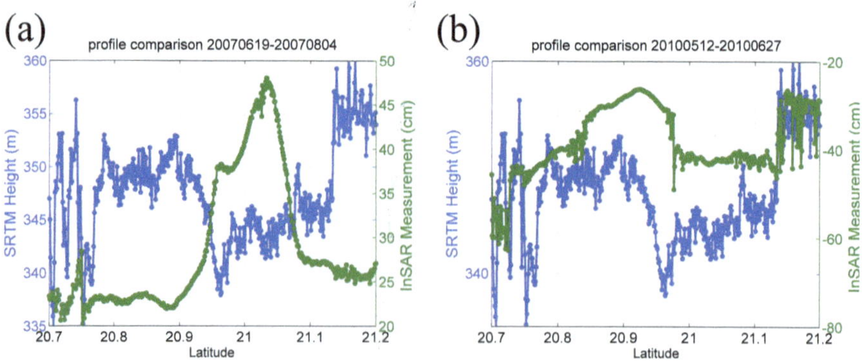

Fig. 4.8 Comparisons between absolute $\partial h/\partial t$ and SRTM topography extracted along the profiles shown as *white* lines in Fig. 4.7c and d

4.5 Conclusion

A stack of PALSAR interferograms was generated over the Congo floodplain near the city of Lisala. Two absolute water level change maps by integrating InSAR-derived relative $\partial h/\partial t$ and Envisat-derived absolute $\partial h/\partial t$ were generated. The proximal floodplain close to the mainstem has greater absolute $\partial h/\partial t$ than the distal floodplain close to the upland. Water level change difference within the floodplains can reach up to 1.2–1.4 m in high water season while the difference is at decimeter level in low water season. The absolute $\partial h/\partial t$ maps have also been used to investigate the water flow hydraulics in the Congo floodplain based on the mass continuity. The absolute $\partial h/\partial t$ maps in low water season suggest that the water flow is not well confined and has broad and diffuse pattern. On the other hand, the absolute $\partial h/\partial t$ maps in high water season show rapid spatial variation indicating water flow from the floodplain toward the mainstem. The relationship between the absolute $\partial h/\partial t$ in low water season and SRTM elevation shows the low topography depression area has the greater water level changes during both the water filling and draining periods.

Acknowledgements This research was supported by grants from NASA's Earth and Space Science Fellowship (NNX15AM70H) and NASA's New Investigator Program (NNX14AI01G). ALOS PALSAR data were provided by Alaska Satellite Facility (ASF) and Japan Aerospace Exploration Agency (JAXA). Envisat altimetry data were provided by European Space Agency (ESA). Some of the figures were prepared using Generic Mapping Tool (GMT) graphics package.

References

Alsdorf DE, Smith LC, Melack JM (2001) Amazon floodplain water level changes measured with interferometric SIR-C radar. IEEE Trans Geosci Remote Sens 39(2):423–431. doi:10.1109/36.905250

Alsdorf D, Bates P, Melack J, Wilson M, Dunne T (2007) Spatial and temporal complexity of the Amazon flood measured from space. Geophys Res Lett 34(L08402):1–5. doi:10.1029/2007GL029447

Alsdorf D, Beighley E, Laraque A, Lee H, Tshimanga R, O'Loughlin F, Mahe G, Bienvenu D, Moukandi G, Spencer RGM (2016) Opportunities for hydrologic research in the Congo Basin. Rev Geophys 54:378–409. doi:10.1002/2016RG000517

Barbier EB (1994) Valuing environmental functions: tropical wetlands. Land Econ 70 (2):155–173. doi:10.2307/3146319

Bayer T, Winter R, Schreier G (1991) Terrain influences in SAR backscatter and attempts to their correction. IEEE Trans Geosci Remote Sens 29(3):451–462. doi:10.1109/36.79436

Esa (2007) ENVISAT RA2/MWR Product Handbook. Report

Frappart F, Calmant S, Cauhope M, Seyler F, Cazenave A (2006) Preliminary results of ENVISAT RA-2-derived water levels validation over the Amazon basin. Remote Sens Environ 100 (2):252–264. doi:10.1016/j.rse.2005.10.027

Freeman A, Durden SL (1998) A three-component scattering model for polarimetric SAR data. IEEE Trans Geosci Remote Sens 36(3):963–973. doi:10.1109/36.673687

Hansen MC, Roy DP, Lindquist E, Adusei B, Justice CO, Altstatt A (2008) A method for integrating MODIS and Landsat data for systematic monitoring of forest cover and change in the Congo Basin. Remote Sens Environ 112(5):2495–2513. doi:10.1016/j.rse.2007.11.012

Hayashi M, Van Der Kamp G, Rudolph DL (1998) Water and solute transport between a prairie wetland and adjacent upland, 1. Water balance. J Hydrol 207:42–55

Hess LL, Melack JM, Melack JM, Filoso S, Wang Y, Wang Y (1995) Delineation of inundated area and vegetation along the Amazon floodplain with the SIR-C synthetic aperture radar. IEEE Trans Geosci Remote Sens 33(4):896–904. doi:10.1109/36.406675

Hess LL, Melack JM, Novo EMLM, Barbosa CCF, Gastil M (1999) Monitoring flooding and vegetation on seasonally inundated floodplains with multi-frequency polarimetric synthetic aperture radar. Remote Sens Environ 87:404–428. doi:10.1016/j.rse.2003.04.001

Hess LL, Melack JM, Novo EMLM, Barbosa CCF, Gastil M (2003) Dual-season mapping of wetland inundation and vegetation for the central Amazon basin. Remote Sens Environ 87:404–428. doi:10.1016/j.rse.2003.04.001

Hong S-H, Wdowinski S, Kim S-W, Won J-S (2010) Multi-temporal monitoring of wetland water levels in the Florida Everglades using interferometric synthetic aperture radar (InSAR). Remote Sens Environ 114(11):2436–2447. doi:10.1016/j.rse.2010.05.019

Hughes RH, Hughes JS (1992) A directory of African wetlands. IUCN, Gland, Switzerland and Cambridge, UK/UNEP, Nairobi, Kenya/WCMC, Cambridge, UK

Jung CH, Hamski J, Durand M, Alsdorf D, Hossain F, Lee H, Azad Hossain AKM, Hasan K, Khan AS, Zeaul Hoque AKM (2010) Characterization of complex fluvial systems using remote sensing of spatial and temporal water level variations in the Amazon, Congo, and Brahmaputra rivers. Earth Surf Process Landf 35(3):294–304. doi:10.1002/esp.1914

Keddy PA, Fraser LH, Solomeshch AI, Junk WJ, Campbell DR, Arroyo MTK, Alho CJR (2009) Wet and Wonderful: the world's largest wetlands are conservation priorities. Bioscience 59 (1):39–51. doi:10.1525/bio.2009.59.1.8

Kim JW, Lu Z, Lee H, Shum CK, Swarzenski CM, Doyle TW, Baek SH (2009) Integrated analysis of PALSAR/Radarsat-1 InSAR and ENVISAT altimeter data for mapping of absolute water level changes in Louisiana wetlands. Remote Sens Environ 113:2356–2365. doi:10.1016/j.rse.2009.06.014

Kim S-W, Wdowinski S, Amelung F, Dixon TH, Won J-S (2013) Interferometric coherence analysis of the Everglades wetlands, South Florida. IEEE Trans Geosci Remote Sens 51 (12):5210–5224

Kim J-W, Lu Z, Jones JW, Shum CK, Lee H, Jia Y (2014) Monitoring Everglades freshwater marsh water level using L-band synthetic aperture radar backscatter. Remote Sens Environ 150:66–81. doi:10.1016/j.rse.2014.03.031

Lee HH, Jung C, Yuan T, Beighley RE, Duan J (2014), Controls of terrestrial water storage changes over the Central Congo Basin determined by integrating PALSAR ScanSAR, Envisat Altimetry, and GRACE data. In: Remote sensing of the terrestrial water cycle. Wiley, Hoboken, NJ, pp 115–129

Lee H, Yuan T, Jung HC, Beighley E (2015) Mapping wetland water depths over the central Congo Basin using PALSAR ScanSAR, Envisat altimetry, and MODIS VCF data. Remote Sens Environ 159:70–79. doi:10.1016/j.rse.2014.11.030

Lehner B, Verdin K, Jarvis A (2008) New global hydrography derived from spaceborne elevation data. Eos Trans Am Geophys Union 89(10):93–94. doi:10.1029/2008EO100001

Lu Z, Kwoun OI (2008) Radarsat-1 and ERS InSAR analysis over southeastern coastal Louisiana: implications for mapping water-level changes beneath swamp forests. IEEE Trans Geosci Remote Sens 46(8):2167–2184. doi:10.1109/TGRS.2008.917271

Lu Z, Kim J-W, Lee H, Shum CK, Duan J, Ibaraki M, Akyilmaz O, Read C-H (2009) Helmand river hydrologic studies using ALOS PALSAR InSAR and ENVISAT altimetry. Mar Geod 32 (3):320–333. doi:10.1080/01490410903094833

Mattia F, Le Toan T, Souyris JC, De Carolis C, Floury N, Posa F, Pasquariello NG (1997) The effect of surface roughness on multifrequency polarimetric SAR data. IEEE Trans Geosci Remote Sens 35(4):954–966. doi:10.1109/36.602537

Mitsch WJ, Gosselink JG (2007) Wetlands, 4th edn. Wiley, New York

Mora O, Lanari R, Mallorqui JJ, Berardino P, Sansosti E (2002) A new algorithm for monitoring localized deformation phenomena based on small baseline differential SAR interferograms. IEEE Int Geosci Remote Sens Symp 2(11):2375–2383. doi:10.1109/IGARSS.2002.1025900

Neue HU, Gaunt JL, Wang ZP, Becker-Heidmann P, Quijano C (1997) Carbon in tropical wetlands. Geoderma 79(1–4):163–185. doi:10.1016/S0016-7061(97)00041-4

O'Loughlin F, Trigg MA, Schumann GJP, Bates PD (2013) Hydraulic characterization of the middle reach of the Congo River. Water Resour Res 49(8):5059–5070. doi:10.1002/wrcr.20398

Pope KO, Rejmankova E, Paris JF, Woodruff R (1997) Detecting seasonal flooding cycles in marshes of the Yucatan Peninsula with SIR-C polarimetric radar imagery. Remote Sens Environ 59(2):157–166. doi:10.1016/S0034-4257(96)00151-4

Rosenqvist A (2008) Mapping of seasonal inundation in the Congo river basin using PALSAR ScanSAR, ALOS PI Symposium, Rhodes, Greece, Nov. 3–7, 2008

Rosenqvist A, Birkett CM (2002) Evaluation of JERS-1 SAR mosaics for hydrological applications in the Congo River Basin. Int J Remote Sens 23(7):1283–1302. doi:10.1080/01431160110092902

Rosenqvist A, Shimada M, Ito N, Watanabe M (2007) ALOS PALSAR: a pathfinder mission for global-scale monitoring of the environment. IEEE Trans Geosci Remote Sens 45(11):3307–3316. doi:10.1109/TGRS.2007.901027

Shimada M, Isoguchi O, Tadono T, Isono K (2009) PALSAR radiometric and geometric calibration. IEEE Trans Geosci Remote Sens 47(12):3915–3932

Wdowinski S, Kim SW, Amelung F, Dixon TH, Miralles-Wilhelm F, Sonenshein R (2008) Space-based detection of wetlands' surface water level changes from L-band SAR interferometry. Remote Sens Environ 112:681–696. doi:10.1016/j.rse.2007.06.008

Werner C, Wegmüller U, Strozzi T, Wiesmann A (2000) Gamma SAR and interferometric processing software. In: Proceedings of the ERS-Envisat Symposium, vol 461. Citeseer, Gothenburg, Sweden

Yuan T, Lee H, Jung H (2015) Toward estimating wetland water level changes based on hydrological sensitivity analysis of PALSAR backscattering coefficients over different vegetation fields. Remote Sens 7(3):3153–3183. doi:10.3390/rs70303153

Chapter 5
Optical and Physical Methods for Mapping Flooding with Satellite Imagery

Jessica Fayne, John Bolten, Venkat Lakshmi, and Aakash Ahamed

5.1 Introduction

Flood and surface water mapping is becoming increasingly necessary, as extreme flooding events worldwide can damage crop yields and contribute to billions of dollars economic damages as well as social effects including fatalities and destroyed communities (Xaio et al. 2004; Kwak et al. 2015; Mueller et al. 2016). Utilizing earth observing satellite data to map standing water from space is indispensable to flood mapping for disaster response, mitigation, prevention, and warning (McFeeters 1996; Brakenridge and Anderson 2006). Since the early 1970s (Landsat, USGS 2013), researchers have been able to remotely sense surface processes such as extreme flood events to help offset some of these problems. Researchers have demonstrated countless methods and modifications of those methods to help increase knowledge of areas at risk and areas that are flooded using remote sensing data from optical and radar systems, as well as free publically available and costly commercial datasets.

In 1972, Landsat 1 also called Earth Resources Technology Satellite-1 (ERTS-1) was launched, prompting an explosion of literature on the ability to map surface

J. Fayne, B.A., M.S. (✉)
Department of Earth and Ocean Sciences, University of South Carolina,
Columbia, SC, USA
e-mail: jessicavfayne@gmail.com

J. Bolten • A. Ahamed, M.Sc.
Hydrological Sciences Laboratory, NASA Goddard Space Flight Center,
Greenbelt, MD, USA

V. Lakshmi
Department of Geophysics, Stanford University, Stanford, CA, USA

School of Earth, Ocean and Environment, University of South Carolina,
Columbia, SC, USA

V. Lakshmi (ed.), *Remote Sensing of Hydrological Extremes*, Springer Remote Sensing/Photogrammetry, DOI 10.1007/978-3-319-43744-6_5

processes from space using wavelengths in the optical and near-infrared spectrum (Irons et al. 2016). Landsat 1 paved the way for the Landsat sensor series (1972–most recent launch of Landsat 8 in 2013), along with many other optical and radar surface monitoring sensors and sensor series, including Advanced Very High Resolution Radiometer (AVHRR); Moderate Resolution Imaging Spectroradiometer (MODIS); Advanced Spaceborne thermal Emission and Reflection Radiometer (ASTER); Visible Infrared Imager Radiometer Suite (VIIRS); Satellite Pour l'Observation de la Terre (SPOT); Sentinel, Advanced Land Observing Satellite (ALOS); Envisat; Radarsat; and Soil Moisture Active Passive (SMAP).

Earth observing satellites are able to capture images of the earth at varying spatial scales, and with different orbital periods, making each satellite sensor and dataset unique. Optical sensors may have the ability to capture surface reflectance data from visible blue, green, and red wavelengths, as well as emissivity data through infrared wavelengths. Though a variety of wavelengths and bandwidths are available, the two bandwidths found in many sensors are centered around the red wavelength (~0.65 nm) and near-infrared wavelength (~0.85 nm) (Berkeley Lab 2016), with varying bandwidths. These two bands are universally used in surface monitoring studies including mapping vegetation phenology, surface water and flooding, snowmelt, and drought monitoring (Rouse et al. 1973; Colwell 1974; Tucker 1979; Song et al. 2004; Pettorelli 2009; Hasan 2011; Gopinath et al. 2014; Abbas 2014). Infrared wavelengths are key for water studies due to the high absorption over water, and relative ease of identification beside dry land (Frazier 2000; Lei 2009).

The availability of Landsat/ERTS-1 and AVHRR-1 data allowed scientists (Rouse et al. 1973; Colwell 1974; Tucker 1979) to develop an algorithm to quantify vegetation change over time, the Normalized Difference Vegetation Index (NDVI). Because the index uses near-infrared and red wavelengths, it can be universally applied to most sensor datasets. In addition, the exploitation of the difference between the near-infrared and red wavelengths allows NDVI to be applied to other studies such as water detection (Lei 2009), and it has inspired similar spectral indices specifically for water mapping such as the Normalized Difference Water Index (NDWI) (Gao 1996; McFeeters 1996), the Vegetation Supply Water Index (VSWI) (Cai et al. 2011; Abbas et al. 2014), and the Normalized Difference Pond Index (NDPI) (Lacaux et al. 2006).

While these indices may be useful for analyzing surface conditions, they are also subject to errors introduced by input datasets. Some sensor-based errors include scan line errors (SLC-Off Products, 2013), scan angle errors (TIRS SSM Anomaly, 2015), and edge of swath pixel-bowtie effects (NOAA-NESDIS VIIRS User's Guide 2013), while atmospheric conditions such as cloud coverage, cloud shadow, haze, pollution, scattered light from ground-based reflective objects, and atmospheric scattering may render many images unsuitable for analysis (Anderson et al. 2007; Holben 2007; Pettorelli et al. 2009). In areas where cloud coverage is pervasive, the use of cloud-penetrating radar is ideal, allowing data on surface texture to be collected instead of reflectance or emissivity (Brivio et al. 2002; Parinussa et al. 2016). When radar data is unavailable, preprocessing methods such as regression (Swets et al. 1999; Zhang et al. 2003), curve fitting (van Dijk et al. 1987), and maximum value (Holben 2007) formulas as well as cloud masking techniques (Fayne et al. 2015) have been implemented alongside atmospheric correction (University of California Berkeley College of Natural Resources) to determine the likely pixel value.

The sensors chosen for flood mapping efforts largely depend on the goals of the individual project, as well as sensor availability and funding. Optical sensors may be limited to surface inundation detection, while radar sensors may to determine water depths (Smith 1997; Hess et al. 2003). The ability to see fine details on the ground, known as spatial resolution, and to get data in a timely manner, known as temporal resolution, are important questions asked in remote sensing research. Studies may choose to sacrifice temporal resolution in favor of spatial resolution to create risk maps (Mueller et al. 2016; Revilla-Romero et al. 2015), while others may create high temporal resolution near real-time mapping products (Nigro et al. 2014). Further, studies may require the use of elevation models (Gallant and Dowling 2003; Guerschman et al. 2011), land cover datasets (Townsend and Walsh 1998; Gallant and Dowling 2003; Sun et al. 2012), or hydrological modeling software (Knebl et al. 2005) to obtain a certain level of precision or accuracy.

This chapter will review methods for mapping floods and open water using spectral formulas and statistical methods commenting on false color composite techniques with optical data, physical models using radar and ancillary datasets such as land cover maps and digital elevation models (DEMs), and will conclude with a look into the future of flood mapping techniques and applications. Some methods may be demonstrated with MODIS Terra 250- and 500-m data (path/row tile H28v07) (NASA-LPDAAC; USGS-EROS 2016) over the Lower Mekong Basin (LMB) to demonstrate visual impacts of the differences over the same study area. The demonstration area is located in Cambodia and Southern Vietnam, and the focus will be around the Tonle Sap Lake and Mekong Delta. This region was chosen because the LMB experiences monsoonal flooding between May and December, the MODIS data and a modified flood extent polygon (UNITAR-UNOSAT 2013) used in the example figures are observed during October 2013 (Fig. 5.1).

5.2 Optical Sensors

Flood mapping research using optical and near-infrared sensors may use a combination of statistics and empirical formulas to measure flood extent, such as spectral indices and single-band thresholds. To understand the capabilities and limitations of spectral indices, researchers must consider atmospheric penetration at certain wavelengths and the product availability of the desired wavelengths.

The United States Geological Survey (USGS) has produced an online Spectral Characteristics Viewer (USGS 2014), graphing the spectral response patterns of nine different minerals, nine vegetation types, four water types (ice, snow, clear,

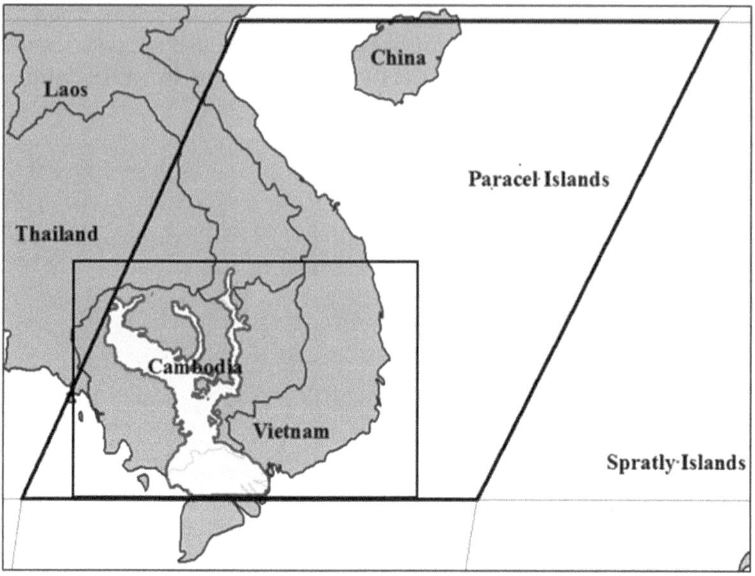

Fig. 5.1 The demonstration region is the Lower Mekong Basin, seen here with MODIS (path/row tile H28v07). The southern region of the MODIS tile is extracted to highlight areas that are commonly flooded along with the modified flood extent

turbid), and three types of "desert varnish," between 0 and 3000 nm (0–3 μm). The spectral viewer also allows users to view the bands relative to the spectral response graph for four Landsat sensors, Earth Observing-1 Advanced Land Imager, (EO-1 ALI), the ASTER and MODIS sensors on the Terra platform, and Sentinel 2A MultiSpectral Instrument (MSI).

Spectral graphs and others (ASTER Spectral Library 2008) like this help to explain how spectral indices and wavelength-based algorithms help scientists identify water features as separate from land. The spectral bands from MODIS show sensitivity to differences in reflectance from lawn grass, dry grass, and water. Note how the lawn grass spectra reflectance increases sharply in the near-infrared, while clear and turbid water show very low reflectance in visible wavelengths, and almost zero reflectance in the infrared wavelengths. The clear and turbid water in blue and dark blue do not reflect light past 1.2 μm, or 1200 nm. This allows for a clear delineation of water and other features using wavelengths beyond 1200 nm, such as bands MODIS 5, 6, and 7. In addition, the contrast between the highly reflective grass and very low reflectivity in clear and turbid water make the use of band 2 particularly useful.

Because water absorbs infrared radiation instead of reflecting (Campbell and Wynne 2011), also evidenced by the Spectral Viewer, many studies have been able to take advantage of the "dark pixel" values that occur as a result of low reflectance. As you can see in Fig. 5.2, the near-infrared shows clear and turbid water having a

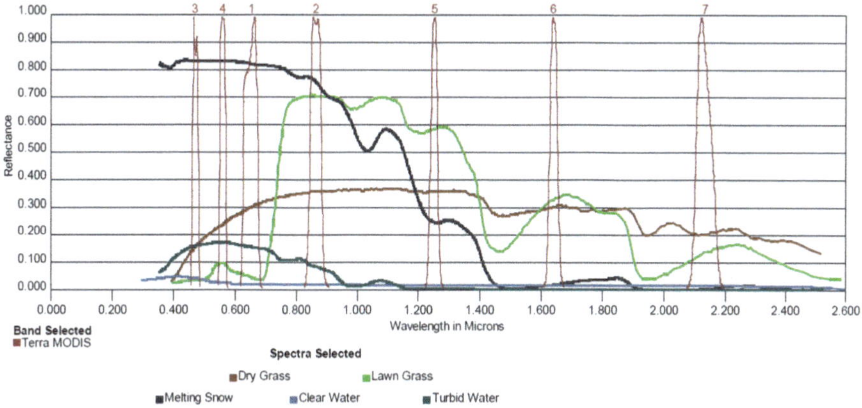

Fig. 5.2 USGS Spectral Viewer with dry and lawn grass, along with melting snow, clear water, and turbid water with Terra MODIS bands 1–7

much lower reflectance than grasses, which will yield an infrared image where both water and grasses are present to show grass as very bright as water as very dark.

5.2.1 Band Thresholding

In 1993, Manavalan et al. applied a density slicing technique to a series of near-infrared (760–900 nm) images to identify the appropriate thresholds for the land–water boundary to monitor reservoir capacity. Density slicing generally involves an iterative process of the arbitrary segmentation of image values into intervals to aid in the visual identification of spectrally dissimilar features. Similarly, Frazier and Page (2000) compared the classification accuracy of density slicing to a more sophisticated maximum-likelihood classification to identify water bodies. In Frazier and Page (2000) the boundaries to extract the water bodies were selected by first identifying 12 different training sites over three water body types (river, lagoon, and dam) and using the maximum and minimum values for all training areas across Landsat-5TM bands 1–7. The study found that the values extracted for band 5 (1550–1750 nm) gave the best visual approximation of the ground truth image, achieving an overall accuracy of 96.9 %, compared to 97.4 %, the overall accuracy of the maximum-likelihood classification, demonstrating that single-band threshold techniques may be equally beneficial for mapping water compared to more data and computation intensive methods (Fig. 5.3).

The density slice technique demonstrated here shows how a grayscale infrared image (a) can be first transformed by applying a color ramp that helps distinguish between different features (b). The image here is in the raw digital number format from MODIS. The areas that reflect very highly may reach 3000, while areas that absorb infrared will be very low. Because water absorbs infrared, it is expected to

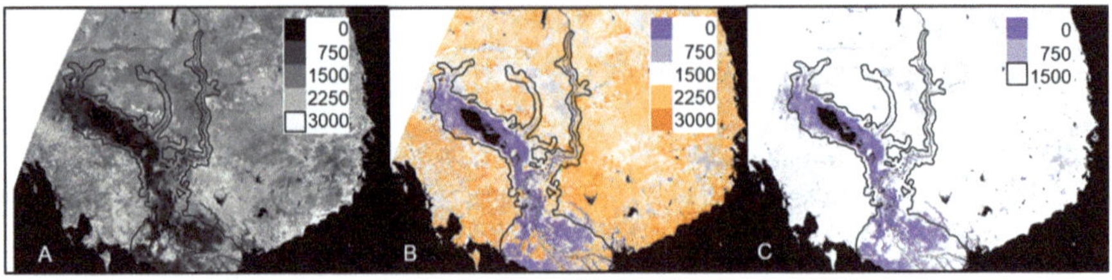

Fig. 5.3 Density slice threshold using Terra MODIS Band 6 (1628–1652 nm) 500 m MOD09A1 8-day composite October 24, 2013

have a very low reflectance in MODIS band 6. Finally, upon visual examination, a threshold such as 1500 in this cause can be selected that best depicts the feature of interest (c).

5.2.2 *Spectral Indices*

Other uses of red and infrared reflectance data combine the two bands into spectral indices. The one such spectral index is the Normalized Difference Vegetation Index (Rouse et al. 1973; Colwell 1974; Tucker 1979). The index was created to identify and measure vegetation health and phenology in ERTS-1, using a normalized scale −1 to 1, where 1 is very healthy vegetation, and values approaching zero are unhealthy vegetation or not vegetation at all (McFeeters 1996). The contrast of the high absorption of the red wavelength and the high reflectance of infrared by plant chlorophyll allows researchers to normalize chlorophyll activity between different plant types and various stages of development (Tucker 1979; Gao 1996). However, the normalization of the infrared and red wavelengths allows researchers to use the negative side of NDVI to focus on water's absorption of infrared to separate land and water (Lacaux et al. 2006; Lei et al. 2009), which may also be useful in flood monitoring studies.

Two similar studies in 1996 created indices particularly for measuring water, the Normalized Difference Water Index (NDWI). One index maximizes the green (Landsat 4MSS 500–600 nm) reflectance of water features, while minimizing the infrared (Landsat 4MSS 800–1100 nm) reflectance, (called NDWIg hereafter) (McFeeters 1996) in order to delineate open water features. A modification of the NDWIg index helped to reduce sensitivity and over-detection of water in urban areas (Xu 2006). The MNDWI (MNDWIg) replaces the near-infrared wavelength (Landsat-5TM 760–900 nm) with mid-wave infrared (Landsat-5TM 1550–1750 nm), as there is greater contrast in the reflectance of lake water, urban areas, and vegetation in the mid-wave band in the study region, compared to infrared reflectance (Fig. 2 seen in Xu 2006).

Researchers compared these spectral indices to help understand their relevance across water coverage fractions and sensor types (Lei 2009). The analysis focused on the available infrared (or SWIR/MWIR) wavelength combinations with green (variations of NDWIg), with Landsat 7 ETM+, SPOT-5, ASTER, and MODIS. The study found that for all sensors excluding MODIS, the short wave bandwidth (~1550 ~ 1750 nm) worked best with the green bandwidths (~500 ~ 600 nm). Still, a bandwidth with a shorter frequency than the other sensors' SWIR, but longer than the NIR is recommended, as evidenced by the choice of MODIS band 5 (1230–1250 nm) over band 6 (1628–1652 nm), which is more similar to the SWIR bandwidths on other sensors in the study. The research further concluded that NDVI is an inappropriate choice for delineating water bodies when short wave and green bands are available (Lei 2009).

The second NDWI focused on vegetation liquid moisture using short wave infrared (AVIRIS 1240 nm) and infrared (AVIRIS 860 nm) (called NDWIs hereafter), where both wavelengths are sensitive to canopy chlorophyll and moisture content (Gao 1996). As NDWIs is sensitive to variations in vegetation moisture, it may be beneficial to identify spatial variations where forested or crop areas are becoming inundated with floodwater in order to delineate open water from vegetation with high water content.

The Normalized Difference Pond Index (NDPI) helps to identify small water bodies (greater than 100 m^2) where vegetation might be present, which may not be detected by other water indices or NDVI, particularly as pixel sizes increase to more coarse resolution (Lacaux et al. 2006). NDPI was developed using the SPOT-5 sensor, utilizing the green (500–590 nm) and shortwave (1580–1750 nm called middle infrared MIR in the text) wavelengths. NDPI identifies standing water when vegetation is present, which may allow for a more rigorous account of shallow water bodies that are not easily distinguishable in other water indices that focus on pure water or turbid water with little vegetation presence, which may be dominant in deeper flooded areas.

While some rice and grain crops may benefit from seasonal flooding, prolonged floods can cause immense damage to agricultural areas by wash away soils and crops. To contribute to agricultural food risk assessment the Modified Land Surface Water Index (MLSWI) was created by comparing combinations of infrared (841–876 nm) and two different shortwave bands (1628–1652 and 2105–2155 nm) (Kwak 2015). The near-infrared MODIS band 2 (841–876 nm) and shortwave infrared MODIS band 7 (2105–2155 nm) were shown to be ideal in Bangladesh. This is a surprising result, compared with the laboratory spectra (Lei 2009), where the shorter wavelength (1230–1250 nm) was an improvement over the longer wavelength (1630–1650 nm) when green (550–570 nm) was the complementary bandwidth.

The surface water indices all show a very similar map of flooding in the Mekong Region seen in Fig. 5.4. The Normalized Difference Water Index (NDWIg), when using a threshold of 0–1, shows the smallest flood extent (a). The Modified Normalized Difference Water Index (mNDWI) shows more flooding, as it was meant to be more sensitive to the effects of water in heterogeneous pixels (b). Finally, the Normalized Difference Pond Index (NDPI) shows the largest flood extent, as it is very sensitive to water turbidity. The variance shown between the

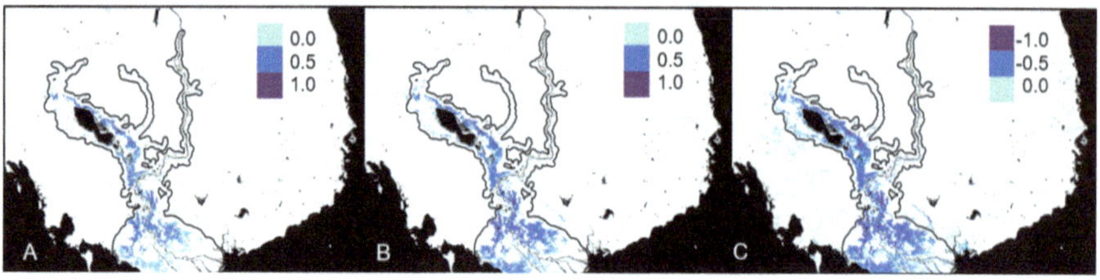

Fig. 5.4 Surface water indices (**a**) NDWIg, (**b**) mNDWIg, (**c**) NDPI

water indices might be due to the nature of water turbidity in the region and the relatively coarse resolution, compared to the finer resolution data used in the original index development.

5.2.3 *Color Composite*

Earlier research (Ali et al. 1989) identified AVHRR bands 1 (580–680 nm) and 2 (725–1100 nm) as suitable for studying water turbidity and land water separation, respectively, prompting a study by Rasid and Pramanik (1990) to use the two bands in a color composite method to delineate flood boundaries and identify areas inundated with deeper water. Researchers have also used near-infrared color composites to determine which pixel might have a mixture of water and land as transition zones can be difficult to distinguish in coarse resolution imagery (Chen 2013). The low reflectance in the near-infrared wavelengths allows researchers to identify turbid water, which may be shallower compared to clear water, which may be deeper. The color composites aid in preliminary visual inspection before continuing onto further studies to incorporate other datasets such as elevation models and land cover datasets, which will be discussed in the next section (Table 5.1).

5.3 Physically Based Models, Additional Input Data

Physical data such as temperature, texture, or elevation have also proven to be reliable methods to mapping floods. Temperature data can be formulated from measurements derived from long wavelength infrared data or brightness temperature conversion from passive microwave sensors, while texture and elevation can be derived from active microwave products and photogrammetric products derived from optical data. Additional sources of data that may be used in flood mapping efforts may be land cover datasets, outputs from hydrological models, and soil moisture and precipitation information.

Table 5.1 Formulas wavelengths and thresholds used in Figs. 5.3 and 5.4

Formula Author	Formula with original bandwidth used Formula with MODIS bands used
NDWIgreen McFeeters (1996)	NDWIgreen = ((500–600 nm)−(700–800 nm))/((500–600 nm) + (700–800 nm)) (band 4 − band 2)/(band 4 + band 2)
mNDWIgreen Xu (2006)	NDWIgreen = ((520–600 nm)−(1550–1750 nm))/((520–600 nm) + (1550–1750 nm)) (band 4 − band 5)/(band 4 + band 5)
NDPI Lacaux et al. (2006)	NDPI = ((1580–1750 nm)−(500–590 nm))/((1580–1750 nm) + (500–590 nm)) (band 6 − band 4)/(band 6 + band 4)
NDVI Rouse et al. (1973)	NDVI = ((700–800 nm)−(600–700 nm))/((700–800 nm) + (600–700 nm)) (band 2 − band 1)/(band 2 + band 1)
NDWIswir Gao (1996)	NDWIswir = (860 nm–1240 nm)/(860 nm + 1240 nm) (band 2 − band 5)/(band 2 + band 5)
VSWI Cai et al. (2011)	VWSI = NDVI/LST_{kelvin}

5.3.1 *Thermal*

As water bodies have a relatively consistent temperature compared to dry land masses (Schaaf and Lakshmi 2000), land surface temperature (LST) can be used to delineate water bodies by identifying temperature contrasts between dry land and water. Researchers have studied this relationship over the United States using the High Resolution Infrared Sounder (HRIS) (Schaaf and Lakshmi 2000) as well as in Australia using MODIS (Parinussa et al. 2016), citing differences in surface temperature recorded between morning and evening satellite overpasses. The sensors used in these studies contain long wave infrared bands ranging from 10,600 to 12,510 nm to measure temperature through the computation of the data recorded from surface radiance and emissivity (Chedin et al. 1984; Wan 1999; Lei et al. 2016).

Cloud coverage during flood events makes the use of infrared imaging for flood detection a difficult task. While infrared bands may be able to penetrate small particle masses such as haze or cirrus clouds, denser stratus clouds pose a problem, preventing clear observations of the surface. In these cases, the use of cloud-penetrating radar is available, as the large wavelengths are able to surpass the relatively small particle sizes of the clouds (Liou 2002; Parinussa 2008). Surface temperature from radar measurements can similarly be used to identify flooding while penetrating cloud cover. Surface temperature can be calculated from brightness temperatures observed from microwave wavelengths. Brightness temperature, the measure of radiation emission from the surface, can be recorded by measuring the temperature at the antenna of the sensor. Surface temperature T_S may be produced with varying accuracy, as determined by wavelength and algorithms used, the general principle of the conversation is given by the relationship with brightness temperature T_B and the emissivity of the surface in kelvin e; r is reflectivity at the surface (Lakshmi 2013)

$$e = 1 - r,$$
$$T_B = eT_S.$$

The Advanced Microwave Scanning Radiometer—Earth Observing System (AMSR-E) and its ancestor AMSR-2 are dual-polarized passive microwave sensors with a wide range of spatial resolutions, and a twice daily revisit at 1:30 p.m. and 1:30 a.m. AMSR-E functioned from June 2002 to October 2011, while AMSR-2 became available beginning July 2012. MODIS is flown along with AMRS-E on the Aqua satellite platform, as a complement of sensor types for hydrologic applications capturing measurements at the same time. Both sensors can derive surface temperature products; however, Parinussa et al. (2008, 2016) found that because of the relative consistency between the datasets and the high accuracy as compared with ground measurements, a combined dataset was proposed, featuring MODIS as the primary data, and filling in cloud cover gaps using AMSR-2. The MODIS/ AMSR-2 combined surface temperature product could produce a daily flood map with no cloud coverage when day and night observations are compared. In contrast, studies have identified that the difference between the vertically and horizontally polarized brightness temperatures observed simultaneously can also be used to identify open water, as large differences signify the presence of strongly polarized signals as are found in open water (Choudhury 1989; Smith 1997).

In addition to identifying flooding through diurnal changes in surface temperature, LST data can be used with NDVI to create the Vegetation Water Supply Index (VWSI), as VWSI = NDVI/LST, which was originally created to monitor drought conditions (Cai et al. 2011), by identifying vegetation stress under arid settings. The inverse of the drought values may be used to identify above average moisture. A modification of the VWSI, known as the Normalized Vegetation Water Supply Index,

$$\mathrm{NVSWI} = \left(\frac{\mathrm{VSWI} - \mathrm{VSWI}_{\mathrm{min}}}{(\mathrm{VSWI}_{\mathrm{max}} - \mathrm{VSWI}_{\mathrm{min}})}\right) * 100$$

normalizes the values over the study period to give more context to the severity of the output values, which are then segmented into equal interval classifications. The drought values are below 60, while wet values are above 80 (Abbas et al. 2014).

The three vegetation indices focus on identifying vegetation health and not water. However, based on the premises that floodwater may obscure vegetation or that floods might destroy healthy vegetation, vegetation indices may be used to help map floods. The Normalized Difference Water Index (NDWIs) identifies water content of vegetation; in Fig. 5.5a, the flooding in the region is shown with high

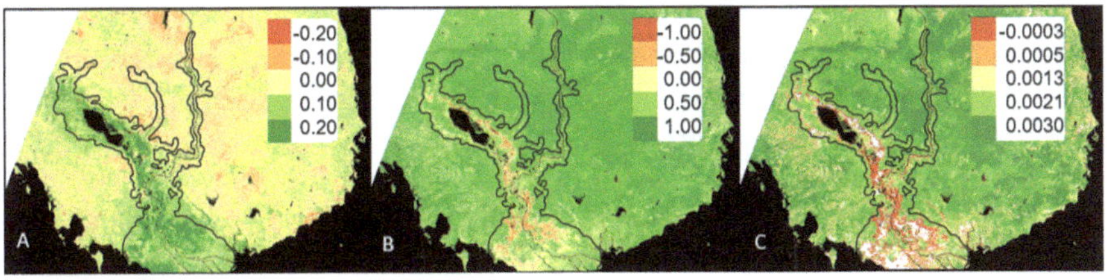

Fig. 5.5 Vegetation indices for water content and vegetation health (**a**) NDWIs, (**b**) NDVI, (**c**) VSWI

water content values as 0–0.20. The Normalized Difference Vegetation Index (NDVI) is generally symbolized as values −1 to 1 with healthy vegetation being above zero; Fig. 5.5b shows the flooded region as negative values. The Vegetation Water Supply Index in Fig. 5.5c is based on NDVI and provides more information than NDVI alone as land surface temperature information is added. Because of surface water, surface temperature is not always able to be recorded from MODIS, leaving no data regions shown in white inside the floodplain.

5.3.2 Radar Imaging

The surface temperature example is not the only combination of radar and optical remote sensing methods to map floods. Radar is able to measure ground texture through backscatter at multiple wavelengths much longer than is found in the optical spectrum. Wavelengths for radar sensors are generally measured in length centimeters (cm) or frequency (MHz, GHz). Ground features are expected to have a coarse texture, giving a speckled appearance in radar imagery, whereas water features are expected to be very flat or specular. Another study showed that while a C-Band (5.6 cm) synthetic aperture radar (SAR) image could penetrate cloud cover to identify surface features, backscatter from wind caused waves reduced the specular nature of the water bodies, preventing the water to be identified by the sensor (Alsdorf et al. 2007). It is then suggested that the L-Band (24-cm wavelength) sensor may be ideal for measuring inland surface water bodies, as it is not as sensitive to the rough texture of water caused by wind or flow turbulence.

Although raw backscatter data can be detrimental to direct observations of water surface when there is wind roughening, the backscatter coefficients of X- and L-band sensors have been used (Rosenqvist and Birkett 2002; Hess et al. 2003) to extract flood inundation extent when surface water is not specular, or is mixed with vegetative features such as in wetlands. To complement SAR systems that have the capability to measure stage height, Smith (1997) devised a method to combine European Space Agency SAR data with optically derived inundation extent from Landsat to obtain elevation extent/discharge rating curve to derive water elevations at the land–water boundary.

Researchers have also presented the case that while satellite imagery in the visible and near-infrared wavelengths is useful for mapping water extents, problems such as canopy cover and emergent vegetation can obscure and mix pixels, respectively, causing classification errors (Töyrä et al. 2002). This study utilized the C-Band on the satellite radar sensor RADARSAT and the NIR (790–890 nm) data from SPOT to create a composite to identify flood boundaries, which was repeated in another study, where visible imagery from Landsat thematic mapper (TM) and Envisat advanced synthetic aperture radar (ASAR) system were used to identify flooded regions (Ramsey et al. 2012). Acquiring flood depth information can also be difficult to using visible imagery with varying vegetation types or regularly flooded marsh areas (Rasid and Pramanik 1990; Ramsey et al. 2012). The Ramsey et al. (2012) study provided a solution to this problem by utilizing SAR and ASAR to identify relative water penetration depths in different marsh areas.

5.3.3 Digital Elevation Models

Incorporating ancillary data such as elevation is an important part of flood mapping. When available, digital elevation models are used in conjunction with optical, radar, and modeled data.

The integration of optical and radar data with digital elevation models using geographic information system (GIS) processing techniques is described in Townsend and Walsh (1998). The Position Above the River Index (PARI) model is an integrative approach creates a potential inundation map based on the river's proximity to other hydrologic features, such as tributaries or streams.

In a study based in northern Italy, researchers found that due to the delay of the satellite overpass from the peak inundation time, only a fraction of the flooded area was observed by the satellite (Brivio et al. 2002). In a technique similar to the PARI model from Townsend and Walsh (1998), a cost-distance matrix was created. Using the areas that were mapped using C-Band SAR after the peak flood, Brivio et al. (2002) created a digital elevation model to create a cost-distance matrix to calculate the difficulty of water traveling from the river to the remaining flooded regions; the matrix was then used to trace the path of the river to the flooded region.

Flood depths provide a useful dynamic to flood maps giving the user specific information about the inundation level and the type of risks that exist in that area. Emergency planners and disaster mitigation teams typically require water depth information in areas other than at gage locations. A method to create water depth grids is identified (Lant 2013) by subtracting the DEM from the inundation extent.

In calibration of coarse resolution mapped flood extents, it was suggested (Fayne and Bolten 2014; Guerschman et al. 2011) that a higher resolution elevation model should be used to remove areas that would be unlikely to flood such as ridge tops or high hillsides. Gallant and Dowling (2003) created a multistep iterative process to categorize digital elevation models as flat valley bottoms or flat ridge tops and areas in between, known as the Multi-Resolution Index of Valley Bottom Flatness (MRVBF). As flooding is expected to occur along valley bottoms, a threshold may be used to mask out values that MRVBF and the complement Multi-Resolution Ridge Top Flatness (MRRTF) consider a high hillside or hilltop.

5.3.4 *Classification Algorithms Using Elevation Data*

Another method of mapping floods using coarse resolution imagery and elevation models is seen in the Open Water Likelihood (OWL) algorithm. OWL uses a logistic regression to incorporate MODIS shortwave infrared reflectance bands, NDVI, NDWIs, and MRVBF to obtain the probability that a fraction of the coarse resolution pixel is inundated (Guerschman 2011). The formula is as follows:

$$\mathrm{OWL} = \left(1 + \exp\left(a_0 + \sum_{i=1}^{5} a_i{}^{*}x_i\right)\right)^{-1}$$

where

a_0—3.41375620
a_1—0.000959735270
a_2—0.00417955330
a_3—14.1927990
a_4—0.430407140
a_5—0.0961932990
x_1—SWIR (1628–1652 nm) MODIS band 6 (reflectance × 1000)
x_2—SWIR (2105–2155 nm) MODIS band 7 (reflectance × 1000)
x_3—NDVI
x_4—NDWIs
x_5—MRVBF

This method was again validated (Chen et al. 2013), as it was applied to 500-m MODIS Daily and 8-day images, and was visually compared against Landsat 5 (Fig. 5.6).

The four elevation-based products shown here are derived from 90-m elevation data (a) collected from the Shuttle Radar Topography Mission and were preprocessed by removing voids and sinks in the data (CGIAR-CSI). The permanent water bodies are overlayed in white. The Multi-resolution Valley Bottom Flatness (b) used the SRTM data as a product input to estimate the flatness of the floodplain, and therefore the likelihood of deposition. The light blue region in b is the flattest, while the white region is the smoothed, flat water surface. The water depth grid (c) was created by subtracting a binary classification of the NDPI (Fig. 5.4c) from the elevation model. Using this method, the region surrounding the lake shows the deepest flooding, while the delta shows shallower flooding.

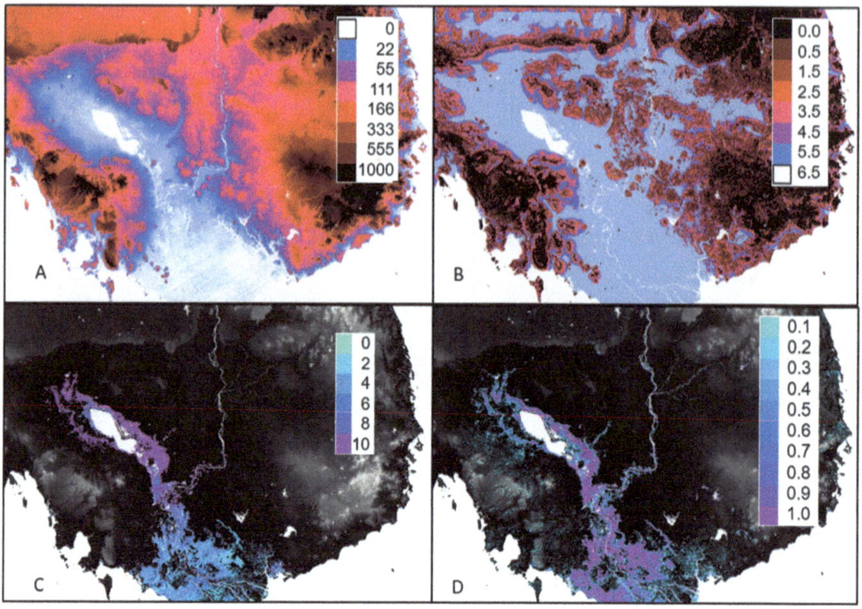

Fig. 5.6 (**a**) SRTM Elevation Model in meters, (**b**) MrVBF, (**c**) Water depth grid in meters, (**d**) OWL Water Likelihood Fraction

The Open Water Likelihood (OWL) algorithm combines inputs from two short-wave infrared bands, NDWIs (Fig. 5.5a) NDVI (Fig. 5.5b), and MrVBF (b) to identify what fraction of the pixel is likely inundated. Unlike other water detection algorithms, OWL is more sensitive to mixed pixels, reducing uncertainty caused by the other algorithms which may use less input variables.

Outputs from the OWL and MRVBF algorithms have also been combined in a decision tree and logistic regression approach with Landsat to create binary classifications of water bodies over time, then were combined to create a map of cumulative observations of surface water from space. Similarly, another study used a regression tree approach to integrate the predictors of water presence to derive a map of water fraction, instead of a binary classification, applied to coarse resolution MODIS imagery.

As many studies have cited problems with cloud and terrain shadow being spectrally similar to the low reflectance of water in the infrared wavelengths (Xu et al. 2006; Sun et al. 2012; Nigro et al. 2014), Li et al. (2015) used a geometric algorithm to identify and remove cloud shadow. This study is particularly relevant to all of the research related to optical data, as many of the spectral indices and thresholds mentioned in section one, use infrared reflectance. The terrain shadow removal study identifies that the root-mean-square (RMS) height, internal and external height difference are good indicators of surface roughness and delineation of water from terrain.

5.3.5 Software Models, Rainfall, and Soil Moisture

It is also worth mentioning that flood inundation maps can also be prepared using computer software, using input data such as gage height, elevation models, and land cover maps. Part of the US Army Corps of Engineers, the Hydrological Engineering Center developed the River Analysis System (HEC-RAS) and other hydrological modeling software, which is commonly used in combination with remotely sensed data (Lant 2013). The software suite enables researchers to create inundation maps, surface profiles, and model flow direction and physics, as well as taking into account topographic features such as surface roughness and slope, or seasonal changes in seasonal vegetation, and anthropogenic factors such as changes in impervious surfaces or crop cycles (United States Army Corps of Engineers HEC-RAS).

The implementation of software for flood modeling is particularly useful compared to, or in conjunction with, remote sensing studies as issues with temporal latency, cloud cover, cloud or terrain shadow, and spatial resolution are reduced or eliminated. Incorporating rainfall estimates is particularly useful for flood forecasting and flash flood analysis (Krajewski and Smith 2002). Many of the studies discussed here focus on slow and persistent flooding; however, one study created a framework to map inundation threats and flash flooding at city and regional scales by integrating the HEC-RAS system with precipitation data (Knebl 2005), while another estimated flood extent by combining precipitation data with a routing model (Wu et al. 2014).

As the surface soil moisture state is key to the infiltration or runoff of precipitation (Entakhabi et al. 2010), the recently launched Soil Moisture Active Passive (SMAP) sensor (NASA-JPL-SMAP) and other soil moisture products can be useful tools in mapping floods and determining flood risk when soil moisture is approaching a saturated state. Analogous to the use of radar backscattering to determine standing water, the unique dielectric properties of water and dry soil allows water to be measured as a fraction of soil to determine volumetric moisture. Therefore, if the soil is approaching saturation, then flooding is likely to occur. While the SMAP data was not available during the demonstration year, the 2015 volumetric soil moisture from the SMAP sensor (L3 SM_P 36 km 2015, National Snow and Ice Data Center NSIDC 2015, 2016) is able to capture increased soil moisture around the areas shown as flooded from the 2013 MODIS data.

The SMAP volumetric soil moisture data is publically available at 36 km although special algorithms and processing may be capable of creating a higher resolution product to 9 or 3 km. The coarse resolution of the 36-km pixels may not be enough to measure flooding independently, but it is clear in Fig. 5.7 that the SMAP sensor is able to capture increased moisture over the target region. While SMAP is not intended to map floods, the ability for SMAP to identify soils that are increasing in saturation before a flood event is very important for flood hazard and damage mitigation.

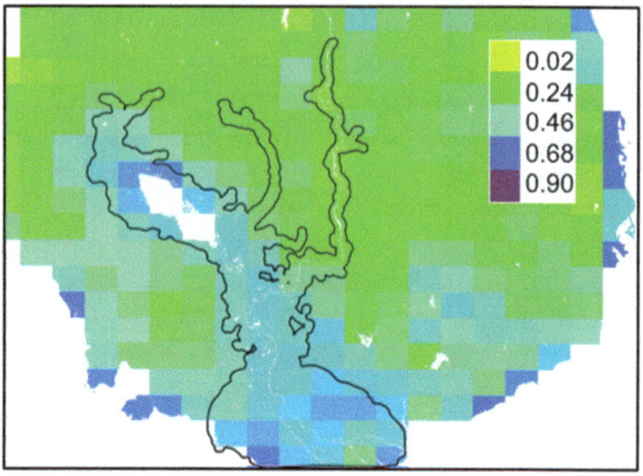

Fig. 5.7 SMAP 36 km Volumetric Soil Moisture Data October 23, 2015

5.3.6 Change Detection Methods

Finally, change detection is a universal method to monitor flood events, as areas that were not previously flooded would appear different spectrally, thermally, and texturally. Simple subtraction between images of different dates also known as differencing (Song et al. 2004), calculating the standard deviation of a baseline of data with the z-score of the newer data (Sarp 2011), and percent change formulas (Hasan and Islam 2011) are all useful tools for identifying surface changes, and can be implemented across sensor types, spatial and temporal resolutions. However, it is important to note that although the algorithms may not be temporally dependent, the latency between compared datasets may skew the validity of the results, as other factors may contribute to the observed change. Trend analysis is one method to monitor flooded areas over time, as individual pixel values or basin averages can be made into a time series to identify when flooding is occurring or may occur in the future when flooding is cyclical. While Nash et al. (2014) used auto-regression techniques on NDVI to predict seasonal variations in vegetation health, another study used the TIMESAT (Eklundh et al. 2009) software to fit a function on the time series of NDVI data for snow-vegetation dynamics (Jönsson et al. 2010).

5.4 Conclusion

The field of remotely sensed flood mapping continues to evolve and improve. The development and improvements of real-time data access systems have allowed scientists to harness power of programming languages such as Python, C++, and R, in order to digest data as soon as it becomes available, and to create output data in a rapid manner. Instead of simply hosting the data in a list format online for users to download, many authors have found it useful to demonstrate their flood products on online web dashboards, such as the Dartmouth Flood Observatory (Brakenridge and Daniel 1996), the MODIS Near Real-Time (NRT) Global Flood Mapping Project (Nigro et al. 2014), and the Near Real-Time Flooding in Southeast Asia Project (Ahamed and Bolten 2016). Both the Global Flood Mapping and the Flooding in Southeast Asia projects rely on methods discussed here using MODIS data, such as infrared band thresholding and spectral index combined with change detection, respectively.

Free publically available and costly commercial Earth observing satellites are able to capture images of the earth at varying spatial scales, and with different orbital periods. Optical sensors may have the ability to capture surface reflectance data from visible blue, green, and red wavelengths, as well as longer wave infrared bands and emissivity data. Radar sensors may focus on brightness temperature or backscattering coefficients to identify moist and saturated soil. Finally, these first-level datasets can be used as inputs to other computational modeling software, time series, or regression algorithms to provide value-added improvements, increasing the spatial or temporal resolution of the input datasets, or creating a wholly different product entirely.

Utilizing earth observing satellite data to map standing water from space is indispensable to flood mapping for disaster response, mitigation, prevention, and warning as extreme flooding events worldwide can damage crop yields and contributing to billions of dollars economic damages as well as social effects including fatalities and destroyed communities. The increase in the quantity and variety of flood mapping techniques using satellite data has allowed broader and less-technical audiences to be able to benefit from flood products. The use of remotely sensed data by diverse audiences increases the general knowledge of flooding in a given area and may help to mitigate pervasive economic and social damages caused by flooding.

References

Abbas S, Nichol JE, Qamaer FM, Xu J (2014) Characterization of drought development through remote sensing: a case study in Central Yunnan, China. Remote Sens 4998–5018

Ahamed A, Bolten JD (2016) Near real time flooding in Southeast Asia. Project Mekong. http://mekongflood.appspot.com/

Ali A, Quadir DA, Huh OK (1989) Study of river flood hydrology in Bangladesh with AVHRR data. Int J Remote Sens 10(12):1873–1891

Alsdorf DE, Rodriguez E, Lettenmaier DP (2007) Measuring surface water from space. Rev Geophys 45:RG2002. doi:10.1029/2006RG000197

Anderson MC, Norman JM, Mecikalski JR, Otkin JA, Kustas WP (2007) A climatological study of evapotranspiration and moisture stress across the continental United States based on thermal remote sensing: 1. Model formulation. J Geophys Res Atmos 112:D10117

Berkeley Lab (2016) Electromagnetic spectrum. http://www2.lbl.gov/MicroWorlds/ALSTool/EMSpec/EMSpec2.html

Brakenridge GR, Anderson E (2006) MODIS-based flood detection, mapping and measurement: the potential for operational hydrological applications. In: Transboundary floods: reducing

risks through flood management. NATO Science Series: Earth and Environmental Sciences, vol 72. pp 1–12
Brakenridge GR, Daniel K (1996) The Dartmouth Flood Observatory: an electronic research tool and electronic archive for investigations of extreme flood events. In: Geological Society of America Annual Meeting, Geoscience Information Society Proceedings
Brivio PA, Colombo R, Maggi M, Tomasoni R (2002) Integration of remote sensing data and GIS for accurate mapping of flooded areas. Int J Remote Sens 23(3):429–441. doi:10.1080/01431160010014729
Cai G, Du M, Liu Y (2011) Regional drought monitoring and analyzing using MODIS data – A case study in Yunnan Province. In Computer and Computing Technologies in Agriculture IV, Nanchang, China, 243–251
California Institute of Technology Jet Propulsion Laboratory (2008) ASTER Spectral Library. California Institute of Technology, Pasadena
Campbell JB, Wynne RH (2011) Introduction to remote sensing, 5th edn. The Guilford Press, New York
Chedin A et al (1984) The improved initialization inversion method: a high resolution physical method for temperature retrievals from satellites of the TIROS-N Series. http://dx.doi.org/10.1175/1520-0450(1985)
Chen Y, Huang C, Ticehurst C, Merrin L, Thew P (2013) An evaluation of MODIS DAILY and 8-day composite products for floodplain and wetland inundation mapping. Wetlands 33(5):823–835
Choudhury BJ (1989) Monitoring global land surface using Nimbus-7 37 GHz data: theory and examples. Int J Remote Sens 10(10):1579–1605
Colwell JE (1974) Vegetation canopy reflectance. Remote Sens Environ 3:175–183
Eklundh L, Johansson T, Solberg S (2009) Mapping insect defoliation in Scots pine with MODIS time-series data. Remote Sens Environ 113:1566–1573
Entakhabi D, Njoku EG, O'Neill PE, Kellogg KH, Crow WT et al (2010) The soil moisture active passive (SMAP) mission. In: Proceedings of the IEEE, vol 98. pp 704–716
Fayne JV, Bolten JD (2014) Validating flood mapping products using a digital elevation model comparison technique. In: American Geophysical Union Fall Meeting Poster Presentation. San Francisco, CA
Fayne JV, Bolten JD, Fuhrmann S, Rice MT (2015) Real-time multi-scale mapping for emergency management. In: International Cartographic Conference. Rio De Janeiro, Brazil
Frazier PS, Page KJ (2000) Water body detection and delineation with Landsat TM data. Photogramm Eng Remote Sens 1461–1467
Gallant JC, Dowling TI (2003) A multiresolution index of valley bottom flatness for mapping depositional areas. Water Resour Res 4:1–14
Gao BC (1996) NDWI—a normalized difference water index for remote sensing of vegetation liquid water from space. Remote Sens Environ 58:257–266
Gopinath G, Ambili GK, Gregory SJ, Anusha CK (2014) Drought risk mapping of South-Western State in the Indian Peninsula—a web based application. J Environ Manage 161:453–459
Guerschman JP, Warren G, Byrne G, Lymburner L, Mueller N, Van Dijk A (2011) MODIS-based standing water detection for flood and large reservoir mapping: algorithm development and applications for the Australian continent. Report, Commonwealth Scientific and Industrial Research Organization
Hess LL, Melack JM, Novo EMLM, Barbosa CCF, Gastil M (2003) Dual season mapping of inundation and vegetation for the Central Amazon Basin. Remote Sens Environ 87 (2003):404–428
Holben BN (2007) Characteristics of maximum-value composite images from temporal AVHRR data. Int J Remote Sens 7:1417–1434
Irons JR, Taylor MP, Laura R (2016) Landsat1. Landsat Science. NASA. Accessed 25 Mar 2016

Jönsson AM, Eklundh L, Hellström M, Bärring L, Jönsson P (2010) Annual changes in MODIS vegetation indices of Swedish coniferous forests in relation to snow dynamics and tree phenology. Remote Sens Environ 114:2719–2730

Knebl MR, Yang Z-L, Hutchinson K, Maidment DR (2005) Regional scale flood modeling using NEXRAD rainfall, GIS, and HEC-HSM/RAS: a case study for the San Antonio River Basin Summer 2002 storm event. J Environ Manage 75:325–336

Krajewski WF, Smith JA (2002) Radar hydrology: rainfall estimation. Adv Water Resour 25:1387–1394

Kwak Y, Arifuzzanman B, Iwami Y (2015) Prompt proxy mapping of flood damaged rice fields using MODIS-derived indices. Remote Sens 7:15969–15988. doi:10.3390/rs71215805

Lacaux JP, Tourre YM, Vignolles C, Ndione JA, Lafaye M (2006) Classification of ponds from high-spatial resolution remote sensing: application to Rift Valley Fever epidemics in Senegal. Remote Sens Environ 106:66–74

Lakshmi V (2013) Remote sensing of soil moisture. ISRN Soil Science. http://dx.doi.org/10.1155/2013/424178

Lant JG (2013) Flood-inundation maps for a 6.5-mile reach of the Kentucky river at Frankfort, Kentucky. Pamphlet to accompany Scientific Investigations Map 3278, USGS

Lei J, Zhang L, Wylie B (2009) Analysis of dynamic thresholds for the normalized difference water index. Photogramm Eng Remote Sens 75(11):1307–1317

Lei S et al (2016) Algorithm development of temperature and humidity profile retrievals for long-term HIRS observations. http://www.mdpi.com/2072-4292/8/4/280

Li S, Sun D, Goldberg ME, Sjoberg B (2015) Object based automatic terrain shadow removal from SNPP/VIIRS flood maps. Int J Remote Sens 36(21):5504–5522

Liou KL (2002) An introduction to atmospheric radiation. Academic, New York

Manavalan P, Sathyanath P, Rajegowda GL (1993) Digital image analysis techniques to estimate waterspread for capacity evaluations of reservoirs. Photogramm Eng Remote Sens 59 (9):1389–1395

McFeeters SK (1996) The use of the normalized difference water index (NDWI) in the delineation of open water features. Int J Remote Sens 17:1425–1432

MODIS Data was Downloaded from LP DAAC Data Pool. http://e4ftl01.cr.usgs.gov/MOLT/MOD09A1.005/2013.10.24/MOD09A1.A2013297.h28v07.005.2013316135122.hdf

MODIS Data was processed using the MODIS Reprojection Tool. https://lpdaac.usgs.gov/tools/modis_reprojection_tool

Mueller N, Lewis A, Roberts D, Ring S, Melrose R, Sixsmith J, Lymburner L, McIntyre A, Tan P, Curnow S, Ip A (2016) Water observations from space: mapping surface water form 25 years of Landsat imagery across Australia. Remote Sens Environ 174:341–352

Hasan M, Saiful Islam AKM (2011) Drought Assessment using remote sensing and GIS in North-West region of Bangladesh. Proceedings of the 3rd International Conference on Water & Flood Management, 797–804

Nash MS, Bradford DF, Wickham JD, Wade TG (2014) Detecting change in landscape greenness over large areas: an example for New Mexico, USA. Remote Sens Environ 150:152–162

National Snow and Ice Data Center NSIDC (2015) SMAP L3 Radiometer Global Daily 36 km EASE-Grid Soil Moisture, Version 3 http://nsidc.org/data/docs/daac/smap/sp_l3_smp/

National Snow and Ice Data Center (2016) NASA Distributed Active Archive Center (DAAC) at NSIDC: SMAP Data. http://nsidc.org/data/SMAP

Nigro J, Slayback D, Policelli F, Brakenridge R (2014) NASA/DFO MODIS near real-time (NRT) global flood mapping product evaluation of flood and permanent water detection. Evaluation, Greenbelt, MD

NOAA-NESDIS VIIRS User's Guide (2013) NOAA-NESDIS Visible Infrared Imaging Radiometer Suite (VIIRS) Sensor Data Record (SDR) User's Guide September 2013. http://www.star.nesdis.noaa.gov/smcd/spb/nsun/snpp/VIIRS/VIIRS_SDR_Users_guide.pdf

NOAA SIS (2013) NOAA Satellite Information System: advanced very high resolution radiometer—AVHRR. http://noaasis.noaa.gov/NOAASIS/ml/avhrr.html. Accessed 26 Nov 2013

Parinussa RM, de Jeu RAM, Holmes TRH, Walker JP (2008) Comparison of microwave and infrared land surface temperature products over the NAFE'06 research sites. IEEE Geosci Remote Sens Lett 5(4):783–787

Parinussa RM, Lakshmi V, Johnson F, Sharma A (2016) Comparing and combining remotely sensed land surface temperature products for improved hydrological applications. Remote Sens 8(2):162. doi:10.3390/rs8020162

Pettorelli N, Vik JO, Mysterud A, Gaillard J-M, Tucker CJ, Stenseth NC (2009) Using the satellite-derived NDVI to assess ecological responses to environmental change. Trends Ecol Evol 20 (9):503–510

Ramsey E III, Werle D, Suzuoki Y, Rangoonwala A, Lu Z (2012) Limitations and potential of satellite imagery to monitor environmental response to coastal flooding. J Coastal Res 28:457–476

Rasid H, Pramanik MAH (1990) Visual interpretation of satellite imagery for monitoring floods in Bangladesh. J Environ Manage 14:815–821

Revilla-Romero B, Hirpa FA, Thielen-del Pozo J, Salamon P, Brakenridge R, Pappenberger F, De Groeve T (2015) On the use of global flood forecasts and satellite-derived inundation maps for flood monitoring in data-sparse regions. Remote Sens 7:15702–15728. doi:10.3390/rs71115702

Rosenqvist A, Birkett CM (2002) Evaluation of JERS-1 SAR mosaics for hydrological applications in the Congo river basin. Int J Remote Sens 23:1283–1302

Rouse JW Jr, Haas RH, Schell JA, Deering DW (1973) Monitoring the vernal advancement and retrogradation (Greenwave effect) of natural vegetation. Type II report. Goddard Space Flight Center, Greenbelt, MD, pp 1–93

Sarp G (2011) Determination of vegetation change using thematic mapper imagery in Afşin-Elbistan Lignite Basin; SE Turkey. Procedia Technol 1:407–411

Schaaf K, Lakshmi V (2000) Analysis of the 1993 midwestern flood using satellite and ground data. IEEE Trans Geosci Remote Sens 39(8)

Smith L (1997) Satellite remote sensing river inundation area, stage, and discharge: a review. Hydrol Process 11:1427–1439

Song X, Saito G, Kodama M, Sawada H (2004) Early detection system of drought in East Asia using NDVI from NOAA/AVHRR data. Int J Remote Sens 25:3105–3111

Sun DL, Yu YY, Zhang R, Li SM, Goldberg MD (2012) Toward operational automatic flood detection using EOS-MODIS. Photogramm Eng Remote Sens 78(6):637–646

Swets DL, Reed BC, Rowland JD, Marko, SE (1999) A weighted least-squares approach to temporal NDVI smoothing. In: ASPRS Annual Conference: From Image to Information, Portland, Oregon, 17–21 May. American Society of Photogrammetry and Remote Sensing, Bethesda, MD

Terra MODIS Surface Reflectance 8-Day L3 Global 500m (MOD09A1). https://lpdaac.usgs.gov/dataset_discovery/modis/modis_products_table/mod09a1

The MODIS data products were retrieved from the online Data Pool, courtesy of the NASA Land Processes Distributed Active Archive Center (LP DAAC), USGS/Earth Resources Observation and Science (EROS) Center, Sioux Falls, South Dakota, https://lpdaac.usgs.gov/data_access/data_pool

Townsend PA, Walsh SJ (1998) Modeling floodplain inundation using an integrated GIS with radar and optical remote sensing. Geomorphology 21:295–312

Töyrä J, Pietroniro A, Martz LW, Prowse TD (2002) A multi-sensor approach to wetland flood monitoring. Hydrol Process 16:1569–1581

Tucker CJ (1979) Red and photographic infrared linear combinations for monitoring vegetation. Remote Sens Environ 8:127–150

United Nations Institute of Training and Research (2013) UNOSAT: Maps and Data. http://www.unitar.org/unosat/maps/KHM

United States Army Corps of Engineers. HEC-RAS. The Hydrologic Engineering Center. http://www.hec.usace.army.mil/

University of California, Berkeley. Remote Sensing and Image Analysis: Atmospheric Correction of Remotely Sensed Data. http://nature.berkeley.edu/~penggong/textbook/chapter5/html/sect52.htm
USGS (2013) Landsat 7. Scan Line Corrector (SLC) -off Products http://landsat.usgs.gov/products_slcoffbackground.php
USGS. 2013. Landsat 1 History http://landsat.usgs.gov/about_landsat1.php
USGS (2014) Spectral Characteristics Viewer. http://landsat.usgs.gov//tools_spectralViewer.php
USGS (2015) Landsat 8 TIRS Scene Select Mechanism (SSM) Anomaly. http://landsat.usgs.gov/calibration_notices.php
van Dijk A, Callis SL, Sakamoto CM, Decker WL (1987) Smoothing vegetation index profiles: an alternative method for reducing radiometric disturbance in NOAA/AVHRR data. Photogramm Eng Remote Sens 53:1059–1067
Wan Z (1999) MODIS Land Surface Temperature Algorithm Theoretical Basis Document (LST ATBD). http://modis.gsfc.nasa.gov/data/atbd/atbd_mod11.pdf
Wu H, Adler RF, Tian Y, Huffman GJ, Li H, Wang J (2014) Real-time global flood estimation using satellite-based precipitation and a coupled land surface and routing model. Water Resour Res 50(3):2693–2717
Xaio X, Boles S, Liu J, Zhuang D, Frolking S, Li C, Salas W, Moore B III (2004) Mapping rice paddy agriculture in southern China using multi-temporal MODIS images. Remote Sens Environ 95:480–492
Xu H (2006) Modification of normalised difference water index (NDWI) to enhance open water features in remotely sensed imagery. Int J Remote Sens 27:3025–3033
Zhang X, Friedl MA, Schaaf CB, Stahler AH, Hodges JCF, Gao F, Reed BC, Huete A (2003) Monitoring vegetation phenology using MODIS. Remote Sens Environ 84:471–475

Chapter 6
Near Real-Time Flood Monitoring and Impact Assessment Systems

Case Study: 2011 Flooding in Southeast Asia

Aakash Ahamed, John Bolten, Colin Doyle, and Jessica Fayne

6.1 Introduction and Background

Floods are the costliest natural disaster (United Nations 2004), causing approximately 6.8 million deaths in the twentieth century alone (Doocy et al. 2013). Worldwide economic flood damage estimates in 2012 exceed $19 Billion USD (Munich Re 2013). Extended duration floods also pose longer term threats to food security, water, sanitation, hygiene, and community livelihoods, particularly in developing countries (Davies et al. 2014).

Projections by the Intergovernmental Panel on Climate Change (IPCC) suggest that precipitation extremes, rainfall intensity, storm intensity, and variability are increasing due to climate change (IPCC 2007). Increasing hydrologic uncertainty will likely lead to unprecedented extreme flood events. As such, there is a vital need to enhance and further develop traditional techniques used to rapidly assess flooding and extend analytical methods to estimate impacted population and infrastructure.

A. Ahamed, M.Sc. (✉)
Hydrological Sciences Laboratory, NASA Goddard
Space Flight Center/Universities Space Research Association, 8800 Greenbelt Road, Building 33, Room G216, Greenbelt, MD 20771, USA
e-mail: Aakash.ahamed@nasa.gov

J. Bolten
Hydrological Sciences Laboratory, NASA Goddard Space Flight Center, 8800 Greenbelt Road, Building 33, Room G216, Greenbelt, MD 20771, USA

C. Doyle
Department of Geography and the Environment, University of Texas at Austin, 305 E. 23rd Street, Austin, TX 78712, USA

J. Fayne, B.A., M.S.
Department of Earth and Ocean Sciences, University of South Carolina at Columbia, 701 Sumter Street, EWS 617, University of South Carolina, Columbia, SC 29208, USA

V. Lakshmi (ed.), *Remote Sensing of Hydrological Extremes*, Springer Remote Sensing/Photogrammetry, DOI 10.1007/978-3-319-43744-6_6

Measuring flood extent in situ is generally impractical, time consuming, and can be inaccurate (Brakenridge and Anderson 2006). Remotely sensed imagery acquired from space-borne and airborne sensors provides a viable platform for consistent and rapid wall-to-wall monitoring of large flood events through time (Kussul et al. 2011). Terabytes of freely available satellite imagery are made available online each day by NASA, ESA, and other international space research institutions. Advances in cloud computing and data storage technologies allow researchers to leverage these satellite data and apply analytical methods at scale.

Repeat-survey earth observations help provide insight about how natural phenomena change through time, including the progression and recession of floodwaters. In recent years, cloud-penetrating radar remote sensing techniques (e.g., Synthetic Aperture Radar) and high temporal resolution imagery platforms (e.g., MODIS and its 1-day return period), along with high performance computing infrastructure, have enabled significant advances in software systems that provide flood warning, assessments, and hazard reduction potential (Brakenridge et al. 2003). By incorporating social and economic data, researchers can develop systems that automatically quantify the socioeconomic impacts resulting from flood disaster events.

6.1.1 Methods, Satellites, and Sensors Used to Determine Flood Extent

MODIS, Landsat, radar sensors, and land-surface models have all been successfully employed to determine flood extent (Nigro et al. 2014; Kuenzer et al. 2013; Wu et al. 2014). The daily orbit cycle of MODIS instruments aboard Aqua and Terra makes these instruments particularly attractive for near real-time studies (Brakenridge and Anderson 2006). Persistent cloud cover can obscure optical sensors like MODIS and Landsat during the monsoon season, but images compositing can broaden spatial coverage (Chen et al. 2013). Land-surface and hydrologic models work regardless of atmospheric conditions, but the accuracy is limited to the resolution of model input data (Orth et al. 2015). Although these models can estimate inundation depth in addition to extent, the results are typically too spatially coarse for community-level disaster response.

Radar sensors such as Sentinel, Radarsat 2, ENVISAT Advanced Synthetic Aperture Radar (ASAR), and TerraSAR-X can penetrate clouds and are unaffected by atmospheric conditions. Radar backscatter is sensitive to water, and data is typically obtained at medium to high resolution, rendering these systems extremely useful for flood detection (e.g., Long and Trong 2001; Kussul et al. 2011). However, radar sensors like Sentinel have longer return times (12 days), and other radar satellites must generally tasked for data acquisition. Further, wall-to-wall coverage of an area is often not practical or a prohibitively lengthy operation. Continuous now-casting and near real-time flood monitoring by any satellite platform is therefore a challenging task for all flood-prone regions.

MODIS instruments aboard Aqua and Terra offer twice-daily multispectral imagery through NASA Land, Atmosphere Near real-time Capability for EOS (LANCE) download systems free of cost. Numerous studies have demonstrated the usefulness of MODIS imagery for flood mapping (Brakenridge et al. 2003; Brakenridge and Anderson 2006; Nigro et al. 2014). Temporal image compositing coupled with the frequent return time of MODIS instruments allows for broad spatial coverage despite intermittent cloud cover. These factors make MODIS a viable alternative to radar imagery and in situ observations for flood detection (Islam et al. 2010).

6.1.2 Near Real-Time Flood Monitoring Systems

The Near Real-time Global Flood Mapping Project (GFMP; http://oas.gsfc.nasa.gov/floodmap/index.html) and Dartmouth Flood Observatory (DFO; Brakenridge and Anderson 2006) apply an algorithm to MODIS near real-time 250-m resolution data (Bands 1, 2, and 7) to detect surface water. The Flood Observatory web site (http://floodobservatory.colorado.edu/) provides recent and historical archives of global flooding, as well as other flood information such as discharge estimates. Accuracy assessment of this product on a global scale is still ongoing (Nigro et al. 2014).

The Global Flood Monitoring System (GFMS) uses near real-time radar precipitation data from the Global Precipitation Measurement Mission (GPM) and historic data from the Tropical Rainfall Measuring Mission (TRMM) coupled with a hydrologic land surface model to estimate runoff and produce a global flood product that updates every 3 h (Wu et al. 2014). The products provided by GFMS have the advantage of being independent of cloud cover since they are radar based, though the spatial resolutions of these products are coarse (Input data is 12 km $\times$ 12 km).

Each flood monitoring system is subject to limitations. The GFMP can experience difficulty monitoring areas with seasonally driven flooding, such as Southeast Asia. Higher resolution radar satellites have long return times or must be tasked for image acquisition. The GFMS relies on relatively coarse resolution radar precipitation inputs, which can exhibit bias and uncertainty (e.g., Falck et al. 2015). These inaccuracies can cascade down the land surface model and skew inundation results. To this end, there is still a need for effective regional flood inundation mapping tools for detecting flooding in most areas of the world.

6.2 Case Study: Near Real-Time Flood Monitoring and Socioeconomic Impact Estimation in Southeast Asia

6.2.1 *Flood Detection*

Researchers at the NASA Goddard Space Flight Center developed regional flood detection software for Southeast Asia (Doyle et al., in prep), consisting of Thailand, Vietnam, Cambodia, and Laos. The software routines use near real-time MODIS data made available at 3-h latency through the NASA Land, Atmosphere Near real-time Capability for EOS (LANCE) servers. MODIS instruments aboard Aqua and Terra satellites pass over the Southeast Asia region around 10:30 a.m. and 1:30 p.m. each day (Nigro et al. 2014). New imagery is ingested, composited spatially and temporally in daily, 4-day, and 8-day increments to reduce cloud cover, and compared to historic conditions to detect near real-time flooding.

The Normalized Difference Vegetation Index (NDVI; Tucker 1979) is a numerical ratio between spectral bands 1 and 2 that is useful for flood detection, land classification, and change detection studies (Brakenridge and Anderson 2006). Figure 6.1 shows annual NDVI (green) and discharge (blue) over time, illustrating the strong seasonality of the Mekong River and Southeast Asia region. The southwest monsoon rain pulse results in flooding and increased surface water from May through October (MRC, 2010). Figure 6.1 shows decreases in NDVI during the summer monsoon season as floodwaters obscure vegetation. Since vegetation has a

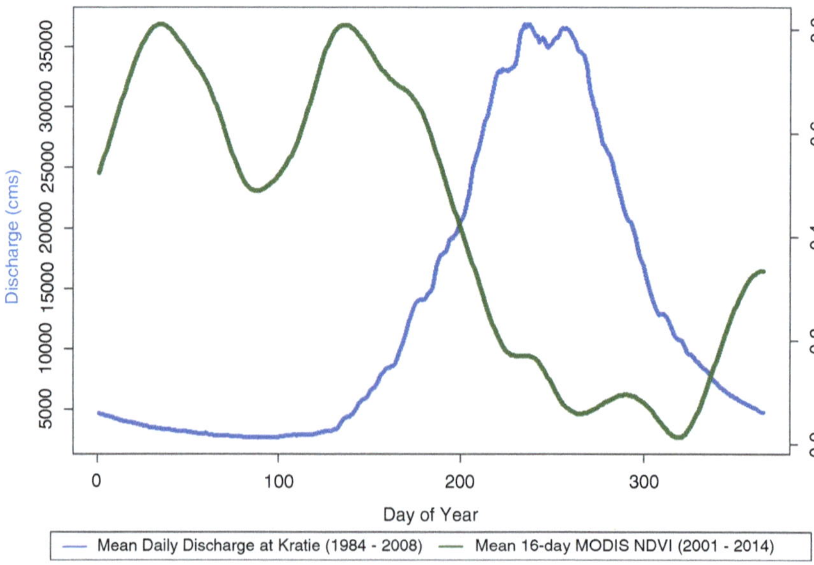

Fig. 6.1 Illustrates variability in regional hydrologic cycle by showing NDVI (*green*) and Discharge (*blue*) as a function of day of year

high NDVI, floodwaters can be detected by analyzing decreases in NDVI relative to historical conditions.

To determine flood extent, near real-time multispectral MODIS imagery is downloaded from NASA LANCE servers. The Normalized Difference Vegetation Index is calculated using the formula:

$$\mathrm{NDVI}_{\mathrm{NRT}} = \left(\frac{\mathrm{NIR} - \mathrm{Red}}{\mathrm{NIR} + \mathrm{Red}}\right) \tag{6.1}$$

where $\mathrm{NDVI}_{\mathrm{NRT}}$ is the near real-time NDVI, NIR is the near infrared spectral band (band 1), and Red is the red band (band 2). $\mathrm{NDVI}_{\mathrm{NRT}}$ is compared to historic $\mathrm{NDVI}_{\mathrm{Dry}}$ conditions that are preprocessed and calculated by averaging the 32-day January NDVI for from 2003 to 2011. Changes between current and historic conditions are detected on a per-pixel basis as a percent difference:

$$\Delta\mathrm{NDVI} = \left(\frac{\mathrm{NDVI}_{\mathrm{Flood}} - \mathrm{NDVI}_{\mathrm{Dry}}}{\mathrm{NDVI}_{\mathrm{Dry}}}\right) \times 100 \tag{6.2}$$

When $\Delta\mathrm{NDVI} < -75$, the pixel is classified as "flooded," and when $\Delta\mathrm{NDVI} > -75\,\%$, the pixel is considered "nonflooded." Permanent water and high slope areas are masked out (Fig. 6.3). This methodology is under refinement and product accuracy assessments are ongoing (Doyle et al., in prep; Fayne et al., in prep). The ΔNDVI images are saved and served to the web using a geoserver (http://geoserver.org).

The flood extent model was tested for accuracy against Landsat imagery (Fig. 6.2) and retrospectively applied to floods that occurred during August–December 2011 (Fig. 6.4a, b). Flood extent derived from MODIS imagery demonstrates good agreement with higher resolution Landsat imagery (Fig. 6.2) as well as Synthetic Aperture Radar (SAR) data (Radarsat 2, Envisat ASAR, TerraSAR-X) acquired from the United Nations Institute for Training and Research Operational Satellite Applications program (UNOSAT) during this time period (Fig. 6.4). The software enables time series analyses and visualizations of flood extent (Fig. 6.5), as well as historic flood extent analysis.

6.2.2 Socioeconomic Impact Estimation

Flood extent data (Figs. 6.3, 6.4 and 6.5) can be fused with socioeconomic data in order to formulate rapid spatial and numerical estimates of population and infrastructure affected by regional flooding. Figure 6.6 illustrates the data processing algorithm to formulate socioeconomic impacts from MODIS-derived flood extent. Static data are preprocessed to match the 250 m resolution of flood extent images (Fig. 6.6, top left and top right). Raster data describing population per grid cell for

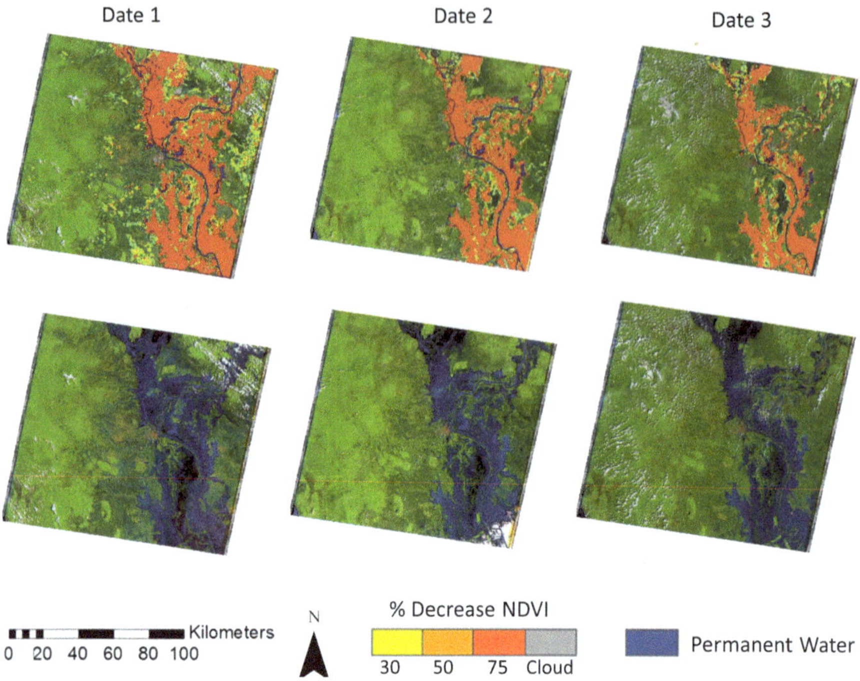

Fig. 6.2 Landsat scenes (*bottom*, Path/Row 126/52) show good agreement for flooded area detection with the corresponding MODIS NDVI change product (*top*) in the Phnom Penh region of Cambodia. Date 1 shows MODIS NDVI change product from 8-day composite acquired 8/29 – 9/5/2006 and Landsat 5 TM acquired 9/3/2006. Date 2 shows MODIS NDVI change product from 8-day composite acquired 10/16/2006 – 10/23/2006 and Landsat 5 TM acquired 10/21/2006. Date 3 shows MODIS NDVI change product from 8-day composite acquired 11/1/2006 – 11/8/2006 and Landsat 5 TM acquired 11/6/2006. Credit: C. Doyle

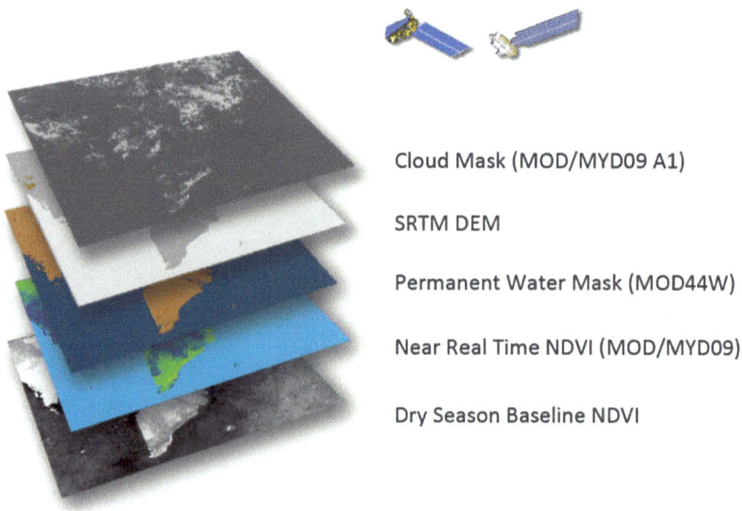

Fig. 6.3 Stack of images used in near real-time flood detection

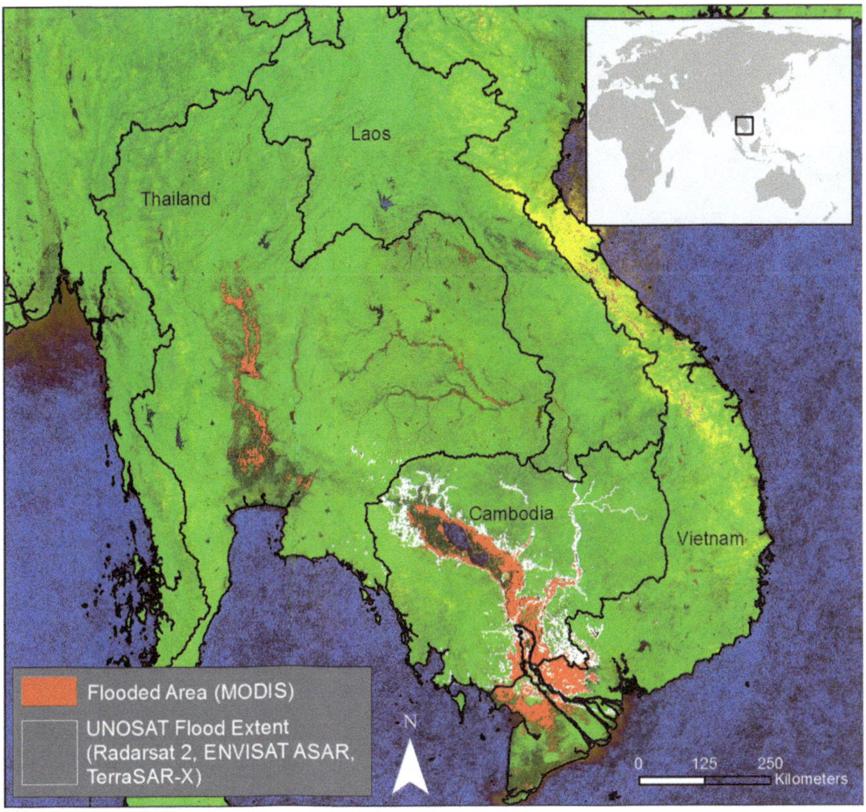

Fig. 6.4 (**a**) Comparison between 2011 flood extent determined from 4-Day composited MODIS imagery obtained on October 16th, 2011 and flood extent determined from the United Nations Institute for Training and Research (UNITAR) Operational Satellite Applications Programme (UNOSAT; white) using Radarsat 2, ENVISAT ASAR, and TerraSAR-X imagery acquired on August 28th, September 21st, 23rd, 30th, October 17th, 18th, and 27th. The background image is a false color image of eight MODIS scenes composited between September 27, 2011 and November 30th, 2011. (**b**) Zoom of the previous image demonstrating agreement between MODIS flood detection method and UNOSAT SAR flood detection

2015 were obtained for Cambodia, Thailand, Laos, and Vietnam from the Worldpop data set (Sorichetta et al. 2015; https://worldpop.org). Worldpop uses a number of covariates and a random forest model (Stevens et al. 2015) to estimate population at 100 m resolution. The data were resampled to 250 m resolution to match MODIS resolution.

Data describing the locations of roads, hospitals, airports, rice paddy, and schools in Cambodia were obtained from sources described in Table 6.1. Point data were converted to rasters matching MODIS resolution with pixel values signifying point density of features within each grid cell. The length of roadways within each grid cell was calculated and a float-type raster image describing roadway length per cell was preprocessed.

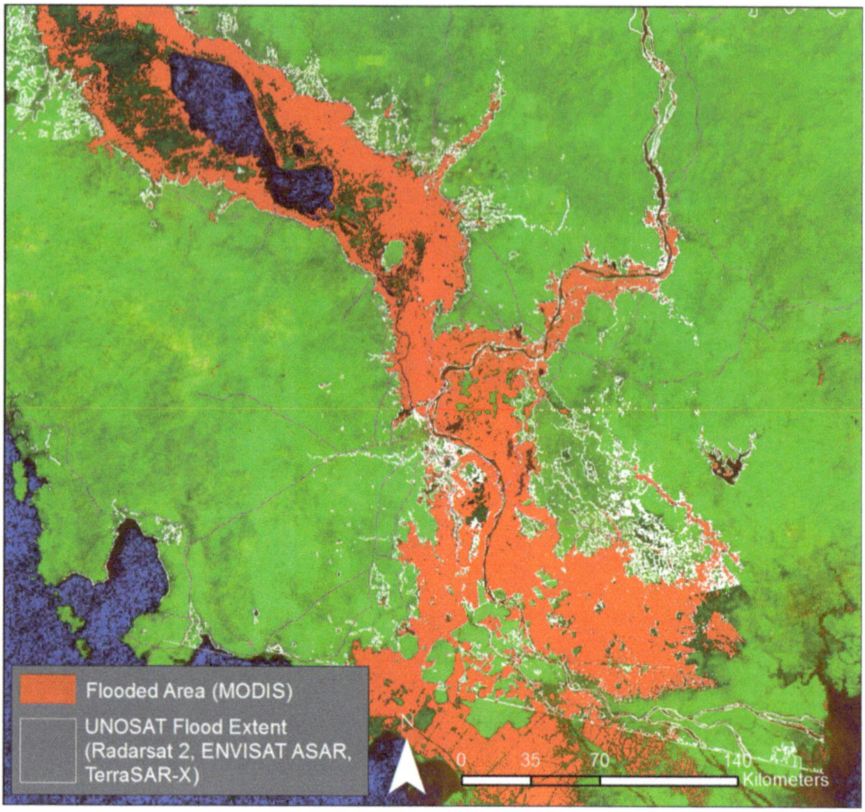

Fig. 6.4 (continued)

Once flood extent algorithms are applied to MODIS data, flood extent data is read as a presence/absence array and socioeconomic data are read as float arrays and multiplied (Fig. 6.6), resulting in estimates of affected population (Fig. 6.7) and infrastructure (Fig. 6.8) per grid cell. These estimates are summed, written to text files, and served live to a flood dashboard as interactive and downloadable graphs (Fig. 6.9). The arrays are also converted to raster imagery so results can be visualized and manipulated in geographic information system (GIS) software (Fig. 6.8).

6.2.3 Comparison with Agency and Government Estimates

Socioeconomic impact estimates (Figs. 6.7 and 6.8) were computed for severe flooding that occurred in 2011 (Figs. 6.3, 6.4 and 6.5) and compared against United Nations or regional government damage estimates (Tables 6.2 and 6.3). Estimates of affected population agree reasonably well with agency estimates for Cambodia

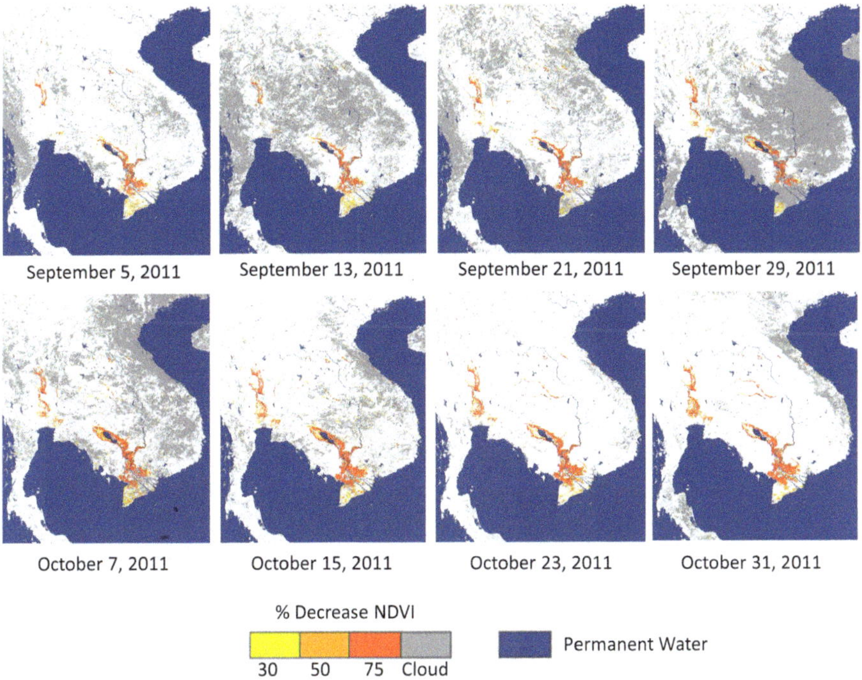

Fig. 6.5 Time series of MODIS NDVI flood extent model applied to MODIS images acquired during widespread flooding occurring September and October 2011 in Southeast Asia. Credit: C. Doyle

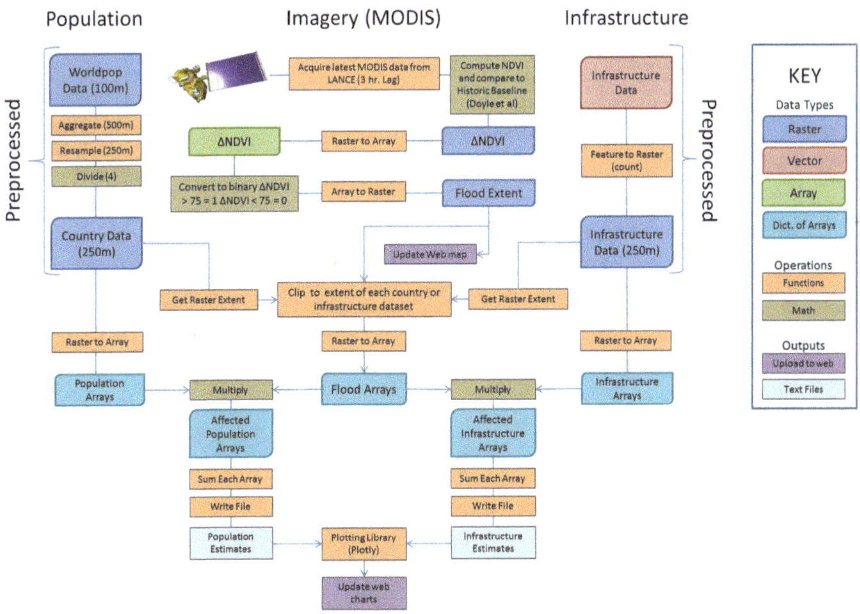

Fig. 6.6 Schematic of data processing workflow for socioeconomic impact estimation model. The python model framework operates in near real time as new MODIS imagery becomes available

Table 6.1 Summary of data sources used in this study

Data metric	Source	Link
Population	Worldpop	http://worldpop.org
Roads	UNOCHA	http://geodata.pdc.org/geodata/unocha/
Hospitals	UNOCHA	http://geodata.pdc.org/geodata/unocha/
Airports	UNOCHA	http://geodata.pdc.org/geodata/unocha/
Schools	Humanitarian Data Exchange	https://data.hdx.rwlabs.org/dataset/cambodia-education
Rice paddy	Open Development Cambodia	http://www.opendevelopmentcambodia.net/download_maps/rice-ecosystem/

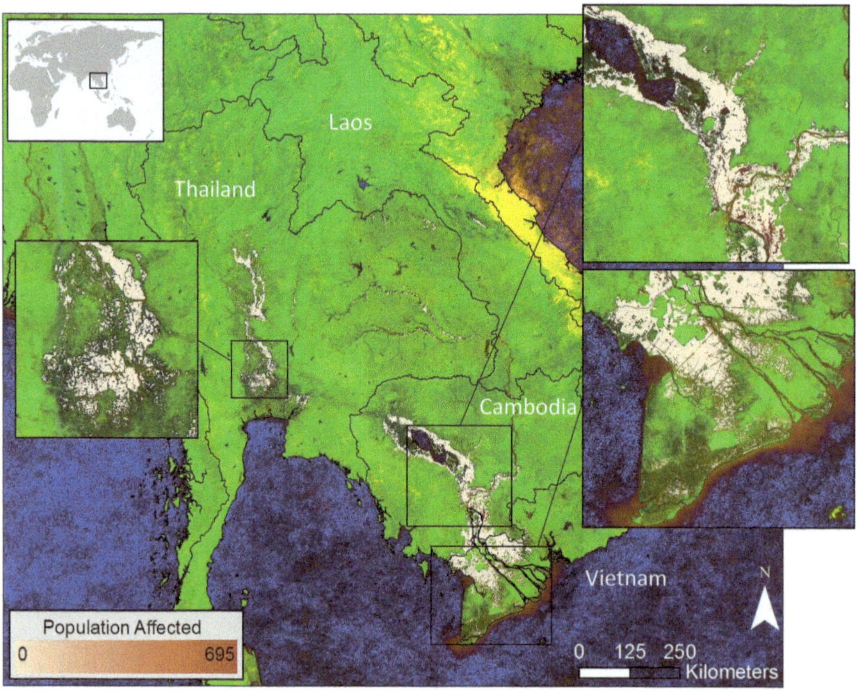

Fig. 6.7 Shows people per grid cell affected by flooding for the 2011 floods around Bangkok, Thailand, near Tonle Sap Lake, Cambodia, and in the Mekong Delta, Vietnam

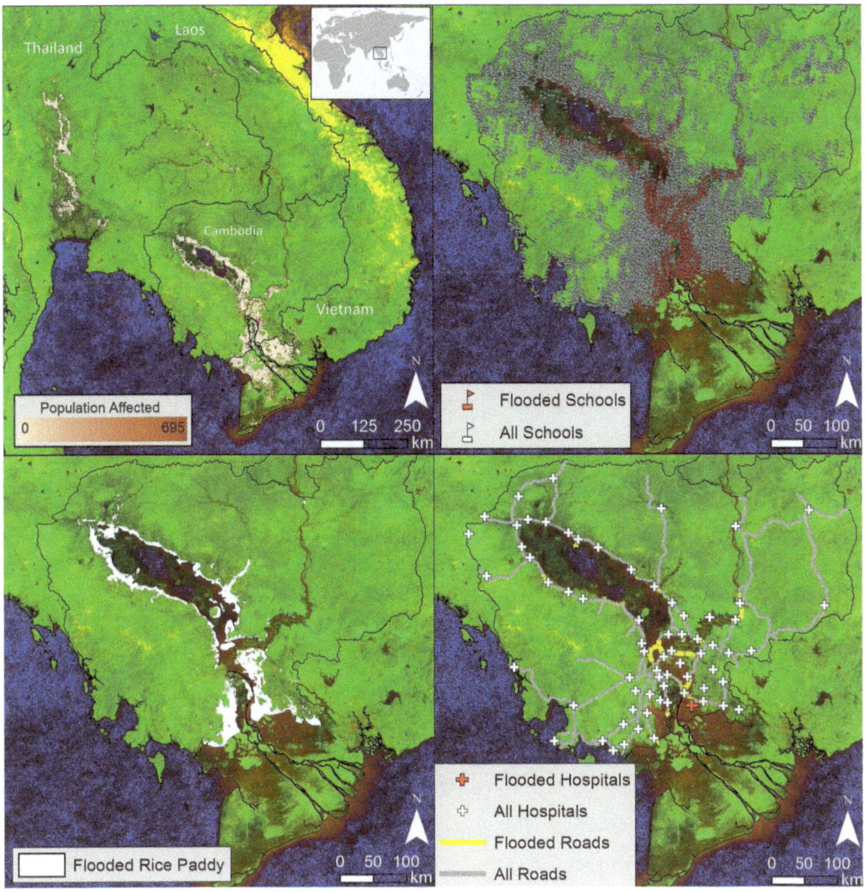

Fig. 6.8 Depicts people per grid cell, (*top left*), schools (*top right*), rice paddy plantation areas, (*bottom left*) and roads and hospitals (*bottom right*) affected by the 2011 flood event for Cambodia. Table 6.3 shows numerical estimates of affected population and infrastructure

and Thailand but are substantially different for Vietnam and Laos (Table 6.2; Fig. 6.7). Potential reasons for this discrepancy include lack of clear agency methods in determining impacted population as well as incomplete flood extent or population data. Estimates of affected infrastructure (roads, rice, schools) tested for the 2011 floods in Cambodia agree well with agency estimates (Table 6.3; Fig. 6.8). These results are archived with each MODIS pass and displayed in interactive graphics on the dashboard (e.g., Fig. 6.9).

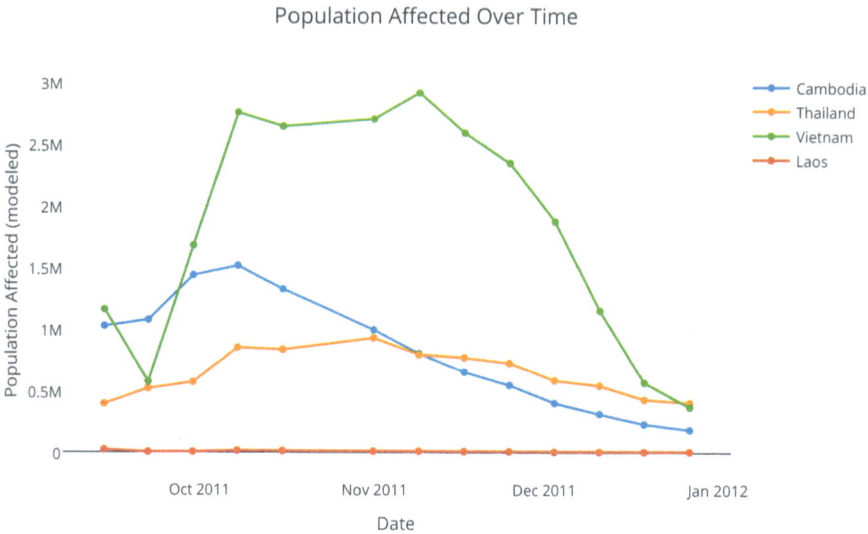

Fig. 6.9 Time series plots generated from 8-day composite images between September and December 2011 showing the rise and decline of population affected by severe flooding

Table 6.2 Estimates of population impacted by 2011 flooding modeled from 4-day image composited on October 16, 2011

	Cambodia	Thailand	Vietnam	Laos
Population affected (modeled)	1,864,988	1,409,507	4,078,799	25,386
Population affected (Agency or Government) estimate	1,500,000	2,100,000	700,000	430,000
Source	USAID (2011)	USAID (2011)	USAID (2011)	USAID (2011)

USAID Source: http://pdf.usaid.gov/pdf_docs/PA00J4C7.pdf

Table 6.3 Results of socioeconomic impact estimation model for Cambodia

Metric	2011 Cambodia flood	Other estimate(s) and source
Persons affected (count)	1,864,988	1,200,000 (UN)
		1,600,000 (Chea and Sharp 2015)
Major roadways (km)	208	180 (NCDM)
Rice paddy (km^2)	4593	3900 (NCDM)
		4870 (Chea and Sharp 2015)
Schools (count)	818	1132 (NCDM)
		1360 (Chea and Sharp 2015)
Hospitals	1	N/A
Airports	0	N/A

Most agency estimates are made available by the National Center for Disaster Management (NCDM)

6.3 Conclusions

During disasters, determining affected population and infrastructure should preclude the allocation of life-saving humanitarian resources such as food, shelter, health supplies, water, sanitation, and hygiene. Automating near real-time flood extent and socioeconomic impact analyses allows for estimates to be tallied for events with sudden or unexpected onset. Systems described here employ open source software and cloud-based infrastructure, promoting easy access and cost-effective solutions that can strengthen the decision-making process with the best and newest available data.

These systems have the potential to save lives, limit infrastructure damage, aid in urban and agricultural planning, and safeguard economic and community livelihoods. Though tailored for floods in Southeast Asia, similar systems may be applicable to various disaster scenarios in other regions of the world.

Acknowledgments Research was funded by a grant from the NASA Applied Sciences Program. We would like to thank NASA DEVELOP National Program and the Southeast Asia Disasters teams for carrying out much of the exploratory research and automation of analysis. Also, we would like to acknowledge the USFS Forecast Mekong project and Jerry Gasser, Lockheed Martin Stennis Operations and James Smoot, CSC who helped process and provide the needed MODIS monthly NDVI data. This work made use of the Open Science Data Cloud (OSDC) which is an Open Cloud Consortium (OCC)-sponsored project. This work made use of Geospatial Data Abstraction Library (GDAL) which is an Open Geospatial Consortium (OGC) project. This work was supported in part by grants from Gordon and Betty Moore Foundation and the National Science Foundation and major contributions from OCC members like the University of Chicago. Thanks to Matt Handy and Anika Cartas for assistance in development of the Southeast Asia Flood Dashboard.

References

Brakenridge GR, Anderson E, Nghiem SV, Caquard S, Shabaneh TB (2003) Flood warnings, flood disaster assessments, and flood hazard reduction: the roles of orbital remote sensing. Jet Propulsion Laboratory, National Aeronautics and Space Administration, Pasadena, p 869

Brakenridge R, Anderson E (2006) MODIS-based flood detection, mapping and measurement: the potential for operational hydrological applications. Transboundary floods: reducing risks through flood management, 1–12

Chen Y, Huang C, Ticehurst C, Merrin L, Thew P (2013) An evaluation of MODIS daily and 8-day composite products for floodplain and wetland inundation mapping. Wetlands 33(5):823–835

Chea and Sharp (2015) Flood Management in Cambodia: case studies of flood in 2009 and 2011. 2015 International Conference on Environment And Civil Engineering

Davies GI, McIver L, Kim Y, Hashizume M, Iddings S, Chan V (2014) Water-borne diseases and extreme weather events in Cambodia: review of impacts and implications of climate change. Int J Environ Res Public Health 12(1):191–213

Doocy S, Daniels A, Murray S, Kirsch TD (2013) The human impact of floods: a historical review of events 1980–2009 and systematic literature review. PLoS Curr 5

Falck AS, Maggioni V, Tomasella J, Vila DA, Diniz FL (2015) Propagation of satellite precipitation uncertainties through a distributed hydrologic model: a case study in the Tocantins–Araguaia basin in Brazil. J Hydrol 527:943–957

IPCC (2007) Climate change 2007: the physical science basis. Cambridge University Press, Cambridge

Islam AS, Bala SK, Haque MA (2010) Flood inundation map of Bangladesh using MODIS time-series images. J Flood Risk Manag 3(3):210–222

Kussul N, Shelestov A, Skakun S (2011) Flood monitoring from SAR data. In: Kogan F, Powell A, Fedorov O (eds) Use of satellite and in situ data to improve sustainability. Springer Netherlands, Heidelberg, pp 19–29

Kuenzer C, Guo H, Huth J, Leinenkugel P, Li X, Dech S (2013) Flood mapping and flood dynamics of the Mekong Delta: ENVISAT-ASAR-WSM based time series analyses. Remote Sens 5(2):687–715

Munich Re (2013) NatCatSERVICE database. Munich Reinsurance Company, Munich

Nigro J, Slayback D, Policelli F, Brakenridge GR (2014) NASA/DFO MODIS Near Real-Time (NRT) global flood mapping product evaluation of flood and permanent water detection, October 2014. http://oas.gsfc.nasa.gov/floodmap/documents/NASAGlobalNRTEvaluationSummary_v4.pdf

Orth R, Staudinger M, Seneviratne SI, Seibert J, Zappa M (2015) Does model performance improve with complexity? A case study with three hydrological models. J Hydrol 523:147–159

Sorichetta A, Hornby GM, Stevens FR, Gaughan AE, Linard C, Tatem AJ (2015) High-resolution gridded population datasets for Latin America and the Caribbean in 2010, 2015, and 2020. Scientific data, 2. doi: 10.1038/sdata.2015.45

Stevens FR, Gaughan AE, Linard C, Tatem AJ (2015) Disaggregating census data for population mapping using Random forests with remotely-sensed and ancillary data. PLoS One 10(2): e0107042

Tucker CJ (1979) Red and photographic infrared linear combinations for monitoring vegetation. Remote Sens Environ 8(2):127–150

United Nations (2004) Guidelines for reducing flood losses. https://sustainabledevelopment.un.org/content/documents/flood_guidelines.pdf

Wu H, Adler RF, Tian Y, Huffman GJ, Li H, Wang J (2014) Real-time global flood estimation using satellite-based precipitation and a coupled land surface and routing model. Water Resour Res 50(3):2693–2717

Part II
Droughts

Chapter 7
Remote Sensing of Drought: Vegetation, Soil Moisture, and Data Assimilation

Ali Ahmadalipour, Hamid Moradkhani, Hongxiang Yan, and Mahkameh Zarekarizi

7.1 Introduction

Drought is a natural recurring hazard with slow onset affecting large areas. It can impose severe impacts on agriculture, hydropower generation, water supply, environment, society, and economy. Drought is a complex phenomenon and is classified into four types including meteorological (lack of precipitation), agricultural (deficit in soil moisture, and vegetation response), hydrological (deficit in runoff, streamflow, or groundwater storage), and socioeconomic (social responses to water supply and demand) droughts (Mishra and Singh 2010). The impacts of drought have aggravated primarily due to the increase in vulnerability of assets, as well as the changes in frequency of drought events due to climate change (Ahmadalipour et al. 2016; Sahoo et al. 2015). Therefore, it is necessary to improve the current drought-monitoring tools and develop reliable methods for prediction and early warning of potential droughts (Svoboda et al. 2002; Wan et al. 2004).

Drought monitoring has been historically carried out mainly using ground-based observations (Svoboda et al. 2002). The monitoring had been more focused on meteorological and agricultural droughts by means of statistical indices (Palmer 1965). However, many regions do not have adequate gauge instruments (e.g., remote regions or agricultural areas) to obtain required precipitation, temperature, relative humidity, and wind speed data that are needed for accurate assessment of drought. Furthermore, gage (point) data do not capture the spatial variability of drought. Therefore, spatial interpolation techniques were used to determine drought condition of unsampled locations.

A. Ahmadalipour, M.S. • H. Moradkhani, Ph.D. (✉) • H. Yan, M.S. • M. Zarekarizi, M.S.
Remote Sensing and Water Resources Lab, Department of Civil and Environmental Engineering, Portland State University, 1930 SW 4th Ave., suite 200, Portland, OR 97201, USA
e-mail: aahmad2@pdx.edu; hamidm@pdx.edu; yan2@pdx.edu; mahkam2@pdx.edu

V. Lakshmi (ed.), *Remote Sensing of Hydrological Extremes*, Springer Remote Sensing/Photogrammetry, DOI 10.1007/978-3-319-43744-6_7

Remote-sensing imagery has been utilized from various aspects in weather and climatology studies. The Television Infrared Observation Satellite (TRIOS-1) was the first meteorological satellite and was launched in 1960. It started a new era of earth monitoring and led to the development of additional satellites such as the Advanced Very High Resolution Radiometer (AVHRR), Landsat, and the Geostationary Operational Environmental Satellites (GOES).

Satellite measurements overcome the limitations of gauge-based meteorological observation. They are cost-effective with high spatial resolution, which provide consistent and continuous monitoring of the Earth environment (Rhee et al. 2010; Zhang and Jia 2013). Remote-sensing imagery has also been used for drought-monitoring purposes. Satellite-based drought indices are capable of characterizing spatial variability of drought, and therefore, they became the most promising tools for monitoring drought at regional scales (Kogan 1997).

Generally, the application of satellite data in drought studies can be classified in two categories: an atmospheric perspective focusing on acquiring drought-related atmospheric variables from satellites (e.g., precipitation or relative humidity), as well as a land surface standpoint which is associated with earth surface. The latter includes variables that are used directly for drought-monitoring purposes (e.g., NDVI), which is capable of determining the ecosystem impacts of drought; as well as the variables used indirectly for assimilating state variables to hydrologic and land surface models (e.g., soil moisture or snow cover) in order to improve the accuracy of hydrologic predictions (DeChant and Moradkhani 2014a, 2015).

During the past decade, a number of satellite instruments have been launched. They provide observation for various environmental variables, some of which are useful for drought-monitoring purposes from different perspectives (e.g., precipitation, soil moisture, snow cover, land surface temperature, terrestrial water storage, evaporation, and vegetation). These progresses along with the analytical/modeling advances have eventuated in many recent remote-sensing tools and products that are applicable for drought-monitoring purposes.

This chapter highlights the advances in earth remote sensing and provides an insight into satellite-based drought monitoring and prediction. First, from a hydrometeorological perspective, the remote-sensing retrievals of different variables are introduced along with their features and attributes. Furthermore, some of the drought indices related to each variable are explained. Then, the satellite instruments acquiring vegetation health data are introduced along with the development and progresses of vegetation health indices. The application of one of the most common vegetation health indices is then investigated for long-term assessment of agricultural drought. Finally, the latest advances in assimilation of satellite data into hydrologic models for drought monitoring and prediction purposes are described.

7.2 Remote Sensing of Hydrologic Variables for Drought Assessment

7.2.1 Precipitation

Precipitation is one of the primary meteorological variables, which has a direct impact on drought. Because of large spatial variability as well as dynamic nature of precipitation, it is difficult to generate an accurate spatial pattern of precipitation from point measurements. Obtaining gridded mean areal precipitation (MAP) is essential since most of the current distributed hydrologic models require gridded input data. Hence, satellite estimation of precipitation has the potential to be used for hydrologic modeling as well as drought-monitoring purposes.

After the launch of TIROS-1 in 1960, satellite observations of weather provided an invaluable tool for weather forecast. This lead to the establishment of the World Weather Watch program in 1963 by the World Meteorological Organization (WMO) to commence the application of a network of operational geostationary (GEO) as well as low Earth orbiting (LEO) meteorological satellites referred to as the Global Observing System (GOS) (Kidd et al. 2011). The remarkable capabilities of satellite observations demanded constant advancement and development of new algorithms and methods for further improving the quality of observations. The main input to all of these methods is the brightness temperature which is acquired from infrared instruments in geostationary orbit satellites and allows them to provide continuous space–time precipitation measurement.

Several satellite precipitation missions have provided data which have been utilized for various applications. Among these datasets are the CPC Merged Analysis of Precipitation (CMAP) (Xie and Arkin 1997), Special Sensor Microwave Imager (SSM/I) on board the Defense Meteorological Satellite Program (DMSP) platform (Ferraro 1997), Global Precipitation Climatology Project (GPCP) (Adler et al. 2003), Climate Predicting Center (CPC) Morphing Technique (CMORPH) (Joyce et al. 2004), Tropical Rainfall Measuring Mission (TRMM) Multi-satellite Precipitation Analysis (TMPA) (Huffman et al. 2007), Precipitation Estimation from Remotely Sensed Information using Artificial Neural Networks (PERSIANN) (Ashouri et al. 2015; Sorooshian et al. 2000); all of which have been widely used for different climate and hydrologic applications. Recently, National Aeronautics and Space Administration (NASA) started a new era in the satellite remote sensing of precipitation through the Global Precipitation Measurement (GPM) mission (Hou et al. 2014) to develop the Integrated Multi-satellitE Retrievals for GPM (IMERG) (Huffman et al. 2015). Table 7.1 provides a summary of characteristics of the above-mentioned satellite precipitation measurements.

Although satellite precipitation retrievals provide useful high-resolution global precipitation estimates, they are associated with high uncertainty and bias (Alemohammad et al. 2015; Sorooshian et al. 2008). It has been shown that the uncertainty in precipitation (input uncertainty) is the major source of uncertainty in hydrologic modeling and ensemble streamflow prediction (Hong et al. 2006;

Table 7.1 Summary of the characteristics of some of the satellite-based precipitation measurement missions

Product abbreviation	Period	Temporal resolution	Spatial resolution	References
CMAP	1979–Present	Monthly	2.5 Deg.	Xie and Arkin 1997
SSM/I	1988–Present	Daily	12.5 km	Ferraro 1997
GPCP	1979–Present	Monthly	2.5 Deg.	Adler et al. 2003
CMORPH	2002–Present	Daily	0.25 Deg.	Joyce et al. 2004
TMPA	1998–Present	3 h	0.25 Deg.	Huffman et al. 2007
PERSIANN	2000–Present	3 h	0.25 Deg.	Sorooshian et al. 2000
PERSIANN-CDR	1983–Present	Daily	0.25 Deg.	Ashouri et al. 2015
IMERG	2014–Present	0.5 h	0.1 Deg.	Huffman et al. 2015

Moradkhani et al. 2005, 2006). Because of indirect measurement of satellite precipitation, they are noisy with lower accuracy than ground measurements, and sometimes limited to space and time.

Satellite-based precipitation products have been extensively evaluated in many studies (Yong et al. 2016). The issue of high uncertainty and large biases in satellite precipitation has urged some researchers quantify and characterize precipitation uncertainty for various applications (Alemohammad et al. 2015), as well as developing post-processing techniques which are capable of bias correcting these precipitation products (Madadgar et al. 2014). Recently, Khajehei (2015) presented a new ensemble post-processing approach to bias correct precipitation data using Copula functions and account for uncertainties through their ensemble generating approach. The method is capable of bias correcting monthly precipitation and generating post-processed ensemble precipitation data. This provides more accurate precipitation forecast eventuating in more reliable meteorological drought characteristics.

Satellite precipitation measurements have been widely utilized for drought-monitoring and characterization purposes (Zhang and Jia 2013). Standardized Precipitation Index (SPI) (McKee et al. 1993) is among the most widely used drought indices that solely requires precipitation data. However, it needs long period of data (about 30 years) for accurate characterization of drought. Therefore, some of the more recent missions (e.g., TMPA), although having acceptable accuracy in many regions, are not yet practical for data-driven drought indices such as SPI. Nevertheless, remotely sensed precipitation has been utilized in some recent studies for drought-monitoring purposes (Sahoo et al. 2015; Zhang and Jia 2013). Some studies have utilized remotely sensed data as the input to land surface models to estimate other hydro-meteorological fluxes such as runoff or soil moisture and analyzed them from a different perspective (Zulkafli et al. 2014). The availability of long-term records for satellite-based precipitation, e.g., GPCP or PERSIANN-CDR, provides the opportunity to assess drought from a climatological perspective.

7.2.2 *Surface Soil Moisture*

Soil moisture generally refers to the amount of water stored in the unsaturated zone. Soil moisture is a key variable in examining the global water-energy cycle, agricultural planning, and water resources management (Seneviratne et al. 2010). In 2010, the Global Climate Observing System (GCOS) included soil moisture in the list of 50 Essential Climate Variables (ECVs) to support the work of the United Nations Framework Convention on Climate Change (UNFCCC) and the Intergovernmental Panel on Climate Change (IPCC) (Bojinski et al. 2014).

Contrary to the meteorological and hydrological drought, the agricultural drought is defined based on soil moisture availability in the root-zone. The soil moisture can be measured either by ground sensors or satellite sensors. Networks of ground sensors are only available for limited regions and cannot represent the spatial heterogeneity of soil moisture. As an alternative, the remotely sensed surface soil moisture retrievals can provide an unprecedented spatial and temporal resolution of soil moisture data across a range of scales. Especially, microwave satellites (C/X/L bands) have shown useful retrievals of surface soil moisture as they are able to penetrate cloud cover and are sensitive to soil moisture (Njoku et al. 2003). The principles of the microwave retrievals are based on the sensitivity of passive brightness temperature and active backscattering to the soil moisture content.

A series of passive/active microwave sensors have been launched since 1978 to measure soil moisture, including the Scanning Multichannel Microwave Radiometer (SMMR), the Special Sensor Microwave Imager (SSM/I), the microwave imager from the Tropical Rainfall Measuring Mission (TRMM), the European Remote Sensing scatterometer, the Advanced Microwave Scanning Radiometer, the Advanced Scatterometer (ASCAT), the Soil Moisture Ocean Salinity Satellite (SMOS), the Advanced Microwave Scanning Radiometer2 (AMSR2), and the Soil Moisture Active Passive (SMAP) (Moradkhani 2008; Entekhabi et al. 2010; McMullan et al. 2008; Wagner et al. 2013). Besides the single sensor soil moisture products, Liu et al. (2011) developed a blended soil moisture product by merging both passive and active soil moisture from multiple sensors. The long-term blended soil moisture products (1978–2014) can be used for agricultural drought monitoring and trend studies. Table 7.2 summarizes the microwave sensors used for soil moisture estimation.

7.2.3 *Evapotranspiration*

Evapotranspiration (ET) is a crucial variable in the hydrologic cycle and one of the major hydrological fluxes (Sima et al. 2013). It has huge impacts on agriculture, irrigation water demand, and consequently crop yield. Similar to precipitation, gauge-based ground observations of ET are not capable of quantifying spatial

Table 7.2 Summary of the microwave sensors for soil moisture estimation

Sensor	Channel	Temporal resolution	Spatial resolution (km)	Data period
Scanning Multichannel Micro-wave Radiometer (SMMR)	Multiple from 6.6 GHz	Daily	140	1978–1987
Special Sensor Microwave Imager (SSM/I)	Multiple from 19.4 GHz	Daily	25	1987–Present
Microwave Imager TRMM	Multiple from 10.7 GHz	Daily	25	1997–2015
Scatterometer ERS	5.3 GHz	3–4 days	50	1992–Present
Advanced Microwave Scanning Radiometer (AMSR-E)	Multiple from 6.9 GHz	Daily	25	2002–2011
Advanced Scatterometer (ASCAT)	5.3 GHz	1–2 days	25	2006–Present
Soil Moisture Ocean Salinity Satellite (SMOS)	1.4 GHz	2–3 days	35	2009–Present
Advanced Microwave Scanning Radiometer2 (AMSR2)	Multiple from 6.9 GHz	Daily	25	2012–Present
Soil Moisture Active Passive (SMAP)	1.4 GHz	2–3 days	3, 9, 36	2015–Present

variations. Meanwhile, remotely sensed data provides valuable tools to estimate ET at large scales.

The application of actual and potential evapotranspiration (ET and PET) for drought analysis has been developed and examined in numerous studies, and several different drought indices have been developed to characterize drought from various perspectives. Palmer Drought Severity Index (PDSI) (Palmer 1965), Crop Water Stress Index (CWSI) (Jackson et al. 1981), Supply-Demand Drought Index (SDDI) (Rind et al. 1990), Reconnaissance Drought Index (RDI) (Tsakiris and Vangelis 2005), Evaporative Drought Index (EDI) (Yao et al. 2010), Standardized Precipitation Evapotranspiration Index (SPEI) (Vicente-Serrano et al. 2010), Evaporative Stress Index (ESI) (Anderson et al. 2011, 2013), Drought Severity Index (DSI) (Mu et al. 2013), Green Water Scarcity Index (GWSI) (Núñez et al. 2013), and Green Water Stress Index (GrWSI) (Wada 2013) are among the drought variables that incorporate evapotranspiration into their assessment.

One of the satellite instruments that is extensively practical for drought analysis from evapotranspiration viewpoint is the Moderate Resolution Imaging Spectroradiometer (MODIS). Yao et al. (2010) proposed Evaporative Drought Index (EDI) to monitor droughts over the Contiguous United States (CONUS). Mu et al. (2007, 2011) developed a technique to use MOD16 products in order to generate a global ET/PET dataset with an algorithm based on Penman-Monteith equation. Then, they developed Drought Severity Index (DSI), which incorporates ET/PET data as well as vegetation status (Mu et al. 2013).

Air temperature, radiation, and air relative humidity are among the primary variables affecting evapotranspiration (Yin et al. 2014). Therefore, satellite missions measuring each of these variables are useful in understanding evapotranspiration and drought. Currently, several remote-sensing platforms provide continuous worldwide measurement of land surface temperature (LST) including Terra/Aqua-MODIS, Terra-Aster, NOAA-AVHRR, Meteosat-MVIRI, Landsat, and the ATSR series of instruments (Westermann et al. 2011). These LST products have been evaluated in several studies at various regions, and results show that they provide accurate measurements with slight positive bias in some cases (Crosman and Horel 2009; Sima et al. 2013). Furthermore, numerous studies have focused on estimating air temperature from land surface temperature measurement of satellite products (Benali et al. 2012; Shen and Leptoukh 2011). Temperature data can also be utilized to calculate radiation, another variable affecting evapotranspiration.

7.2.4 Terrestrial Water Storage and Groundwater

Since 2002, regional estimates of changes in the terrestrial water storage (TWS), i.e., water content both on and below the land surface, have been provided by the Gravity Recovery and Climate Experiment (GRACE) mission on a monthly timescale (Reager and Famiglietti 2009). Changes in TWS include the changes in surface water storage, soil moisture storage, and groundwater storage. GRACE consists of two twin satellites that orbit the Earth with about 220 km distance at an altitude of about 500 km. Changes in Earth's gravity field alter the distance between the two satellites, and precisely acquiring the distance, makes it possible to determine the TWS anomalies. The distance between the twin satellites is determined using a highly accurate inter-satellite K-Band microwave ranging system. GRACE tracks the variations in gravity field of an area, which is then translated into estimates of TWS (Rodell and Famiglietti 2002).

GRACE data provides a valuable tool for observing the water availability (Velicogna et al. 2012). It has been utilized to study regional climate variability and drought (Chen et al. 2013; Zhang et al. 2015b), assess the impacts of water availability on vegetation growth (Velicogna et al. 2015), and assimilate into land surface models (Reager et al. 2015). Many studies have evaluated GRACE data across various regions (Feng et al. 2013). Some studies have tried to utilize GRACE data to estimate different hydrological variables. Syed et al. (2005) developed an approach to estimate basin discharge using GRACE data. Rodell et al. (2004) employed GRACE data to estimate basin-scale evapotranspiration.

One of the limitations of GRACE is that it provides data at a spatial resolution of more than 150,000 km^2 per grid for TWS, which makes it impractical for regional drought assessments. Furthermore, currently the GRACE TWS timeseries contains about 14 years of data, which is not sufficient for drought assessment from a climatological perspective. Therefore, researchers have utilized statistical methods, e.g., Artificial Neural Networks, to extend and reconstruct TWS data for previous years (Zhang et al. 2015a), as well as downscaling GRACE TWS data to a finer spatial resolution (Zaitchik et al. 2008). Information offered by GRACE TWS data can be employed for investigating the impacts of drought and estimating the amount of precipitation required for drought recovery.

7.2.5 Snow Cover

Snow cover is an important characteristic of land surface as it can impact surface radiation, energy, and hydrologic cycle (Kongoli et al. 2012). It is a natural reservoir of water resources and a great source of soil moisture supply for agricultural crops. Snowmelt contributes to a considerable fraction of runoff in many mountainous regions. A shortage in winter snowpack can yield to a hydrological drought in summer, and a deficit in soil moisture during winter can result in agricultural drought. Therefore, monitoring snow is an important task for water management and it provides an essential tool for drought prediction in many regions.

Snowpack measurements that are important in hydrology are the Snow Water Equivalent (SWE), Snow Depth (SD), Snow Covered Area (SCA), and Snow Albedo (SA) (Kongoli et al. 2012). SWE indicates the amount of water contained in the snowpack (which is calculated by multiplying SD by the effective snow cover density). It is the most important parameter for runoff during the snowmelt season. In the western US and Alaska, an automated SWE monitoring system is maintained by the U.S. Department of Agriculture (USDA), which is called the SNOwpack TELemetry (SNOTEL) (Molotch and Bales 2006).

Remote sensing of snow can be classified into three methods of optical, microwave (MW), and the combination of optical and MW. Optical measurements can only provide estimate for SCA, whereas MW and combined methods can measure SCA, SC, and SWE. Snow has an intense spectral gradient in reflectance from a high albedo in visible ranges to low reflectance in the middle infrared wavelengths (Dozier et al. 2009; Lettenmaier et al. 2015). Therefore, the visible reflectance (R_{VIS}) and the middle IR reflectance (R_{mIR}) can be utilized to monitor snow (Romanov et al. 2000). The Normalized Difference Snow Index (NDSI) is defined as $(R_{\mathrm{VIS}} - R_{\mathrm{mIR}})/(R_{\mathrm{VIS}} + R_{\mathrm{mIR}})$ and was developed to detect snow using reflectance information (Dozier 1989; Hall et al. 2002). Former to this attempt, several algorithms were developed for optical-based remote sensing of snow from AVHRR satellite data.

The accuracy and availability of optical-based snow estimates are affected by the presence of clouds (Lettenmaier et al. 2015). Therefore, microwave measurement of snow can provide a practical alternative for temporally continuous snow monitoring. Microwave sensors are available onboard polar-orbiting satellites, e.g., MODIS and AMSR-E, which provide less frequent observations compared to optical sensors onboard geostationary satellites. Thus, many studies have tried to improve the accuracy of snow-related measurements. In some studies, snow retrieval algorithms are introduced that merge snow observations from both optical and microwave sensors (Foster et al. 2011). Remotely sensed snow observations have also been evaluated for various regions (Moser et al. 2011).

Snow data has been used in numerous studies for drought monitoring and prediction purposes (Guan et al. 2013). Many studies have focused on assimilating snow observations into land surface models to improve the accuracy of streamflow simulations, and consequently, hydrological drought prediction (Dong et al. 2007). Since snow does not directly relate to drought and because there is a spatial and temporal lag between the snow occurrence and its impact on surface and sub-surface water availability, snow-based indicators of drought are not developed yet.

7.3 Remote Sensing of Vegetation for Drought Analysis

Agricultural production, especially in less developed regions, is dependent on weather conditions and rainfall to a great extent (Zhang et al. 2016). The changes in climate along with increase in food demand due to the increase in population have resulted in substantial aggravation of the agricultural drought risk over the past decades (Naumann et al. 2015). Studies have shown that climate change and severe drought events have robust negative impacts on agricultural production (Lobell et al. 2014).

Vegetation abundance and vegetation health information are strongly related to rainfall and can be utilized for drought assessments (Tucker 1979). Among various remote sensing-based indices, the Normalized Difference Vegetation Index (NDVI) has been extensively utilized for drought-monitoring purposes. Studies have found a significant relationship between NDVI and soil moisture and precipitation (Wang et al. 2001). In 1979, the first AVHRR instrument was launched, which revolutionized remote sensing of drought by providing high resolution data for monitoring vegetation condition. In fact, the application of time-series satellite observations for drought monitoring began in 1980s after the availability of AVHRR NDVI data (Tucker and Choudhury 1987). NDVI is defined as the difference between reflected near-infrared (NIR) and red (R) radiances divided by their sum; NDVI $= (\mathrm{NIR} - \mathrm{R})/(\mathrm{NIR} + \mathrm{R})$ (Rouse et al. 1974). Studies have shown that NDVI is related to the leaf area index (LAI) and to the photosynthetic activity of green vegetation (Rojas et al. 2011). Therefore, NDVI can be used during the growing season as a means to monitor vegetation condition and drought.

Several drought indicators have been introduced which consider vegetation condition, surface temperature, and a combination of both. The application of remotely sensed data for drought detection was generalized through the development of the Vegetation Condition Index (VCI) defined as $\mathrm{VCI} = (\mathrm{NDVI} - \mathrm{NDVI}_{\mathrm{min}}) \times 100/(\mathrm{NDVI}_{\mathrm{max}} + \mathrm{NDVI}_{\mathrm{min}})$ (Kogan 1995). The inter-annual variations of NDVI can be due to the changes in weather or ecological components; however, VCI discriminates between these two components. Some of the other vegetation indicators (VIs) that take NDVI into consideration are the Anomaly Vegetation Index (AVI) (Weiying et al. 1994), Standardized Vegetation Index (SVI) (Peters et al. 2002), Normalized Difference Water Index Anomaly (NDWIA) (Gu et al. 2007), and Percent of Average Seasonal Greenness (PASG) (Brown et al. 2008).

Other studies have attempted to monitor drought using remotely sensed brightness temperature acquired from the thermal channels of various instruments, e.g., AVHRR, MODIS, and TM. The Temperature Condition Index (TCI) is among the earliest indices that is based on thermal infrared (TIR) observations for characterizing vegetation stress (Kogan 1995). It is defined as $TCI = (BT_{max} - BT) \times 100/(BT_{max} - BT_{min})$, where BT, BT_{max}, and BT_{min} are the smoothed weekly brightness temperature and its multiple-year maximum and minimum values, respectively. It is demonstrated that the TCI outperforms NDVI and VCI, especially when excessive soil moisture is available due to heavy rainfall or persistent cloudiness, so that NDVI is depressed and VCI indicates low values eventuating in erroneous interpretation of drought (Kogan 1995).

Although NDVI provides valuable information about drought, it is affected by some limitations, e.g., effects of soil humidity and surface anisotropy. Therefore, NDVI alone cannot describe drought condition, and some researchers have developed indices that combine NDVI and temperature information; such as the Vegetation Health Index (VHI) (Kogan 1995), Temperature Vegetation Index (TVI) (McVicar and Jupp 1998), Vegetation Supply Water Index (VSWI) (McVicar and Jupp 1998), Temperature Vegetation Drought Index (TVDI) (Sandholt et al. 2002), and Vegetation Temperature Condition Index (VTCI) (Wan et al. 2004). The VHI is one of the most commonly used indices developed from the additive combination of VCI and TCI, and it provides a powerful tool for monitoring vegetation stress and drought.

Building on the concept of NDVI, several other indices have been developed to monitor vegetation condition, such as the Soil-Adjusted Vegetation Index (SAVI) (Huete 1988), Normalized Difference Infrared Index (NDII) (Hunt and Rock 1989), Normalized Ratio Vegetation Index (NRVI) (Baret and Guyot 1991), Normalized Difference Water Index (NDWI) (Gao 1996), Global Vegetation Monitoring Index (GVMI) (Ceccato et al. 2002a, b), Shortwave Infrared Water Stress Index (SIWSI) (Fensholt and Sandholt 2003), Normalized Difference Drought Index (NDDI) (Gu et al. 2007), Perpendicular Drought Index (PDI) (Ghulam et al. 2007), Normalized Multiband Drought Index (NMDI) (Wang and Qu 2007), and Scaled Drought Condition Index (SDCI) (Rhee et al. 2010).

Currently, satellite-derived vegetation indices can be acquired from various instruments at a global scale with high frequencies. These include AVHRR (Advanced Very High Resolution Radiometer), SPOT (Satellite Pour l'Observation de la Terre) VEGETATION, and Moderate Resolution Imaging Spectroradiometer (MODIS). AVHRR has the longest data period, but has a relatively coarser resolution than others. However, AVHRR offers the advantage of a thermal channel (Rojas et al. 2011). Although MODIS data are available at a finer resolution with more advanced instruments, the delay in their data processing makes it less favorable for real-time drought monitoring (Anyamba and Tucker 2012). Recently, the Global Monitoring for Environment and Security (GMES) Sentinel-3 mission was designed to provide accurate and timely data for environmental management and better understanding of the effects of climate change (Donlon et al. 2012).

7.3.1 Application of VHI for Long-Term Assessment of Agricultural Drought Across the CONUS

The Vegetation Health Index (VHI) has been widely applied for analyzing agricultural drought and vegetation health at different regions (Rhee et al. 2010; Rojas et al. 2011). VHI is calculated by coupling VCI and TCI, i.e., $\mathrm{VHI} = \alpha\mathrm{VCI} + (1 - \alpha)\mathrm{TCI}$, where α depends on the relative contribution of moisture and temperature components. Since the contribution of moisture and temperature during the crop cycle is not known, it is assumed that $\alpha = 0.5$.

VHI acquired from AVHRR has the great advantage of covering a long duration of data, i.e., from 1981 until present, which can even be utilized to investigate the impacts of climate change. VHI is freely available at the Center for Satellite Applications and Research (STAR) of the NOAA Satellite and Information Service (NESDIS) on a weekly timescale at 16 km spatial resolution over the globe. VHI is able to detect drought at any season. However, since crop yield is immensely affected by agricultural drought during the growing season, a lot of studies have focused on the application of VHI during the growing season (Rhee et al. 2010; Rojas et al. 2011). Furthermore, different types of vegetation respond differently to drought and moisture deficit. Therefore, land cover classifications are employed to assess VHI for each land cover types separately.

Here, weekly VHI is utilized across the contiguous United States (CONUS) for long-term assessment of agricultural drought and vegetation stress during the growing season in different land cover types. The spatial resolution of VHI is 16 km, and the study period is from 1982 to 2015 at a weekly temporal scale. Growing season months of May through October are used for analyzing the VHI. In order to classify land cover, MOD12Q1 data from MODIS instrument onboard Terra satellite is acquired with 500 m resolution and aggregated to 16 km. Then, using UMD Land Cover Classification (Hansen et al. 1981), VHI is analyzed for three categories, i.e., grasslands, croplands, and forests. Figure 7.1 represents the coverage of each land cover type over the CONUS. Because forests cover a large area, it is divided into three parts; western, southeastern, and northeastern forests. The number of 16 km grids in each class are as follows: 9777 grids for croplands, 6390 grids for grasslands, 4675 grids for western forests, 7565 grids for northeastern forests, and 5614 grids for southern forests; totaling over 34,000 grids covering the entire CONUS.

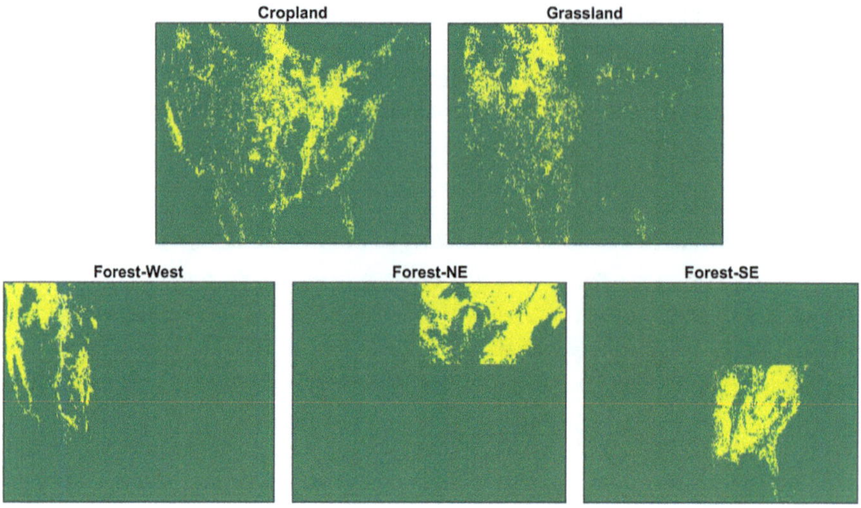

Fig. 7.1 Land cover classification and the sub-regions considered

7.3.1.1 Spatial Extent of Agricultural Drought

A VHI value of 35 or less is translated as drought condition (Rembold et al. 2015; Rojas et al. 2011). Therefore, the spatial extent of drought (grids with VHI < 35) is calculated for each of the five land cover regions (shown in Fig. 7.1) during the growing season of 1982–2015 on a weekly temporal scale, and the results are shown in Fig. 7.2. According to the results, grasslands (which are mostly located in the western U.S.) show the highest increase in drought extent. On the other hand, eastern forests indicate an increase in the spatial extent of drought over the past three decades.

To better understand how the drought extent has changed over time, an inter-decadal assessment has been done, and the drought extent for the period of 1982–1998 is compared to that of 1999–2015. The assessment is done using cumulative distribution function (CDF) of drought extent at each region. Figure 7.3 represents the results of inter-decadal assessment. A shift to the right in the CDF indicates an increase in the value of drought extent. Therefore, CDF plots of Fig. 7.3 clearly highlight the increase in drought extent of cropland, grassland, and western forests.

7.3.1.2 Trend of VHI (Changes in the Intensity of Drought)

As mentioned earlier, a persistent VHI value below 35 is assumed as drought. However, lower values of VHI indicate more severe agricultural drought and vegetation stress. Therefore, it is of great importance to study the time-series of VHI in order to investigate for any significant changes in the intensity of

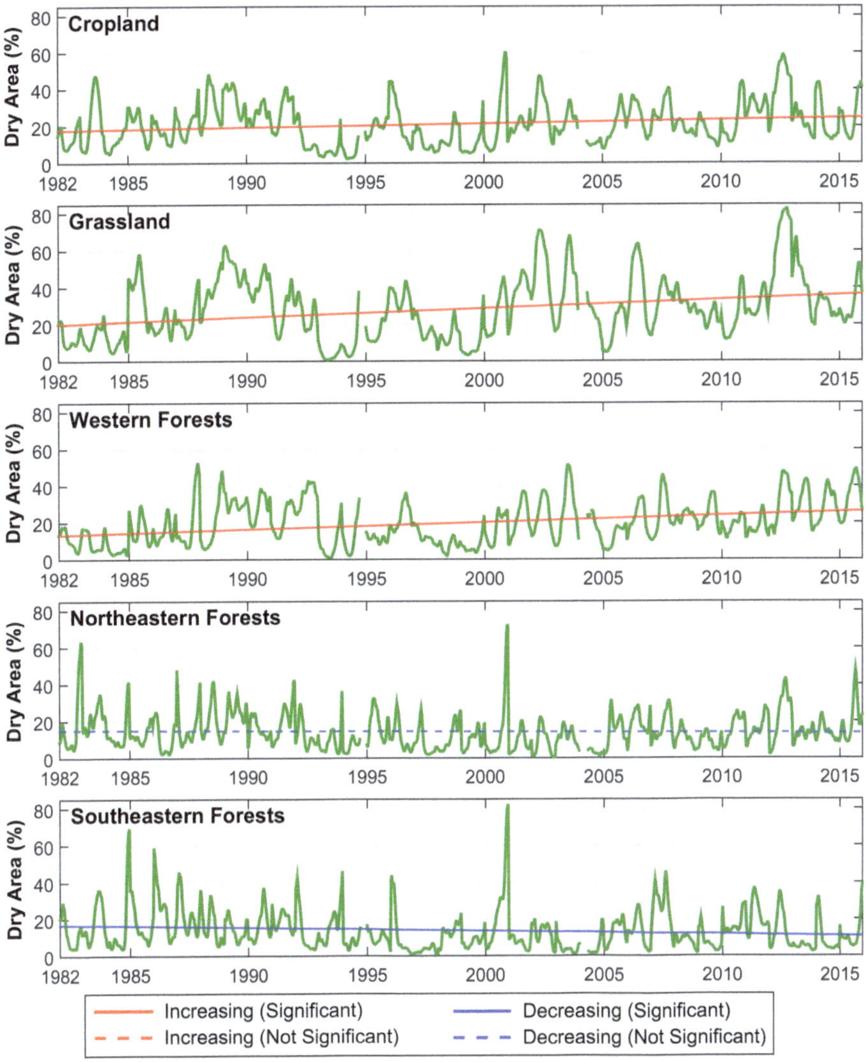

Fig. 7.2 Spatial extent of agricultural drought according to VHI for the growing season of 1982–2015

agricultural drought and vegetation stress. Furthermore, the spatial pattern of VHI changes can be useful for assessing the effects of large-scale climate signals on agricultural drought. Figure 7.4 shows the linear trend of VHI during the growing season of 1982–2015. A negative trend (red color) indicates decreasing VHI, which corresponds to an increase in the severity of drought and vegetation stress. The figure shows that most of the western regions indicate a decrease in VHI value, meaning an increasing trend in the intensity of agricultural droughts at western regions. For instance, more than 60 % of western forests indicate increasing severity

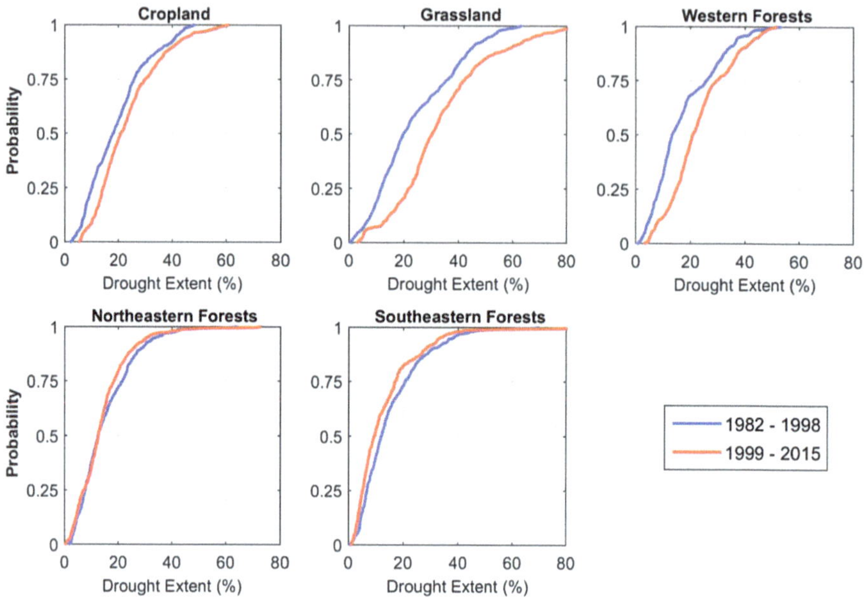

Fig. 7.3 CDF of drought extent based on VHI for two time frames in the historical period. A shift to the right in CDF represents an increase in drought extent

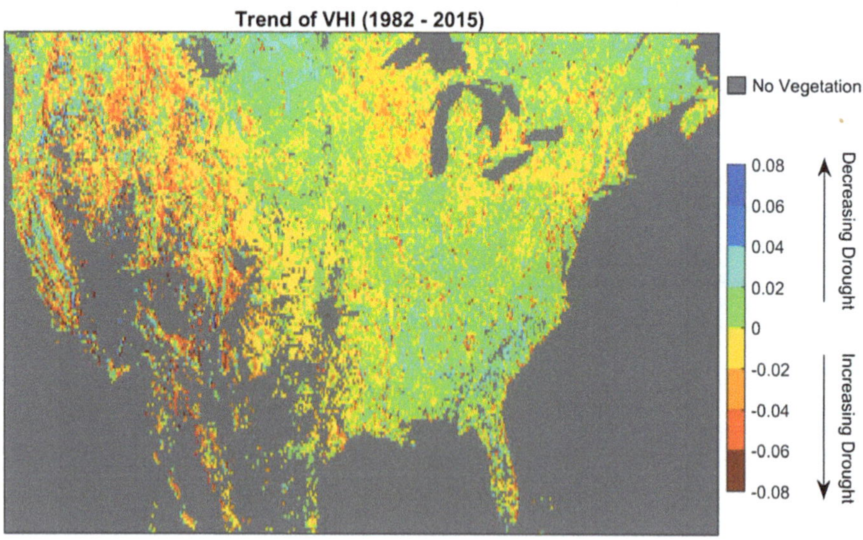

Fig. 7.4 Linear Trend of VHI during the growing season of 1982–2015

of drought. The case is similar for grasslands, with over 60 % of area showing increasing trend in drought intensity. On the other hand, southeastern forests show less areas with increasing drought intensity, with about 25 % of area showing intensifying drought over the past three decades.

7.4 Drought Monitoring with Data Assimilation

7.4.1 The Challenges in Remote Sensing of Drought

Drought monitoring is based on drought indicators computed using hydro-meteorological variables. In general, the remotely sensed observations are not currently used for operational drought monitoring. One of the main limitations is the short length of record (AghaKouchak et al. 2015; Lettenmaier et al. 2015). For instance, a decade of soil moisture data from AMSR-E may not be sufficient to monitor drought. However, the remotely sensed observations still provide valuable information of the hydrologic variables at an unprecedented spatial and temporal resolution. For instance, the L-band soil moisture from the SMAP mission provides immense opportunities to aid agricultural drought detection and improve early agricultural drought warning. In the meantime, current and future satellite missions offer the opportunities to generate higher-quality and long-term data records by merging the satellite observations with model simulations; the method which is referred to as data assimilation (DA). In this chapter, the remotely sensed soil moisture is used as an example to illustrate the agricultural drought-monitoring framework through DA. Other remotely sensed hydrologic variables, such as precipitation from GPM or ground water component from GRACE, can also be assimilated into hydrologic models in the same way for meteorological and hydrologic drought monitoring, respectively.

7.4.2 Data Assimilation in Agricultural Drought Monitoring

Agricultural drought monitoring is based on root-zone soil moisture. The root-zone soil moisture drought indicator has been used in different approaches including the soil moisture percentile, soil moisture anomaly, and standardized soil moisture index (Hao and AghaKouchak 2013; Luo and Wood 2007; Mao et al. 2015). In U.S., ground-based measurements of soil moisture are only available at the point scale at a limited number of sites, while the remotely sensed retrievals can provide unprecedented spatial and temporal resolution of soil moisture. However, microwaves only penetrate the top few centimeters, retrieving only the surface soil moisture rather than the root-zone soil moisture. Hydrologic model simulations can estimate long-term root-zone soil moisture across a range of scales, but the simulation results are potentially biased due to the errors in forcing data, parameters, and the deficiencies in model physics (DeChant and Moradkhani 2014a, b; Parrish et al. 2012).

A plausible approach to improve the root-zone soil moisture estimation is to assimilate the satellite surface soil moisture observations to a dynamic hydrologic model to update the model states (root-zone soil moisture storages) (Moradkhani 2008). Currently, the majority of satellite soil moisture DA studies are relying on the ensemble Kalman filter (EnKF) (Crow et al. 2008; Kumar et al. 2014; Reichle et al. 2002, 2008; Ryu et al. 2009). However, for the highly nonlinear hydrologic models, the assumption of Gaussian distribution of errors in the EnKF may not be valid and adjusting states using the linear updating rule violates the conservation of mass (Dechant and Moradkhani 2012). The effectiveness of EnKF also depends on the cross covariance between the surface and root-zone soil moisture. The weak and highly nonlinear linkage between the surface and root-zone layers prevents the EnKF from updating the root-zone soil moisture effectively when the satellite surface soil moisture is assimilated (Brocca et al. 2012; Massari et al. 2015). As an alternative, the particle filter (PF) method can relax the Gaussian assumption and does not violate the conservation of mass (Moradkhani et al. 2012). The PF also does not depend on the cross covariance between the surface and root-zone layers, rather it seeks good performance for the two layers at the same time. Therefore, the PF can achieve better performance on root-zone soil moisture estimation than the EnKF method (Yan et al. 2015). The newly developed PF-Markov Chain Monte Carlo (MCMC) (Moradkhani et al. 2012) method, with a combination of PF and MCMC, can further improve the root-zone soil moisture estimations. Yan and Moradkhani (2016) suggested that with the assimilation of satellite surface soil moisture using PF-MCMC, the root-zone soil moisture estimation can be improved resulting in improvement of the drought-monitoring skill (Fig. 7.5). In addition, current operational drought-monitoring system is a deterministic system and the root-zone soil moisture uncertainty is not considered (Mao et al. 2015; Sheffield et al. 2004; Wang et al. 2011). In DA framework, the root-zone soil moisture uncertainty can be explicitly quantified within an ensemble framework that leads to increasingly accurate estimate of root-zone soil moisture states for agricultural drought monitoring.

7.4.3 Synthetic Study Results

To objectively assess the potential benefit of assimilation of satellite soil moisture, a synthetic study is first conducted through a procedure called observing system simulation experiment (OSSE) (Moradkhani 2008). The synthetic study includes the following four steps: (1) a "truth" run of hydrologic model with the pre-calibrated model parameters; (2) simulated satellite surface soil moisture observations, which are generated from the truth run by incorporating the observation errors; (3) open-loop (OL) ensemble run with perturbed forcing data without

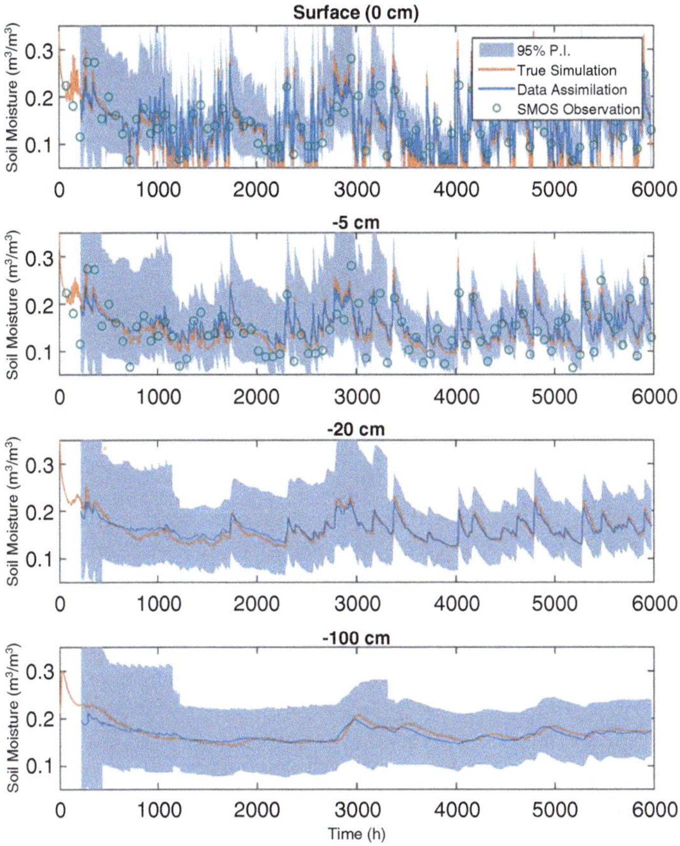

Fig. 7.5 Assimilation of synthetic SMOS soil moisture (temporal resolution: 3 days) for surface and root-zone soil moisture prediction using the PF-MCMC method

DA; and (4) the DA step that assimilates the simulated surface soil moisture observations from step 2 to the model. Then the OL and DA results are compared against the truth simulation to evaluate the impact of satellite soil moisture assimilation.

In this study, the Precipitation-Runoff Modeling System (PRMS) (Leavesley et al. 1983) is used to model the root-zone soil moisture. The meteorological forcing data, precipitation, maximum, and minimum temperature (January 1, 1979 to December 31, 2015) were acquired from the Phase 2 of the North American Land Data Assimilation Systems (NLDAS-2) (Xia et al. 2014). The blended microwave soil moisture climate change initiative (CCI) products v02.2 released on February 2016 (Liu et al. 2012) and the Advanced Microwave Scanning Radiometer2 (AMSR2) soil moisture products (Imaoka et al. 2010) were used in this study. Calibration of the PRMS was performed on a daily timescale utilizing a combination of unregulated U.S. Geological Survey (USGS) streamflow data and No

Regulation No Irrigation (NRNI) streamflow data provided by Bonneville Power Administration (BPA). The results are assessed over the Columbia River Basin (CRB) in the Pacific Northwest U.S. The PF is performed by assimilation of the synthetic satellite surface soil moisture for the period of October 1, 2014 to September 30, 2015. The hindcast focuses on the drought events in the year 2015, since the CRB received historically low snowpack conditions in this year and drought emergencies were declared in Oregon (OR), Washington (WA), and Idaho (ID) States. An ensemble size of 200 is used in this study to obtain the soil moisture posteriors.

The improvement or degradation of DA on soil moisture predictions are assessed using the normalized information contribution (NIC) (Kumar et al. 2014). The NIC for root-mean-square-error (RMSE) is defined as:

$$NIC = \frac{RMSE_{OL}\text{-}RMSE_{DA}}{RMSE_{OL}} \tag{7.1}$$

where the $RMSE_{OL}$ indicates the RMSE values between the OL and synthetic truth, $RMSE_{DA}$ indicates the RMSE values between DA and synthetic truth. If NIC > 0, the DA improves the OL skill; if NIC = 0, the DA does not add any skill; if NIC < 0, the DA degrades the OL skill; and if NIC = 1, the DA achieves the maximum skill. Figure 7.6 presented the NIC values in the surface and root-zone soil moisture and their spatial distributions across the CRB. The majority of the grid cells show the positive NIC values indicating the added-value of the DA. Generally, the improvements in the surface soil moisture field are consistent with the improvements in the root-zone soil moisture field. For surface soil moisture, the daily domain-averaged RMSE (m^3/m^3) for the OL is 0.021, and it decreases to 0.011with DA. Similarly, the daily domain-averaged root-zone soil moisture RMSE value decreases from 0.019 in the OL to 0.012 after DA.

In this study, drought is characterized with the soil moisture percentile and drought intensity is classified based on the National Drought Mitigation Center (NDMC) United States Drought Monitoring (USDM) (Svoboda et al. 2002). Five categories are defined: D0 (abnormally dry, percentile $\leq$ 30 %), D1 (moderate drought, percentile $\leq$ 20 %), D2 (severe drought, percentile $\leq$ 10 %), D3 (extreme drought, percentile $\leq$ 5 %), and D4 (exceptional drought, percentile $\leq$ 2 %). The root-zone soil moisture generated from the OL and DA seasonal integrations was compared against the corresponding synthetic truth percentiles. It is noted that the percentiles from the OL and DA in these comparisons were generated using the ensemble mean estimates. Figures 7.7 and 7.8 present the spatial distribution of drought intensities and the drought extent bias (%) for five different drought (D0–D4) categories. The four seasons in the PNW (fall: OND; winter: JFM; spring: AMJ; summer: JAS) are considered for the period of October 1, 2014 to September 30, 2015. The severe drought events in spring and summer of 2015 across the PNW can be seen in the synthetic truth. In all comparisons, the DA estimates show systematic improvements over the OL estimates. For fall 2014, the OL underestimates the intensity of drought across WA and ID states, whereas DA improves

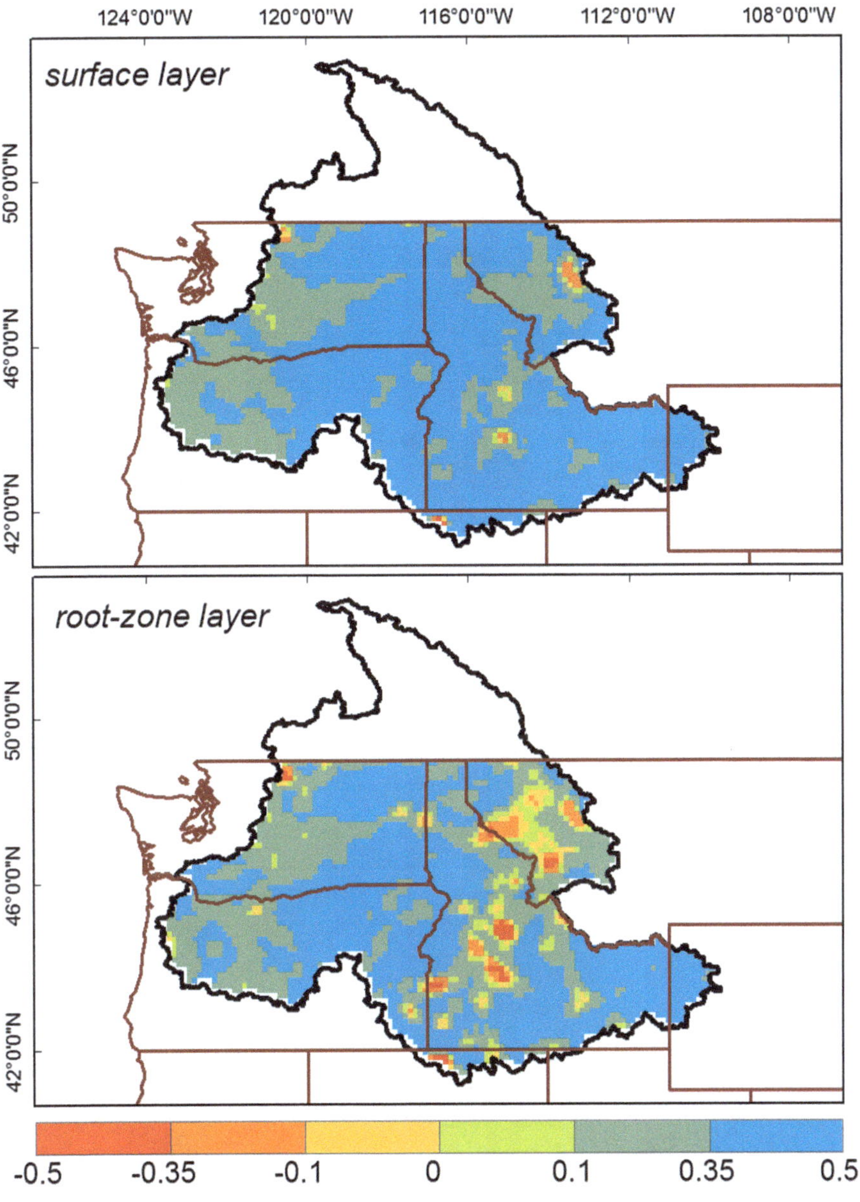

Fig. 7.6 The normalized information contribution (NIC) value between the OL and DA (Eq. 7.1). The positive value indicates that the DA improves soil moisture prediction against OL; negative value indicates the degradation over the OL

these representations. The drought extent bias (%) for D0–D4 between the OL and synthetic truth is 4.33 %, and it decreases to 1.24 % with DA. For winter and spring 2015, the OL underestimates the intensity of drought across PNW, and DA help to reduce these large biases. Similarly, drought extent biases (%) decrease from

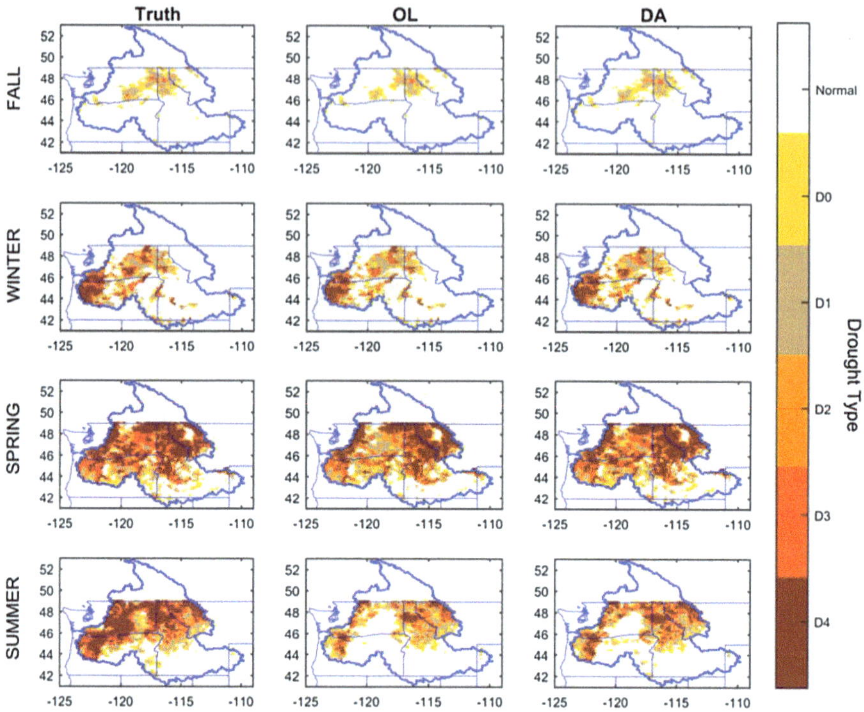

Fig. 7.7 Comparison of the drought-monitoring skill between the OL and DA for fall 2014 and winter/spring/summer 2015

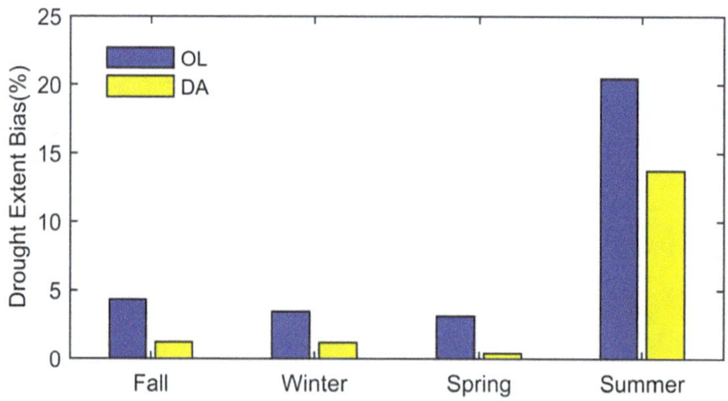

Fig. 7.8 The absolute bias of drought extent (%) against the synthetic truth in the U.S. portion of the CRB

3.46 % and 3.13 % in the OL to 1.20 % and 0.42 % in the DA, respectively. Although the majority of grid cells in WA and OR states show relatively high root-zone soil moisture values in summer 2015, the DA still helps reduce these biases. The drought extent bias (%) decrease from 20.13 to 13.71 %. These results are consistent with the trends in Fig. 7.6, which show the improvements obtained by data assimilation.

7.4.4 Real Case Study Results

Since no "truth" data exist for real case study, the state drought declaration was used as the references to assess the drought-monitoring skill (Kumar et al. 2014; Shukla et al. 2011). Two case studies are presented here to indicate the added-value of soil moisture DA for improving drought-monitoring skill. In spring 2013, drought declarations were issued for nine counties in the southern Idaho (ID). Three months later, a total of 19 counties in ID issued drought emergency. In winter 2015, the PNW received historically low snowpack conditions. In June 2015, the WA Governor declared the statewide drought; the OR Governor declared drought emergencies for 19 out of 36 Oregon counties (about 80 % of the state's landmass); and more than 90 % of Idaho had either declared drought or was bordering drought.

Figures 7.9 and 7.10 present the drought conditions in spring 2013 and winter 2015 and the detected drought areas with OL and DA. For 2013 spring, the OL missed the drought events in parts of southeast ID; however, the DA helped to correct these biases. The detected drought areas of CRB for OL and DA are 73.17 %

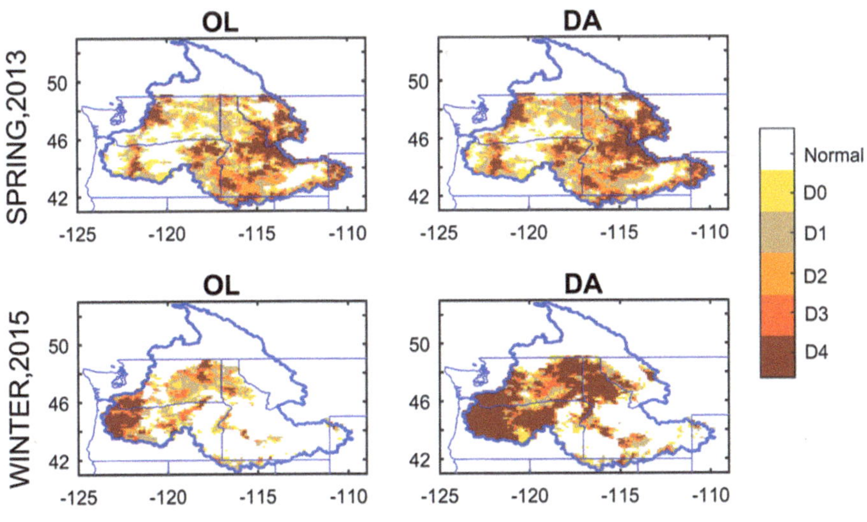

Fig. 7.9 Spatial assessment of drought for 2013 and 2015 before and after DA

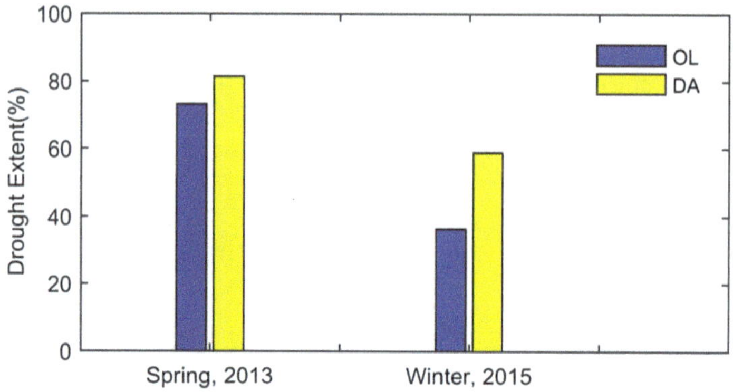

Fig. 7.10 Drought extent for spring 2013 and winter 2015 drought events before and after data assimilation

and 81.43 %, respectively. For winter 2015, the OL underestimated the drought events in parts of OR and ID, whereas DA improved these representations. The detected drought areas increased from 36.37 % in the OL to 58.85 % after the DA. The DA provides a more accurate estimate of drought areas, more consistent with the state drought conditions. In summary, compared with the OL, the DA improves the drought-monitoring skill for both 2013 and 2015. These results demonstrate the added-value of DA to facilitate the state drought preparation and declaration, at least 3 month before the official state drought declaration.

7.5 Summary and Conclusion

Satellite observations have provided invaluable information about many hydroclimatic and environmental variables. This chapter highlights the evolution and progresses of satellite remote sensing of the earth, practical for drought-monitoring purposes from various perspectives. Sections 7.1–7.3 explain the instruments and missions for retrieving hydrological and vegetation data from different satellite platforms and discuss their attributes and characteristics. In Sect. 7.4, data assimilation is explained as a useful method to improve the drought-monitoring skill. We examine the impacts of assimilating remotely sensed surface soil moisture for improving surface and root-zone soil moisture estimates and their subsequent contributions toward an improved assessment of agricultural droughts. Similar to what has been explained here, other satellite data, e.g., precipitation and total water storage, can also be assimilated in the same way to improve the estimations of meteorological and hydrologic droughts.

Acknowledgment Partial Financial supports for this work were provided by NOAA-MAPP program, grant NA140AR4310234 and National Science Foundation (NSF) (Grant No. CCF-1539605).

References

Adler RF, Huffman GJ, Chang A, Ferraro R, Xie P-P, Janowiak J, Rudolf B, Schneider U, Curtis S, Bolvin D (2003) The version-2 global precipitation climatology project (GPCP) monthly precipitation analysis (1979-present). J Hydrometeorol 4(6):1147–1167

AghaKouchak A, Farahmand A, Melton FS, Teixeira J, Anderson MC, Wardlow BD, Hain CR (2015) Remote sensing of drought: progress, challenges and opportunities. Rev Geophys 53 (2):452–480

Ahmadalipour A, Moradkhani H, Svoboda M (2016) Centennial drought outlook over the CONUS using NASA-NEX downscaled climate ensemble. Int J Climatol. doi:10.1002/joc.4859

Alemohammad SH, McLaughlin DB, Entekhabi D (2015) Quantifying precipitation uncertainty for land data assimilation applications. Mon Weather Rev 143(8):3276–3299

Anderson MC, Hain C, Wardlow B, Pimstein A, Mecikalski JR, Kustas WP (2011) Evaluation of drought indices based on thermal remote sensing of evapotranspiration over the continental United States. J Clim 24(8):2025–2044. doi:10.1175/2010JCLI3812.1

Anderson MC, Hain C, Otkin J, Zhan X, Mo K, Svoboda M, Wardlow B, Pimstein A (2013) An intercomparison of drought indicators based on thermal remote sensing and NLDAS-2 simulations with US Drought Monitor classifications. J Hydrometeorol 14(4):1035–1056

Anyamba A, Tucker CJ (2012) Historical perspective of AVHRR NDVI and vegetation drought monitoring. Remote Sens Drought Innov Monit Approach 23

Ashouri H, Hsu K-L, Sorooshian S, Braithwaite DK, Knapp KR, Cecil LD, Nelson BR, Prat OP (2015) PERSIANN-CDR: daily precipitation climate data record from multisatellite observations for hydrological and climate studies. Bull Am Meteorol Soc 96(1):69–83

Baret F, Guyot G (1991) Potentials and limits of vegetation indices for LAI and APAR assessment. Remote Sens Environ 35(2):161–173

Benali A, Carvalho AC, Nunes JP, Carvalhais N, Santos A (2012) Estimating air surface temperature in Portugal using MODIS LST data. Remote Sens Environ 124:108–121

Bojinski S, Verstraete M, Peterson TC, Richter C, Simmons A, Zemp M (2014) The concept of Essential Climate Variables in support of climate research, applications, and policy. Bull Am Meteorol Soc 95(9):1431–1443

Brocca L, Moramarco T, Melone F, Wagner W, Hasenauer S, Hahn S (2012) Assimilation of surface- and root-zone ASCAT soil moisture products into rainfall-runoff modeling. IEEE Trans Geosci Remote Sens 50(7):2542–2555. doi:10.1109/TGRS.2011.2177468

Brown JF, Wardlow BD, Tadesse T, Hayes MJ, Reed BC (2008) The Vegetation Drought Response Index (VegDRI): A new integrated approach for monitoring drought stress in vegetation. GISci Remote Sens 45(1):16–46

Ceccato P, Flasse S, Grégoire J-M (2002a) Designing a spectral index to estimate vegetation water content from remote sensing data: part 2. Validation and applications. Remote Sens Environ 82 (2):198–207

Ceccato P, Gobron N, Flasse S, Pinty B, Tarantola S (2002b) Designing a spectral index to estimate vegetation water content from remote sensing data: part 1: theoretical approach. Remote Sens Environ 82(2):188–197

Chen Y, Velicogna I, Famiglietti JS, Randerson JT (2013) Satellite observations of terrestrial water storage provide early warning information about drought and fire season severity in the Amazon. J Geophys Res Biogeosci 118(2):495–504

Crosman ET, Horel JD (2009) MODIS-derived surface temperature of the Great Salt Lake. Remote Sens Environ 113(1):73–81
Crow WT, Kustas WP, Prueger JH (2008) Monitoring root-zone soil moisture through the assimilation of a thermal remote sensing-based soil moisture proxy into a water balance model. Remote Sens Environ 112(4):1268–1281. doi:10.1016/j.rse.2006.11.033
Dechant CM, Moradkhani H (2012) Examining the effectiveness and robustness of sequential data assimilation methods for quantification of uncertainty in hydrologic forecasting. Water Resour Res 48(4):1–15. doi:10.1029/2011WR011011
DeChant CM, Moradkhani H (2014a) Toward a reliable prediction of seasonal forecast uncertainty: addressing model and initial condition uncertainty with ensemble data assimilation and sequential bayesian combination. J Hydrol 519:2967–2977
DeChant CM, Moradkhani H (2014b) Hydrologic prediction and uncertainty quantification, handbook of engineering hydrology, modeling, climate change and variability. CRC press, Taylor & Francis Group, Boca Raton, pp 387–414
DeChant CM, Moradkhani H (2015) Analyzing the sensitivity of drought recovery forecasts to land surface initial conditions. J Hydrol 526:89–100
Dong J, Walker JP, Houser PR, Sun C (2007) Scanning multichannel microwave radiometer snow water equivalent assimilation. J Geophys Res Atmos 112(D7):D07108
Donlon C, Berruti B, Buongiorno A, Ferreira M-H, Féménias P, Frerick J, Goryl P, Klein U, Laur H, Mavrocordatos C (2012) The global monitoring for environment and security (GMES) sentinel-3 mission. Remote Sens Environ 120:37–57
Dozier J (1989) Spectral signature of alpine snow cover from the Landsat Thematic Mapper. Remote Sens Environ 28:9–22
Dozier J, Green RO, Nolin AW, Painter TH (2009) Interpretation of snow properties from imaging spectrometry. Remote Sens Environ 113:S25–S37
Entekhabi D, Njoku EG, Neill PEO, Kellogg KH, Crow WT, Edelstein WN, Entin JK, Goodman SD, Jackson TJ, Johnson J (2010) The soil moisture active passive (SMAP) mission. Proc IEEE 98(5):704–716
Feng W, Zhong M, Lemoine J, Biancale R, Hsu H, Xia J (2013) Evaluation of groundwater depletion in North China using the Gravity Recovery and Climate Experiment (GRACE) data and ground-based measurements. Water Resour Res 49(4):2110–2118
Fensholt R, Sandholt I (2003) Derivation of a shortwave infrared water stress index from MODIS near-and shortwave infrared data in a semiarid environment. Remote Sens Environ 87 (1):111–121
Ferraro RR (1997) Special sensor microwave imager derived global rainfall estimates for climatological applications. J Geophys Res Atmos 102(D14):16715–16735
Foster JL, Hall DK, Eylander JB, Riggs GA, Nghiem SV, Tedesco M, Kim E, Montesano PM, Kelly REJ, Casey KA (2011) A blended global snow product using visible, passive microwave and scatterometer satellite data. Int J Remote Sens 32(5):1371–1395
Gao B-C (1996) NDWI—a normalized difference water index for remote sensing of vegetation liquid water from space. Remote Sens Environ 58(3):257–266
Ghulam A, Qin Q, Zhan Z (2007) Designing of the perpendicular drought index. Environ Geol 52 (6):1045–1052
Gu Y, Brown JF, Verdin JP, Wardlow B (2007) A five-year analysis of MODIS NDVI and NDWI for grassland drought assessment over the central Great Plains of the United States. Geophys Res Lett 34(6):L06407
Guan B, Molotch NP, Waliser DE, Jepsen SM, Painter TH, Dozier J (2013) Snow water equivalent in the Sierra Nevada: Blending snow sensor observations with snowmelt model simulations. Water Resour Res 49(8):5029–5046
Hall DK, Riggs GA, Salomonson VV, DiGirolamo NE, Bayr KJ (2002) MODIS snow-cover products. Remote Sens Environ 83(1):181–194
Hansen M, DeFries R, Townshend JRG, Sohlberg R (1981) UMD global land cover classification, 1 kilometer, 1.0. Department of Geography, University of Maryland, College Park, 1994: 1998

Hao Z, AghaKouchak A (2013) Multivariate standardized drought index: a parametric multi-index model. Adv Water Resour 57:12–18
Hong Y, Hsu K, Moradkhani H, Sorooshian S (2006) Uncertainty quantification of satellite precipitation estimation and Monte Carlo assessment of the error propagation into hydrologic response. Water Resour Res 42(8):W08421
Hou AY, Kakar RK, Neeck S, Azarbarzin AA, Kummerow CD, Kojima M, Oki R, Nakamura K, Iguchi T (2014) The global precipitation measurement mission. Bull Am Meteorol Soc 95 (5):701–722
Huete AR (1988) A soil-adjusted vegetation index (SAVI). Remote Sens Environ 25(3):295–309
Huffman GJ, Bolvin DT, Nelkin EJ, Wolff DB, Adler RF, Gu G, Hong Y, Bowman KP, Stocker EF (2007) The TRMM multisatellite precipitation analysis (TMPA): quasi-global, multiyear, combined-sensor precipitation estimates at fine scales. J Hydrometeorol 8(1):38–55
Huffman GJ, Bolvin DT, Braithwaite D, Hsu K, Joyce R, Xie P, Yoo S-H (2015) NASA global precipitation measurement (GPM) integrated multi-satellite retrievals for GPM (IMERG). In: Algorithm theoretical basis document, version 4.1. NASA
Hunt ER, Rock BN (1989) Detection of changes in leaf water content using near-and middle-infrared reflectances. Remote Sens Environ 30(1):43–54
Imaoka K, Kachi M, Kasahara M, Ito N, Nakagawa K, Oki T (2010) Instrument performance and calibration of AMSR-E and AMSR2. Int Arch Photogramm Remote Sens Spat Inf Sci 38:13–16
Jackson RD, Idso SB, Reginato RJ, Pinter PJ (1981) Canopy temperature as a crop water stress indicator. Water Resour Res 17(4):1133–1138
Joyce RJ, Janowiak JE, Arkin PA, Xie P (2004) CMORPH: a method that produces global precipitation estimates from passive microwave and infrared data at high spatial and temporal resolution. J Hydrometeorol 5(3):487–503
Khajehei S (2015) A multivariate modeling approach for generating ensemble climatology forcing for hydrologic applications. Portland State University, Portland, OR
Kidd C, Levizzani V, Laviola S (2011) Quantitative precipitation estimation from Earth observation satellites. In: Rainfall: state of the science. Wiley Online Library, p 127–158
Kogan FN (1995) Application of vegetation index and brightness temperature for drought detection. Adv Space Res 15(11):91–100
Kogan FN (1997) Global drought watch from space. Bull Am Meteorol Soc 78(4):621–636
Kongoli C, Romanov P, Ferraro R (2012) Snow cover monitoring from remote sensing satellites: possibilities for drought assessment. In: Remote sensing of drought: innovative monitoring approaches. p 359–386
Kumar SV, Peters-Lidard CD, Mocko D, Reichle R, Liu Y, Arsenault KR, Xia Y, Ek M, Riggs G, Livneh B, Cosh M (2014) Assimilation of remotely sensed soil moisture and snow depth retrievals for drought estimation. J Hydrometeorol 15(6):2446–2469. doi:10.1175/JHM-D-13-0132.1
Leavesley GH, Lichty RW, Thoutman BM, Saindon LG (1983) Precipitation-runoff modeling system: user's manual. US Geological Survey, Colorado
Lettenmaier DP, Alsdorf D, Dozier J, Huffman GJ, Pan M, Wood EF (2015) Inroads of remote sensing into hydrologic science during the WRR era. Water Resour Res 51(9):7309–7342. doi:10.1002/2015WR017616
Liu YY, Parinussa RM, Dorigo WA, De Jeu RAM, Wagner W, Van Dijk A, McCabe MF, Evans JP (2011) Developing an improved soil moisture dataset by blending passive and active microwave satellite-based retrievals. Hydrol Earth Syst Sci 15(2):425–436
Liu YY, Dorigo WA, Parinussa RM, De Jeu RAM, Wagner W, McCabe MF, Evans JP, Van Dijk AIJM (2012) Trend-preserving blending of passive and active microwave soil moisture retrievals. Remote Sens Environ 123:280–297. doi:10.1016/j.rse.2012.03.014
Lobell DB, Roberts MJ, Schlenker W, Braun N, Little BB, Rejesus RM, Hammer GL (2014) Greater sensitivity to drought accompanies maize yield increase in the US Midwest. Science 344(6183):516–519

Luo L, Wood EF (2007) Monitoring and predicting the 2007 U.S. drought. Geophys Res Lett 34 (22):L22702. doi:10.1029/2007GL031673
Madadgar S, Moradkhani H, Garen D (2014) Towards improved post-processing of hydrologic forecast ensembles. Hydrol Process 28(1):104–122
Mao Y, Nijssen B, Lettenmaier DP (2015) Is climate change implicated in the 2013–2014 California drought? A hydrologic perspective. Geophys Res Lett 42(8):2805–2813
Massari C, Brocca L, Tarpanelli A, Moramarco T (2015) Data assimilation of satellite soil moisture into rainfall-runoff modelling: a complex recipe? Remote Sens 7(9):11403–11433. doi:10.3390/rs70911403
McKee TB, Doeskin NJ, Kleist J (1993) The relationship of drought frequency and duration to time scales. In: 8th conference on applied climatology, American Meteorological Society, Anaheim, p 179–184
McMullan KD, Brown MA, Martín-Neira M, Rits W, Ekholm S, Marti J, Lemanczyk J (2008) SMOS: the payload. IEEE Trans Geosci Remote Sens 46(3):594–605
McVicar TR, Jupp DLB (1998) The current and potential operational uses of remote sensing to aid decisions on drought exceptional circumstances in Australia: a review. Agric Syst 57 (3):399–468
Mishra AK, Singh VP (2010) A review of drought concepts. J Hydrol 391(1-2):202–216. doi:10.1016/j.jhydrol.2010.07.012
Molotch NP, Bales RC (2006) Comparison of ground-based and airborne snow surface albedo parameterizations in an alpine watershed: impact on snowpack mass balance. Water Resour Res 42(5):W05410
Moradkhani H (2008) Hydrologic remote sensing and land surface data assimilation. Sensors 8 (5):2986–3004
Moradkhani H, Hsu K, Gupta H, Sorooshian S (2005) Uncertainty assessment of hydrologic model states and parameters: sequential data assimilation using the particle filter. Water Resour Res 41(5):W05012
Moradkhani H, Hsu K, Hong Y, Sorooshian S (2006) Investigating the impact of remotely sensed precipitation and hydrologic model uncertainties on the ensemble streamflow forecasting. Geophys Res Lett 33(12). doi:10.1029/2006GL026855
Moradkhani H, DeChant CM, Sorooshian S (2012) Evolution of ensemble data assimilation for uncertainty quantification using the particle filter-Markov chain Monte Carlo method. Water Resour Res 48(12):W12520. doi:10.1029/2012WR012144
Moser CL, Tootle GA, Oubeidillah AA, Lakshmi V (2011) A comparison of SNOTEL and AMSR-E snow water equivalent data sets in western US watersheds. Int J Remote Sens 32 (21):6611–6629
Mu Q, Heinsch FA, Zhao M, Running SW (2007) Development of a global evapotranspiration algorithm based on MODIS and global meteorology data. Remote Sens Environ 111 (4):519–536
Mu Q, Zhao M, Running SW (2011) Improvements to a MODIS global terrestrial evapotranspiration algorithm. Remote Sens Environ 115(8):1781–1800. doi:10.1016/j.rse.2011.02.019
Mu Q, Zhao M, Kimball JS, McDowell NG, Running SW (2013) A remotely sensed global terrestrial drought severity index. Bull Am Meteorol Soc 94(1):83–98. doi:10.1175/BAMS-D-11-00213.1
Naumann G, Spinoni J, Vogt JV, Barbosa P (2015) Assessment of drought damages and their uncertainties in Europe. Environ Res Lett 10(12):124013
Njoku EG, Jackson TJ, Lakshmi V, Chan TK, Nghiem SV (2003) Soil moisture retrieval from AMSR-E. IEEE Trans Geosci Remote Sens 41(2):215–229
Núñez M, Pfister S, Antón A, Muñoz P, Hellweg S, Koehler A, Rieradevall J (2013) Assessing the environmental impact of water consumption by energy crops grown in Spain. J Ind Ecol 17 (1):90–102
Palmer WC (1965) Meteorological drought. US Department of Commerce, Weather Bureau Washington, DC

Parrish MA, Moradkhani H, Dechant CM (2012) Toward reduction of model uncertainty: integration of Bayesian model averaging and data assimilation. Water Resour Res 48:W03519. doi:10.1029/2011WR011116
Peters AJ, Walter-Shea EA, Ji L, Vina A, Hayes M, Svoboda MD (2002) Drought monitoring with NDVI-based standardized vegetation index. Photogramm Eng Remote Sens 68(1):71–75
Reager JT, Famiglietti JS (2009) Global terrestrial water storage capacity and flood potential using GRACE. Geophys Res Lett 36(23):L23402
Reager JT, Thomas AC, Sproles EA, Rodell M, Beaudoing HK, Li B, Famiglietti JS (2015) Assimilation of GRACE terrestrial water storage observations into a land surface model for the assessment of regional flood potential. Remote Sens 7(11):14663–14679
Reichle R, McLaughlin DB, Entekhabi D (2002) Hydrologic data assimilation with the ensemble Kalman filter. Mon Weather Rev 130(1):103–114. doi:10.1175/1520-0493(2002)130<0103:HDAWTE>2.0.CO;2
Reichle RH, Crow WT, Keppenne CL (2008) An adaptive ensemble Kalman filter for soil moisture data assimilation. Water Resour Res 44(3):W03423. doi:10.1029/2007WR006357
Rembold F, Meroni M, Rojas O (2015) Agricultural drought monitoring using space-derived vegetation and biophysical products: a global perspective. In: Thenkabail PS (ed) Remote sensing of water resources, disasters, and urban studies. CRC Press, Boca Raton, pp 349–365
Rhee J, Im J, Carbone GJ (2010) Monitoring agricultural drought for arid and humid regions using multi-sensor remote sensing data. Remote Sens Environ 114(12):2875–2887
Rind D, Hansen J, Goldberg R, Rosenzweig C, Ruedy R (1990) Potential evapotranspiration and the likelihood of future drought. J Geophys Res 95(D7):9983–10004
Rodell M, Famiglietti JS (2002) The potential for satellite-based monitoring of groundwater storage changes using GRACE: the High Plains aquifer, Central US. J Hydrol 263(1):245–256
Rodell M, Famiglietti JS, Chen J, Seneviratne SI, Viterbo P, Holl S, Wilson CR (2004) Basin scale estimates of evapotranspiration using GRACE and other observations. Geophys Res Lett 31 (20):20504
Rojas O, Vrieling A, Rembold F (2011) Assessing drought probability for agricultural areas in Africa with coarse resolution remote sensing imagery. Remote Sens Environ 115(2):343–352
Romanov P, Gutman G, Csiszar I (2000) Automated monitoring of snow cover over North America with multispectral satellite data. J Appl Meteorol 39(11):1866–1880
Rouse JW, Haas RH, Schell JA, Deering DW, Harlan JC (1974) Monitoring the vernal advancement of retrogradation of natural vegetation, NASA/GSFC, Type III, Final Report, Greenbelt, MD
Ryu D, Crow WT, Zhan X, Jackson TJ (2009) Correcting unintended perturbation biases in hydrologic data assimilation. J Hydrometeorol 10:734–750. doi:10.1175/2008JHM1038.1
Sahoo AK, Sheffield J, Pan M, Wood EF (2015) Evaluation of the Tropical Rainfall Measuring Mission Multi-Satellite Precipitation Analysis (TMPA) for assessment of large-scale meteorological drought. Remote Sens Environ 159:181–193. doi:10.1016/j.rse.2014.11.032
Sandholt I, Rasmussen K, Andersen J (2002) A simple interpretation of the surface temperature/vegetation index space for assessment of surface moisture status. Remote Sens Environ 79 (2):213–224
Seneviratne SI, Corti T, Davin EL, Hirschi M, Jaeger EB, Lehner I, Orlowsky B, Teuling AJ (2010) Investigating soil moisture-climate interactions in a changing climate: a review. Earth Sci Rev 99(3-4):125–161. doi:10.1016/j.earscirev.2010.02.004
Sheffield J, Goteti G, Wen F, Wood EF (2004) A simulated soil moisture based drought analysis for the United States. J Geophys Res D Atmos 109(24):1–19. doi:10.1029/2004JD005182
Shen S, Leptoukh GG (2011) Estimation of surface air temperature over central and eastern Eurasia from MODIS land surface temperature. Environ Res Lett 6(4):45206
Shukla S, Steinemann AC, Lettenmaier DP (2011) Drought monitoring for Washington State: indicators and applications. J Hydrometeorol 12:66–83. doi:10.1175/2010JHM1307.1

Sima S, Ahmadalipour A, Tajrishy M (2013) Mapping surface temperature in a hyper-saline lake and investigating the effect of temperature distribution on the lake evaporation. Remote Sens Environ 136:374–385

Sorooshian S, Hsu K-L, Gao X, Gupta HV, Imam B, Braithwaite D (2000) Evaluation of PERSIANN system satellite-based estimates of tropical rainfall. Bull Am Meteorol Soc 81 (9):2035–2046

Sorooshian S, Hsu K, Coppola E, Tomassetti B, Verdecchia M, Visconti G (2008) Hydrological modelling and the water cycle: coupling the atmospheric and hydrological models. Springer Science & Business Media, Berlin

Svoboda M, LeComte D, Hayes M, Heim R, Gleason K, Angel J, Rippey B, Tinker R, Palecki M, Stooksbury D, Miskus D, Stephens S (2002) The drought monitor. Bull Am Meteorol Soc 83 (8):1181–1190. http://dx.doi.org/10.1175/1520-0477(2002)083<1181:TDM>2.3.CO;2

Syed TH, Famiglietti JS, Chen J, Rodell M, Seneviratne SI, Viterbo P, Wilson CR (2005) Total basin discharge for the Amazon and Mississippi River basins from GRACE and a land-atmosphere water balance. Geophys Res Lett 32(24):L24404

Tsakiris G, Vangelis H (2005) Establishing a drought index incorporating evapotranspiration. Eur Water 9–10:1–9

Tucker CJ (1979) Red and photographic infrared linear combinations for monitoring vegetation. Remote Sens Environ 8(2):127–150

Tucker CJ, Choudhury BJ (1987) Satellite remote sensing of drought conditions. Remote Sens Environ 23(2):243–251

Velicogna I, Tong J, Zhang T, Kimball JS (2012) Increasing subsurface water storage in discontinuous permafrost areas of the Lena River basin, Eurasia, detected from GRACE. Geophys Res Lett 39(9):L09403

Velicogna I, Kimball JS, Kim Y (2015) Impact of changes in GRACE derived terrestrial water storage on vegetation growth in Eurasia. Environ Res Lett 10(12):124024

Vicente-Serrano SM, Beguería S, López-Moreno JI (2010) A multiscalar drought index sensitive to global warming: the standardized precipitation evapotranspiration index. J Clim 23 (7):1696–1718. doi:10.1175/2009JCLI2909.1

Wada Y (2013) Human and climate impacts on global water resources. Utrecht University, Utrecht

Wagner W, Hahn S, Kidd R, Melzer T, Bartalis Z, Hasenauer S, Figa-Saldaña J, de Rosnay P, Jann A, Schneider S (2013) The ASCAT soil moisture product: a review of its specifications, validation results, and emerging applications. Meteorol Z 22(1):5–33

Wan Z, Wang P, Li X (2004) Using MODIS land surface temperature and normalized difference vegetation index products for monitoring drought in the southern Great Plains, USA. Int J Remote Sens 25(1):61–72

Wang L, Qu JJ (2007) NMDI: a normalized multi-band drought index for monitoring soil and vegetation moisture with satellite remote sensing. Geophys Res Lett 34(20):L20405

Wang J, Price KP, Rich PM (2001) Spatial patterns of NDVI in response to precipitation and temperature in the central Great Plains. Int J Remote Sens 22(18):3827–3844

Wang A, Lettenmaier DP, Sheffield J (2011) Soil moisture drought in China, 1950–2006. J Clim 24(13):3257–3271. doi:10.1175/2011JCLI3733.1

Weiying C, Qianguang X, Yongwei S (1994) Application of the anomaly vegetation index to monitoring heavy drought in 1992. Remote Sens Environ 9(2):106–112

Westermann S, Langer M, Boike J (2011) Spatial and temporal variations of summer surface temperatures of high-arctic tundra on Svalbard—implications for MODIS LST based permafrost monitoring. Remote Sens Environ 115(3):908–922

Xia Y, Sheffield J, Ek MB, Dong J, Chaney N, Wei H, Meng J, Wood EF (2014) Evaluation of multi-model simulated soil moisture in NLDAS-2. J Hydrol 512:107–125. doi:10.1016/j.jhydrol.2014.02.027

Xie P, Arkin PA (1997) Global precipitation: a 17-year monthly analysis based on gauge observations, satellite estimates, and numerical model outputs. Bull Am Meteorol Soc 78 (11):2539–2558

Yan H, Moradkhani H (2016) Combined assimilation of streamflow and satellite soil moisture with the particle filter and geostatistical modeling. Adv Water Resour 94:364–378. doi:10.1016/j.advwatres.2016.06.002

Yan H, DeChant CM, Moradkhani H (2015) Improving soil moisture profile prediction with the particle filter-Markov chain Monte Carlo method. IEEE Trans Geosci Remote Sens 53 (11):6134–6147

Yao Y, Liang S, Qin Q, Wang K (2010) Monitoring drought over the conterminous United States using MODIS and NCEP Reanalysis-2 data. J Appl Meteorol Climatol 49(8):1665–1680

Yin D, Roderick ML, Leech G, Sun F, Huang Y (2014) The contribution of reduction in evaporative cooling to higher surface air temperatures during drought. Geophys Res Lett 41 (22):7891–7897

Yong B, Wang J, Ren L, You Y, Xie P, Hong Y (2016) Evaluating four multisatellite precipitation estimates over the Diaoyu Islands during Typhoon seasons. J Hydrometeorol 17(5):1623–1641

Zaitchik BF, Rodell M, Reichle RH (2008) Assimilation of GRACE terrestrial water storage data into a land surface model: results for the Mississippi River basin. J Hydrometeorol 9 (3):535–548

Zhang A, Jia G (2013) Monitoring meteorological drought in semiarid regions using multi-sensor microwave remote sensing data. Remote Sens Environ 134:12–23

Zhang D, Zhang Q, Werner AD, Liu X (2015a) GRACE-based hydrological drought evaluation of the Yangtze River Basin, China. J Hydrometeorol 17(3), 811–828

Zhang Z, Chao BF, Chen J, Wilson CR (2015b) Terrestrial water storage anomalies of Yangtze River Basin droughts observed by GRACE and connections with ENSO. Glob Planet Chang 126:35–45

Zhang J, Mu Q, Huang J (2016) Assessing the remotely sensed Drought Severity Index for agricultural drought monitoring and impact analysis in North China. Ecol Indic 63:296–309

Zulkafli Z, Buytaert W, Onof C, Manz B, Tarnavsky E, Lavado W, Guyot J-L (2014) A comparative performance analysis of TRMM 3B42 (TMPA) versions 6 and 7 for hydrological applications over Andean–Amazon river basins. J Hydrometeorol 15(2):581–592

Chapter 8
Drought Monitoring and Assessment Using Remote Sensing

Z. Su, Y. He, X. Dong, and L. Wang

8.1 Introduction

Drought has wreaked havoc to human societies throughout history. Impacts of drought include devastated crops, famine and conflicts and wars. Serious and severe droughts have occurred on every continent throughout history (Heffernan 2013), including the global mega drought of 4200 years ago that is linked to demise of Akkadian Empire (Kerr 1998) and civilisations in Greece, Egypt and the Indus Valley of Pakistan, the great famine in 1876–1878 which resulted in more than 5 million deaths in India and 30 million in total, the federation drought in 1901 in Australian, the US dust bowl in the 1930s, the central European drought in the 1940s and the Sahel drought in 1970s and 1980s when famine led to 600,000 deaths in 1972–1975, and again in 1984–1985. These droughts have all been associated with climatic shifts that caused low rainfall and as such climate change is now an accepted powerful causal agent in the evolution of civilisation. In many regions, climate change is expected to increase the amount of land at risk from drought and heat and will threaten more arable areas. Timely assessment and monitoring of drought will increase the drought preparedness, relief and mitigation and reduce the damage of drought impacts to the environment, economy and society.

Z. Su, Ph.D. (✉) • L. Wang, M.Sc.
Faculty of Geo-Information Science and Earth Observation (ITC), University of Twente, Hengelosestraat 99, P.O. Box 217, Enschede 7500 AE, The Netherlands
e-mail: z.su@utwente.nl; l.wang-3@utwente.nl

Y. He, Ph.D.
National Meteorological Center, University Road No. 8, Beijing 100081, P.R. China
e-mail: yanbohe@cma.gov.cn

X. Dong, Ph.D.
China Three Gorges University, University Road No. 8, Yichang 443002, Hubei, P.R. China
e-mail: xhdong24@hotmail.com

V. Lakshmi (ed.), *Remote Sensing of Hydrological Extremes*, Springer Remote Sensing/Photogrammetry, DOI 10.1007/978-3-319-43744-6_8

Drought may be defined as the lack of water of a certain location in a certain period compared to a climatic average. From a climatic perspective, we may distinguish meteorological, soil moisture and hydrological drought. Meteorological drought refers to a shortage of precipitation compared to a climatic average; soil moisture drought can be caused by a shortage of precipitation, excessive evaporation and transpiration due to dry weathers and lack of irrigation, while hydrological drought is caused by a combination of lack of precipitation and excessive use of available water resources. When the impact of drought is also taken into consideration, four types of droughts are usually defined such as meteorological drought, agricultural drought, hydrological drought and socioeconomic drought.

As remote sensing provides real-time spatial observations of several atmospheric and land surface variables that can be used to estimate precipitation, evapotranspiration, soil moisture and vegetation conditions, such data can be used for assessment and monitoring of drought characteristics: its intensity, duration and spatial extent. When combined with modelling and forecasting of the water cycle, information on future drought can also be generated for drought preparedness.

In next section, we will briefly review some commonly used drought indices with a focus on the use of remote sensing. A unified framework for drought monitoring and assessment is discussed in Sect. 8.3, and Sect. 8.4 presents several practical examples. Conclusions and recommendations are presented in the last section.

8.2 Drought Indices

Due to the complexity of the drought phenomena and the needs for their descriptions in applications, a panoply of indices has been developed and many recent studies have evaluated their usefulness and consistencies in describing drought events (e.g. Zargar et al. 2011; Eden 2012; van Hoek 2016 among others). In particular, Zargar et al. (2011) reviewed 74 such indices and mentioned that around 150 had been developed in the past. Eden (2012) assessed the droughts in Twente in the Netherlands from 2003 to 2012 by mapping evapotranspiration using the Water Cycle Multi-mission Observation Strategy (WACMOS) methodology (Su et al. 2014) in applying the Surface Energy Balance System (SEBS) (Su 2002). van Hoek (2016) developed a web-based open source platform for global drought monitoring using eight indices (see later for more details).

From a process point of view, meteorological drought occurs when a consistent decline of precipitation reduces water available on land in snow packs, ice sheets, lakes and rivers, therefore a relevant index should primarily describe the decline of precipitation, the Standardised Precipitation Index (SPI) (McKee et al. 1993, 1995; Guttman 1999) can be used for this purpose, although the Palmer Drought Severity

Index (PDSI) (Palmer 1965, 1968) is also widely used but its estimation is more involved than SPI.

Soil moisture drought (or commonly referred to as agricultural drought but this latter gives the impression concerning only agricultural lands and ignoring other natural surfaces) occurs when soil dries out due to evaporation into the atmosphere, drainage into deeper layers that cannot be accessed any more by vegetation or due to excessive human extraction of aquifer storage. The most relevant indices should be those reflecting the soil moisture status in the rooting zone. The PDSI may serve also this purpose. Other relevant indices are the Drought Severity Index (DSI) derived from surface energy balance to infer water balance in the rooting zone (Su et al. 2003a, b) and the Evapotranspiration Deficit Index (ETDI) derived from a hydrological model of the unsaturated zone (Narasimhan and Srinivasan 2005) as well as the Standardised Precipitation and Evaporation Index (SPEI) (Vicente-Serrano et al. 2010a).

Hydrological drought occurs when water reserves in aquifers, lakes and reservoirs fall below averages due to high human demand or low rainfall supplies. The most relevant indices for this should reflect the changes in total storage in an area; the PDSI can be used in part as it considers the accumulation of precipitation, but better is the changes derived from water cycle budget for a certain area which will be described in Sect. 8.3.

From an impact point of view, many remotely sensed indices have been developed in the past decades to describe different aspects of droughts, despite the fact that they mostly reflect the surface conditions instead of the amount of water in the different storages. The following indices are often used: Normalised Difference Vegetation Index (NDVI), Anomaly Vegetation Index, Vegetation Condition Index, Normalised Difference Water Index, Normalised Difference Drought Index and Vegetation Supplication Water Index based on visible, near infrared, shortwave reflectance data for partly and fully covered surface, Temperature Vegetation Dryness Index, Temperature Condition Index, Crop Water Stress Index and Water Deficit Index based on thermal infrared remote sensing data for partly and fully covered surface. A detailed review of these indices can be found in McVicar and Jupp (1998).

In the following, we briefly review some commonly used drought indices focusing on the use of remote sensing as well as reanalysis type of gridded data that use remote sensing data as input or assimilate such data as state variables.

8.2.1 PDSI

The procedure to estimate PDSI is as follows: (1) Carry out a hydrologic accounting by months for a long series of years. (2) Summarise the results to obtain certain constants or coefficients which are dependent on the climate of the area being analysed. (3) Reanalyse the series using the derived coefficients to determine the amount of moisture required for "normal" weather during each month. (4) Convert the departures to indices of moisture anomaly. (5) Analyse the index series to develop: (a) Criteria for determining the beginning and ending of drought periods. (b) A formula for determining drought severity.

As the estimation of PDSI relies on a hydrological accounting, the challenges in using this approach are in the definition of the two soil layers with the corresponding field capacity which requires some knowledge of soil hydraulic properties and the estimation of the potential recharge, loss, precipitation, evaporation and runoff as well as the associated coefficients (i.e. evapotranspiration, recharge, runoff, and loss). A comprehensive land surface hydrological modelling may easily accomplish this task given adequate forcings.

8.2.2 SPI

SPI (McKee et al. 1993) is based only on precipitation data, thus provides a rapid drought index once precipitation data is available. SPI compares precipitation with its climatic mean estimated as multiyear average. To estimate SPI, the precipitation record is first fitted to a gamma distribution, which is then transformed into a normal distribution using an equal probability transformation. After standardisation, i.e. $\mathrm{SPI} = (X_i - m[X])/\mathrm{std}[X]$, where X_i is the precipitation variable, $m[X]$ the average (typically over a period of 30 years) and $\mathrm{std}[X]$ the standard deviation, SPI values above zero indicate wet periods and values below zero indicate dry periods.

For any given event, its SPI score represents how many standard deviations its cumulative precipitation deficit deviates from the normalised average. If a value of less than zero is consistently observed and it reaches a value of -1 or less, a drought is said to have occurred (McKee et al. 1993), thus the onset of a drought can be defined. SPI can be calculated for different time periods, but typically it is applied for the 3, 6, 12, 24 and 48-month periods. Because over time precipitation deficit gradually and variably affects different water resources (e.g. snowpack, stream flow and groundwater), the multitude of SPI durations can be used to reflect change in different drought features.

8.2.3 NDVI, VCI, TCI and VHI

NDVI (Kriegler et al. 1969; Rouse et al. 1973; Tucker 1979) has been used as a surrogate for detecting the condition of green vegetation, as it is defined as $\mathrm{NDVI} = (\mathrm{NIR} - \mathrm{VIS})/(\mathrm{NIR} + \mathrm{VIS})$, where VIS and NIR stand for the spectral reflectance measurements acquired in the visible (red) and near-infrared regions of the spectrum. Because the pigment in plant leaves (chlorophyll) strongly absorbs visible light (from 0.4 to 0.7 μm) for use in photosynthesis and the cell structure of the leaves strongly reflects near-infrared light (from 0.7 to 1.1 μm), the more green leaves a plant has, the more these wavelengths of light are affected, respectively, and hence the higher the NDVI. As such NDVI has also been used for detecting drought effects on vegetation and used as a base index for a number of remote sensing indices that similarly measure vegetation conditions.

Among these the most popular ones are the Vegetation Condition Index (VCI) (Kogan 1990), defined as $VCI = (NDVI - NDVI_{min})/(NDVI_{max} - NDVI_{min})$ for a certain period (week/month), NDVI is the NDVI for the period under study, $NDVI_{max}$ and $NDVI_{min}$ are the maximum and minimum NDVIs, respectively, in the record for the specific period; and the Temperature Condition Index (TCI) (Kogan 1995), defined as $TCI = (BT_{max} - BT)/(BT_{max} - BT_{min})$ which describes the deviation of the brightness temperature (BT) in a period (e.g. a month) from recorded maximum assuming the higher the temperature deviation, the higher the drought. BT, BT_{max} and BT_{min} are the actual, maximum and minimum BTs, respectively, in the record for the specific period. Combining the VCI and TCI using a weight factor α for the contributions of VCI and TCI, the Vegetation Health Index (VHI) (Kogan 1995) can be defined as $VHI = \alpha VCI + (1 - \alpha)TCI$ and often α is set to 0.5 and is found more effective than other vegetative drought indices (Kogan 2001).

By replacing the respective minimum or maximum values in the definition of VCI, TCI with the mean values and make proper adjustment in the equations, Jia et al. (2012) defined the Normalised Vegetation Anomaly Index (NVCI), the Normalised Temperature Anomaly Index (NTCI) and the Normalised Drought Anomaly Index (NDAI), respectively, as, $NVAI = (NDVI - m[NDVI])/(NDVI_{max} - NDVI_{min})$, $NTAI = (BT - m[BT])/(BT_{max} - BT_{min})$, $NDAI = \alpha\ NVCI - (1 - \alpha)\ NTAI$, where $m[.]$ is the respectively corresponding mean value.

van Hoek (2016) has evaluated these indices with ground observations in China and concluded that the normalised values to be more stable than the original indices, while also pointing out difficulties to validate the satellite-derived drought maps with the national hazard database of China Meteorological Administration.

8.2.4 DSI, ETDI and SPEI

The Drought Severity Index (DSI) (Su et al. 2003a) was developed to infer rooting zone soil moisture from surface energy balance consideration. By considering only the vertical movement of water, which is a valid approximation in the absence of heavy rainfall and irrigation and under deep groundwater conditions, it is possible to derive a physical relationship between the temporal changes of the evaporation and the changes of the soil water content. Using the mass conservation principle for an infinitesimal soil layer in the vertical direction, we obtain

$$\frac{\partial\theta(z,t)}{\partial t} = -\frac{\partial q(z,t)}{\partial z} \tag{8.1}$$

where θ, q, t, z are volumetric soil moisture content, soil water flux density (water amount per unit area per unit time) across a vertical plane, time and vertical distance increment, respectively. Integrating (8.1) from z_1 to z_2 in depth and from t_1 to t_2 in time and assuming that there is no source/sink between z_1 and z_2, we obtain the water balance equation,

$$\int_{z_1}^{z_2} \theta(z,t_2)\mathrm{d}z - \int_{z_1}^{z_2} \theta(z,t_1)\mathrm{d}z = Q(z_1) - Q(z_2) \tag{8.2}$$

Applying (8.2) to the active rooting zone with the boundary conditions $Q(z_1) = P_0 + I_0 - E$ at the soil surface and $Q(z_2) = I_\mathrm{c}$ at the bottom of the rooting zone, the change of total soil water content in this zone can be written as

$$\Theta(t_2) - \Theta(t_1) = P_0 + I_0 + I_\mathrm{c} - E \tag{8.3}$$

where Θ is the volumetric soil water content in the rooting zone, P_0, I_0, I_c, E are, respectively, precipitation, irrigation, capillary flux and evapotranspiration amount from t_1 to t_2.

Similar to the determination of relative evaporation as used in SEBS (Su 2002), the water balance at limiting cases can be considered. Without losing generality, we assume that the soil in the rooting zone was completely saturated at t_1, i.e. $\Theta(t_1) = \Theta_\mathrm{wet}$ (i.e. at the wet limit) after a sufficient precipitation or irrigation event. During the drying-down process (i.e. there is no precipitation or irrigation taking place during this period), the rooting zone will eventually reach the dry limit $\Theta(t_2) = \Theta_\mathrm{dry}$, i.e. the evaporation becomes zero due to the limitation of the soil moisture (which drops to below the wilting point), from (8.3) we have,

$$\Theta_\mathrm{wet} - \Theta_\mathrm{dry} = I_\mathrm{c,wet} - E_\mathrm{wet} \tag{8.4}$$

Similarly, for any time between the wet-limit and the dry-limit cases, we obtain from (8.3)

$$\Theta - \Theta_\mathrm{dry} = I_\mathrm{c} - E \tag{8.5}$$

Manipulating (8.4) and (8.5) results in

$$\frac{\Theta - \Theta_\mathrm{dry}}{\Theta_\mathrm{wet} - \Theta_\mathrm{dry}} = \frac{E - I_\mathrm{c}}{E_\mathrm{wet} - I_\mathrm{c,wet}} \tag{8.6}$$

Assuming the capillary flux is only related to soil texture and is small compared to root water uptake which can be quantified by the transpiration flux, i.e. $I_\mathrm{c} = I_\mathrm{c,wet} << E$ as observed by Aydin (1994) who reported that the capillary flux was very small when compared with root water uptake, even in clay soils. Subsequently by defining $R_\theta = \Theta/\Theta_{\text{wet}}$ as the relative soil water content with respect to the porosity and using (8.6) (and multiplying the right-hand side by λ, the latent heat of vaporisation), we arrive at

$$R_\theta = \frac{\Theta}{\Theta_{\text{wet}}} = \frac{\lambda E}{\lambda E_{\text{wet}}} \tag{8.7}$$

Equation (8.7) shows that the relative soil water content is directly related to the relative evapotranspiration. As such if defining the Drought Severity Index as $\text{DSI} = 1 - R_\theta$, i.e. the relative soil water deficit in the rooting zone, one obtains from the SEBS relationship (Su 2002),

$$\text{DSI} = \frac{H - H_{\text{wet}}}{H_{\text{dry}} - H_{\text{wet}}} \tag{8.8}$$

Equation (8.8) is the required quantitative measure for soil water deficit in the rooting depth and can be directly derived from SEBS calculation (Su 2002). When DSI is high, soil water content is low and vice versa.

By relating DSI to soil hydraulic characteristics (i.e. the porosity) and to crop phenology (i.e. the rooting depth), location-specific soil water deficit in the rooting zone can be determined for a particular soil, thus leading to a quantitative assessment of drought severity for the concerned crop at the particular location and time. This can be done as follows by defining the relative evapotranspiration as in SEBS (Su 2002) and linking it to (8.6) as

$$\text{DSI} = 1 - \Lambda_r = 1 - s\left(\frac{\Theta - \Theta_{\text{dry}}}{\Theta_{\text{wet}} - \Theta_{\text{dry}}}\right) \tag{8.9}$$

A general form with a sigmoid function, $y = \frac{1}{1+e^{-x^a}}$, can then be fitted to the soil moisture function $s(.)$ to account for the local soil properties. By relating the factor a (the exponent) to different soil hydraulic properties all conditions can be taken into account as shown in Fig. 8.1.

8.2.5 Case Studies

The Evapotranspiration Deficit Index (ETDI) was defined (Narasimhan and Srinivasan 2005) by using simulated data with the SWAT model (Arnold et al. 1998; Neitsch et al. 2002). The water stress anomaly (WSA) is calculated as case A: if $\text{WS}_{i,j} <= \text{MWS}_j$, then $\text{WSA}_{i,j} = (\text{MWS}_j - \text{WS}_{i,j})/(\text{MWS}_j - \text{minWS}_j)$ $*100$, or case B: If $\text{WS}_{i,j} > \text{MWS}_j$, then $\text{WSA}_{i,j} = (\text{MWS}_j - \text{WS}_{i,j})/$ $(\text{maxMWS}_j - \text{NWS}_j)*100$ (Note: it was specified in Narasimhan and Srinivasan

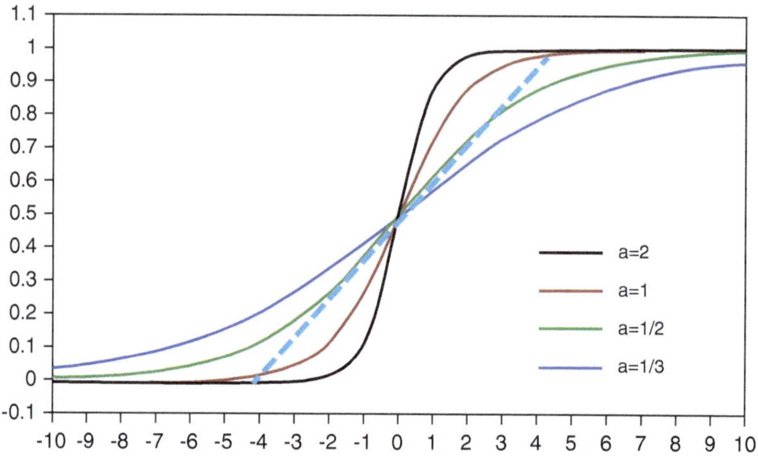

Fig. 8.1 Generalised form $y = \frac{1}{1+e^{-x^a}}$ between relative evaporation (y, ordinate) and normalised soil moisture (x, abscissa), the factor a (the exponent) can be used to fit different soil properties. The *dashed line* indicates the simple linear relationship of (8.7)

2005 that A holds for $WS_{i,j} = MWS_j$, but this is likely a typo, as it would make no difference for A or B if $WS_{i,j} = MWS_j$), and in these expressions, WSA is the weekly water stress anomaly, MWS_j is the long-term median water stress of week j, $MaxWS_j$, $MinWS_j$ are the long-term maximum and minimum water stress of week j, and WS is the weekly water stress ratio, defined as $WS = (ET_0 - ET)/ET_0$, where ET_0 is the weekly reference crop evapotranspiration, and ET is the weekly actual transpiration. Eden (2012) evaluated the usefulness of ETDI for regional drought analysis and compared it to SPI from a meteorological station operated by the Royal Meteorological Institute of the Netherlands (KNMI) from 2003 to 2010 and concluded that the ETDI can be used to estimate the onset, intensity and duration of spatially regional drought.

The Standardised Precipitation Evapotranspiration Index (SPEI) (Vicente-Serrano et al. 2010a) is defined using precipitation and potential evapotranspiration (PET) similar to the PDSI. First, the climatic water balance, i.e. the difference between precipitation and ET_0, is estimated as:

$$D = P - ET_0 \tag{8.10}$$

where P is the monthly precipitation (mm) and ET_0 (mm) is estimated with the method of Thornthwaite (1948), using only mean monthly temperature and the geographical location of the region of interest. Then the calculated D values were aggregated at various time scales:

$$D_n^k = \sum_{i=0}^{k-1} (P_{n-i} - ET_{n-i}), \quad n \geq k \tag{8.11}$$

where k (months) is the time scale of the aggregation and n is the calculation number. The D values are undefined for $k > n$. Finally a log-logistic probability distribution function was then fitted to the data series of D, as it adapts very well to all time scales (Vicente-Serrano et al. 2010b).

The SPEI combines the sensitivity of the PDSI to changes in evaporation demand (caused by temperature fluctuations and trends) with the multitemporal nature of the SPI and the main advantage of the new index lies in its multiscalar character, which allows discrimination between different types of drought.

Different time scales are needed for monitoring drought conditions in different hydrological subsystems, as short time scales are mainly related to soil water content and river discharge in headwater areas, medium time scales to reservoir storages and discharge in the medium course of the rivers, and long time scales to variations in groundwater storage. For regional drought assessment and mitigation, spatial data of high resolution (i.e. kilometre scale or field scale) will be of great benefit; therefore, in next section we propose a unified framework for multiscale temporal and spatial drought assessment, monitoring and analysis as well as prediction.

8.3 A Unified Framework for Drought Monitoring and Assessment

As discussed in Sect. 8.2, many indices have been proposed for drought assessment, each of these indices addresses a specific aspect of drought. It is therefore desirable to have one or a set of consistent indices to describe all aspects of a drought, its onset, severity and duration, as well as their spatial distribution and simultaneously addressing meteorological, soil moisture and hydrological drought. Recently, van Hoek (2016) has evaluated different indices and proposed the use of eight indices for drought monitoring in China in a web-GIS-based system. The used indices are as follows: Precipitation Anomaly Percentage (PAP), TCI, NTAI, VCI, NVAI, VHI, NDAI and the Evapotranspiration Deficit Index (EDI).

On the basis of the review presented in Sect. 8.2 and the studies by Eden (2012) and van Hoek (2016), a drought monitoring and analysis system should take into account the meteorological, soil moisture as well as hydrological characteristics, thus a system using SPI, DSI-ETDI, SPEI and VCI should be able to describe surface characteristics of droughts (i.e. meteorological and soil moisture drought), and as for hydrological drought a water cycle closure (or the simplified water budget closure for a catchment) approach (Fig. 8.2) is then needed.

In order to be able to compare these indices, the same standardisation procedure can be used for other indices and variables providing a uniform descriptor for different aspects of the same event. We shall name this index as: $\mathrm{SXI} = (X_i - m[X])/\mathrm{std}[X]$, where X_i is the variable in question and $m[X]$ is the average (typically over a period of 30 years), std[X] is the standard deviation, and when X is replaced by a certain variable we derive a specific index for that variable (e.g. precipitation,

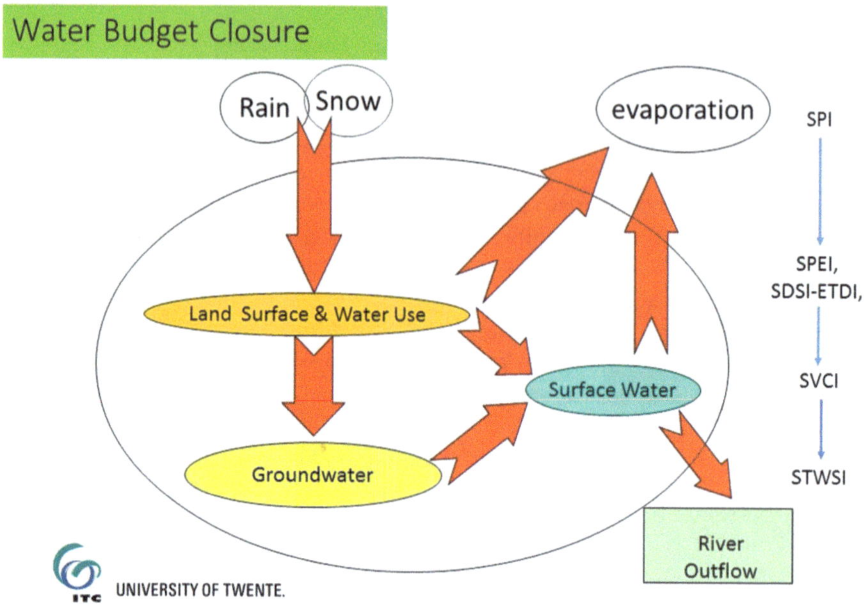

Fig. 8.2 Concept of water budget closure. *SPI* is the Standardised Precipitation Index, *SPEI* is standardised Precipitation Evaporation Index, *SDSI-ETDI* is the Standardised Drought Severity Index-Evapotranspiration Deficit Index, *SVCI* is Standardised Vegetation Condition Index, *STWSI* is the Standardised Terrestrial Water Storage Index, all calculated on weekly or monthly time interval from 1 to *N* time intervals (e.g. for 48 months) (i.e. as anomalies and cumulative anomalies)

evaporation/transpiration, soil moisture, vegetation condition, etc.). As such this SXI is defined in the same manner as standardised anomaly and will be used interchangeably.

For water budget closure of a catchment, the following equation can be used,

$$\frac{\partial S_{i,j}}{\partial t} = P_{i,j} - E_{i,j} - R_{\mathrm{Obs}} {}^{*} f\left(P_{i,j}, E_{i,j}\right) \tag{8.11}$$

The left-hand side of the equation represents the change of total water storage and the right-hand side states that the total water storage change equals to precipitation less evaporation and discharge. The function *f*(.) is used to scale the aggregated discharge at the gauge station to runoff generation at each grid.

In the calculation to be presented in Sect. 8.4, the precipitation is from satellite observation (e.g. Adler et al. 2003), the evaporation is calculated by SEBS (Su 2002; Chen et al. 2014) and the discharge is from in situ observation. This equation also expresses the consistency validation among in situ observation datasets, thus different datasets for precipitation, evaporation and runoff can be used and their consistency evaluated.

In addition to satellite and in situ observation datasets, reanalysis data, e.g. the ERA-Interim, provide these three variables as well, so ERA-interim results will be applied with this equation to calculate the total water storage change as well. On the other hand, the total water storage can be measured by using the GRACE satellite, thus these different data sources can be assessed in one consistent framework.

A set of sequential standardised indices can then be used to describe the different aspects of droughts, SPI for precipitation, SPEI for precipitation and potential evaporation and SDSI-ETDI for evapotranspiration, SVCI for vegetation condition and STWSI for the terrestrial water storage in the catchment, all calculated on weekly or monthly time interval from 1 to N time intervals (e.g. for 48 months) (i.e. as standardised anomalies and cumulative anomalies) (Fig. 8.2).

For the determination of the drought intensity, a similarly uniform approach could be proposed for the different indices, and as they are standardised a normal distribution can be fitted to them (or after some transformation), as such it is sufficient to use the standard deviation of the indices to define their departure from the mean. The corresponding drought intensity classes are listed in Table 8.1.

The utilities of these indices and definitions will be illustrated with some practical examples in next section as executed in the Dragon drought monitoring project supported by the European Space Agency and the Ministry of Science and Technology of China.

8.4 Drought Monitoring and Prediction over Continental China: A Case Study

The conflict between supply and demand of water resources constitutes the biggest problem for food security of an increasing world population. Since 1970s, some progresses have been made in drought quantification and research. However, the methods employed are mostly based on data collected on local meteorological variables and cannot quantify real-time, large-scale actual drought process, i.e. its intensity, duration and its actual extent. It has also often happened that a drought is not recognised while it is in fact already underway of its development. As a result of this inability, no adequate measure can be deployed to effectively fight any big drought disaster often causing great hunger and social instability. Most recently, China Meteorological Administration has installed nation-wide automatic soil moisture monitoring stations providing important in situ observation of the status of soil moisture, such that information derived from spatial remote sensing data can be validated and improved for real-time drought monitoring.

8.4.1 Satellite Earth Observation of Land Surface Processes

The project strategy is to derive land surface parameters and heat fluxes using satellite earth observation techniques and to apply these parameters and fluxes in a hydrometeorological model as initial conditions and constraints at the

Table 8.1 Drought intensity defined as departure from the mean (i.e. number of standard deviations), the suggested category and the return period in number of intervals (e.g. once in how many years) as well as the suggested colour scheme for maps

Value range	Category	Return period (1 in …)
$F <= \mu-3\sigma$	Extremely dry	15787
$\mu-3\sigma < F <= \mu-2\sigma$	Severely dry	370
$\mu-2\sigma < F <= \mu-1\sigma$	Moderately dry	22
$\mu-1\sigma < F < \mu+1\sigma$	Near normal	3
$\mu+1\sigma <= F < \mu+2\sigma$	Moderately wet	22
$\mu+2\sigma <= F < \mu+3\sigma$	Severely wet	370
$\mu+3\sigma <= F$	Extremely wet	15787

land–atmosphere interface. Further data assimilation allows controlling the model drifts. Since a hydrological model treats, in reality, heterogeneous land surfaces by means of effective parameters, and there is no any other adequate means, except satellite earth observation techniques, to derive the actual distribution of these parameters over any large area, this proposed strategy makes a major progress to monitor and predict drought on a quantitative way. It is also well known that both type of models treat soil water content, which controls the energy and water partition at the land–atmosphere interface, as a prognostic variable or put it plainly as a residual of the mass balance equation. Any prediction of soil water content using these models, without other actual physical information, is simply a collection of residual errors of other terms and does not provide any reliable indication of the actual soil water content. The latter as a matter of fact is the most important variable sorted after for drought problems. On the other hand, since satellite earth observation provides actual measurements of land surface parameters and models enable to make predictions, it becomes fairly clear that only a combined approach will provide the definitive opportunities to adequately solve the drought problem over a large area.

Radiometric observations aboard earth observation satellites are an attractive source of spatial observations of land surface processes and can be used to determine the required model parameters. In the Dragon drought monitoring project, we have considered the following variables: sensible and latent heat fluxes; evaporative fraction and actual evaporation; and root zone soil water content index, all important to hydrological modelling. Different satellite data were used to derive land surface physical variables as given by Su (2001).

The state-of-the-art remote sensing algorithm package SEBS (Surface Energy Balance System) (Su 2001, 2002) for mapping heat fluxes for heterogeneous land surfaces was used to derive sensible and latent heat fluxes. The main input data to SEBS are satellite images and meteorological variables. Spatial inputs to SEBS are land surface parameters such as incoming global radiation, albedo, vegetation index for fractional coverage and roughness determination, surface temperature, emissivity and roughness length due to both topography and vegetation effects. The advantage of the system is that no a priori knowledge of heat fluxes is required and no site-specific calibration is needed. Data of high or low spatial resolution from all sensors in the visible, near-infrared and thermal infrared frequency ranges can be used in the system.

After obtaining the sensible and latent heat fluxes, the evaporative fraction (EF) can be calculated as the ratio of the latent heat flux to the total available energy. Further, by making the assumption that the derived evaporative fraction represents the daily average, the actual evaporation can be calculated with an estimate of daily net radiation (Su 2001; Li 2001).

Since it has been demonstrated that EF is often conservative during daytime, and from its definition given earlier (see Sect. 8.2), it represents physically the amount of energy used for evaporation during daytime. As it has long been established that evaporation is strongly regulated by soil water availability under the same atmospheric forcing, i.e. from potential when the soil is saturated to nearly zero when it dries out, EF can be used to quantify the soil water availability. Compared to surface soil moisture as can be determined using both active microwave and passive microwave techniques (e.g. Su et al. 1997), use of EF gives a good indication of the water availability in the whole rooting zone.

Because of cloud contamination and missing data in satellite data, the derived land surface parameters will contain missing values and gaps causing undesirable uncertainty in drought monitoring and prediction. The Harmonic Analysis of Time Series (HANTS) algorithm (Verhoef 2000) can be used to replace the wrong or missing data. The advantages of HANTS include: (a) screening and removing cloud-affected observations; (b) temporal interpolation of remaining observations to reconstruct images at any prescribed time. Roerink et al. (2000) have demonstrated the capability of HANTS in reconstructing cloud-free vegetation index images. Using HANTS reconstructed monthly evaporative fraction and surface albedo in combination with observed monthly global radiation provided by meteorological stations, it has been shown that this strategy is successful in deriving monthly actual evaporation in Northwest China (Li 2001).

8.4.2 Hydrometeorological Modelling

Ideally a large-scale hydrological model should be operated for real-time tracking of water budget and such a model should include precipitation, snow melt, evaporation/transpiration, surface runoff, unsaturated flow and groundwater flow, flow routing, reservoir and lakes regulation. When the model is embedded in a GIS (Geographic Information System) environment and operated on a predefined grid scale, a better compatibility can be achieved with both satellite earth observation products and the atmospheric models, for prediction purposes.

As large hydrometeorological fields are generated by ECMWF (European Centre for Medium-range Weather Forecast) Integrated Forecast System over global scale as well as the global data assimilation system for land (GLDAS) operated by the National Aeronautics and Space Administration (NASA), they can be converted into standardised indices or anomalies for drought analysis. The reliability of the modelling and prediction for China can be established by comparing historical predictions, especially for drought prone areas, to national statistics. Surface observations of precipitation from the meteorological networks in China (around 600 stations) are used to validate and update the consistency of the different fields.

In the following, we show some analyses conducted for several river basins and for the whole China when its southwest region suffered a heavy drought lasting for several months in 2010.

To do so, the country is divided into various river basins by means of a digital elevation model in the Integrated Land and Water Information System (ILWIS) (https://www.itc.nl/Pub/research_programme/Research-review-and-output/ILWIS_-_Remote_Sensing_and_GIS_software.html) (Fig. 8.3).

Then the calculations are done as follows for each river basin using the standardised indices.

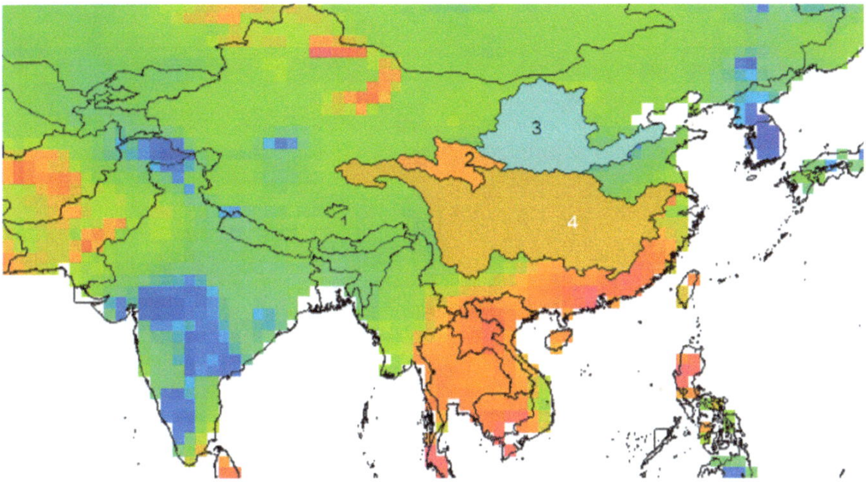

Fig. 8.3 River basins delineated in a GIS system, No. 1—Namco basin, No. 2—The upper Yellow River Basin, No. 3—the whole Yellow River Basin (incl. the upper part), No. 4—the Yangtze river basin

- GLDAS_STWS2I: GLDAS-derived Standardised total water storage index, generated by the following equation:

  GLDAS_STWS2I = ΔTWS2/GLDAS_TWS2 (std)
  GLDAS_TWS2 = $P - \mathrm{ET} - R$
  GLDAS_TWS2: GLDAS-derived total water storage of a specific month and year
  GLDAS_TWS2(mean) = mean long-term GLDAS_TWS2 of a specific month
  ΔTWS2 = GLDAS_TWS2 − GLDAS_TWS2(mean)
  GLDAS_TWS2 (std) = standard deviation of long-term GLDAS_TWS2 of a specific month

- TRMM_SPI

  The Standardised precipitation index from TRMM
  SPI = $(P - P(\text{mean}))/P(\text{std})$
  P = precipitation of a specific month and year
  P(mean) = mean long-term precipitation of a specific month
  P(std) = standard deviation of long-term precipitation of a specific month

- GLDAS_SRI

  The Standardised runoff index from GLDAS
  SRI = $(R - R(\text{mean}))/R(\text{std})$
  R = GLDAS-derived runoff of a specific month and year
  R(mean) = mean long-term GLDAS-derived runoff of a specific month
  R(std) = standard deviation of long-term runoff of a specific month

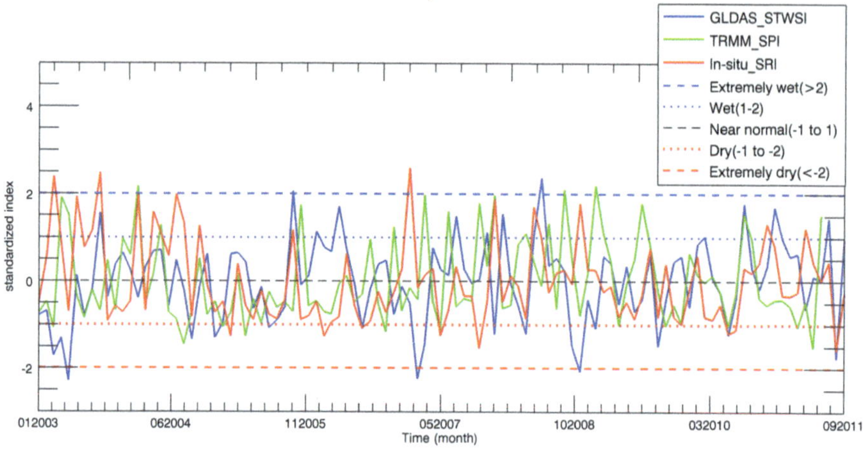

Fig. 8.4 Standardised index in Namco basin (*Note*: the definition of drought categories here follows those by CMA and differs from Table 8.1 in that only severely situations (beyond two standard deviations) were observed but indicated as extremely situations for consistency)

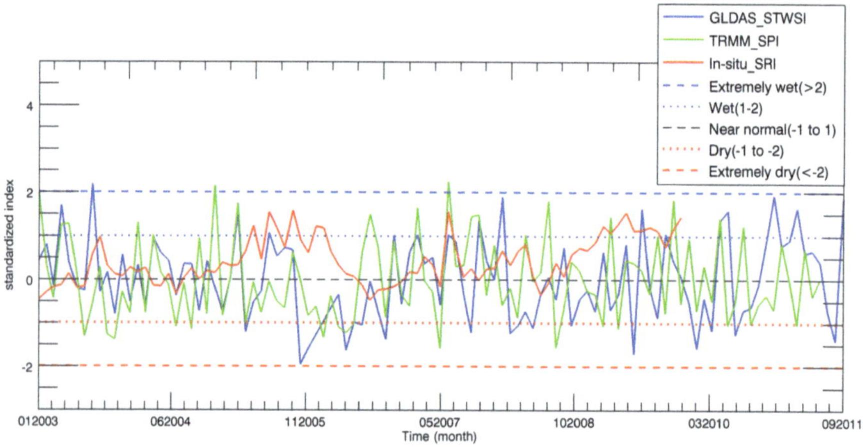

Fig. 8.5 Standardised index in upper Yellow river basin

As the observed runoff is only available for upper yellow river basin, GLDAS-derived runoff is used for TWS2 calculation.

Figures 8.4, 8.5, 8.6 and 8.7 show the calculated indices for the four river basins, and by comparing them to the drought categories, specific drought situation in the basin can be identified.

In Fig. 8.8, we show the Standardised Terrestrial Water Storage Index derived from GLDAS simulation which is generated as follows:

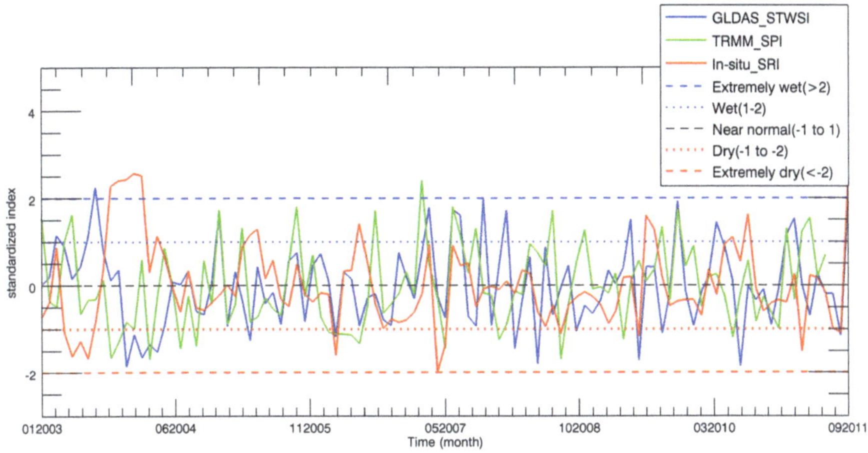

Fig. 8.6 Standardised index in whole Yellow river basin

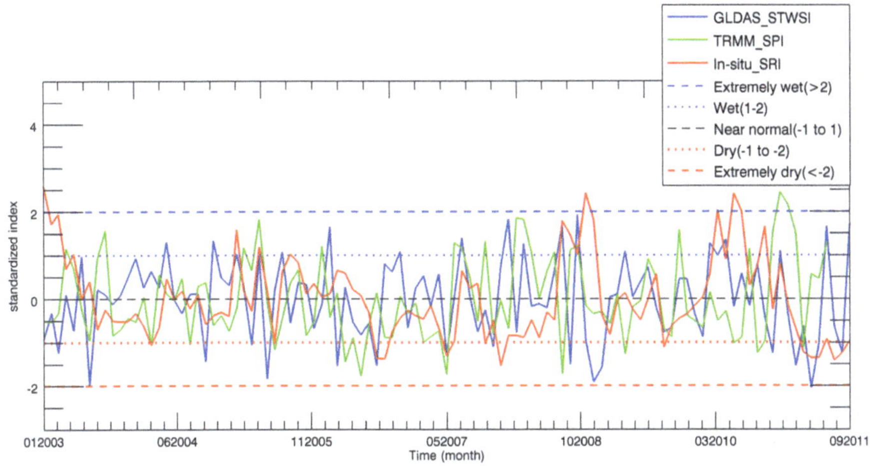

Fig. 8.7 Standardised index in Yangtze basin

$GLDAS\_STWS1I = \Delta TWS1/std(GLDAS\_TWS1)$

GLDAS-derived total water storage of a specific month and year is given as:

$GLDAS\_TWS1 = SM + SWE + CWS$

mean(GLDAS_TWS1) = mean of long-term GLDAS_TWS1 of a specific month

$\Delta TWS1 = GLDAS\_TWS1 - mean(GLDAS\_TWS1)$

std(GLDAS_TWS1) = standard deviation of long-term GLDAS_TWS1 of a specific month

Figure 8.8 shows that the drought affected areas in southwest China persisted from January till June 2010 and affected large area with extremely dry conditions.

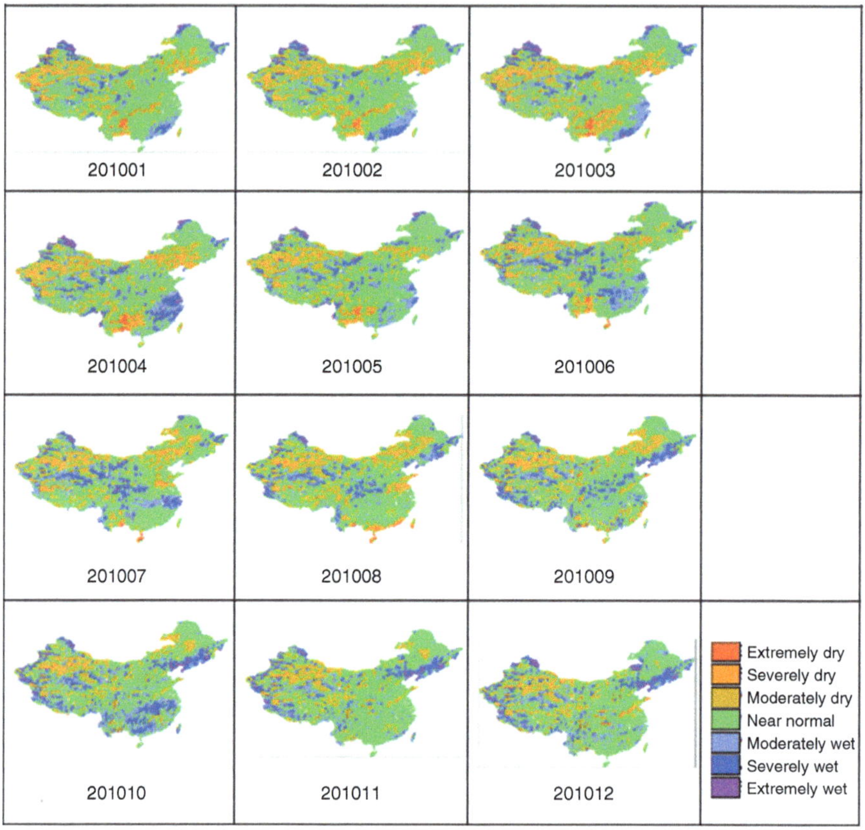

Fig. 8.8 GLDAS-derived Standardised terrestrial water storage index expressed in drought intensity as defined in Table 8.1

To further show the complex character of the drought phenomenon, we display in Fig. 8.9 the spatial drought severity maps derived for southwest China and compare them to in situ observation of soil moisture and drought reports.

The challenges in interpreting these maps and comparing them to both soil moisture observation and drought reports are very obvious. From the satellite-derived SEBS-DSI values, it can be concluded that the regional spatiotemporal variability is very big within the studied region in southwest China. The drought in northwest of Chengdu appeared most severe and persistent for months from beginning of January till beginning of May, but unfortunately there are no in situ soil moisture observations, nor drought reports available. Near Chongqing, the in situ soil moisture stations report large wet conditions, while SEBS-DSI have often missing values due to cloud contamination, but implicitly the consistency of the two datasets is confirmed. Near Guiyang, the information provided by in situ soil moisture observation and drought reports are not consistent while droughts are reported for all the months, the soil moisture observations indicate wetter

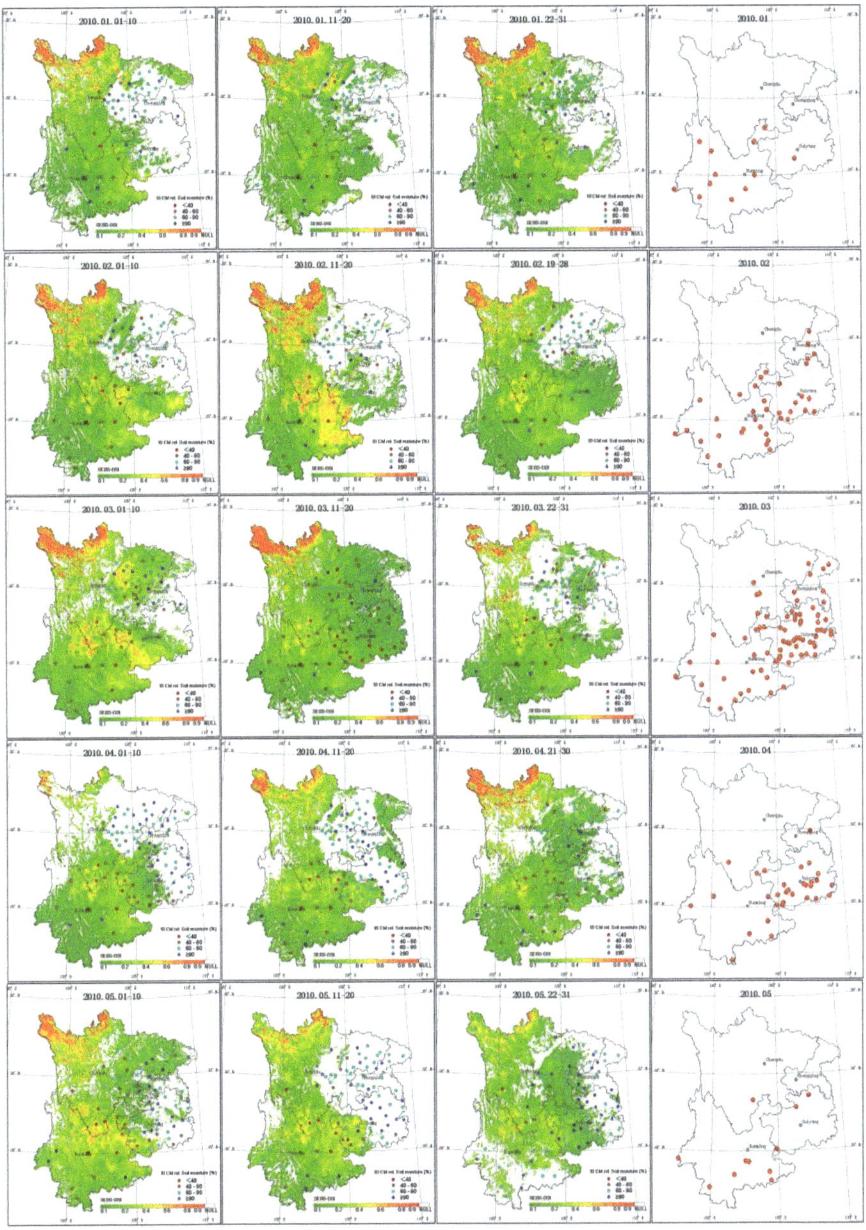

Fig. 8.9 SEBS-DSI with measured 10 cm relative soil moisture overlaid and reported drought situation, January–May 2010 (*from left to right*: first, second, third decade and reported droughts)

conditions. Ideally the SEBS-DSI should be compared to in situ soil moisture observations to establish the accuracy of the calculations, but the real challenges lay in the very local nature of the in situ observations where both very dry and very wet observations are reported within short distance of each other, which may be influenced by local agricultural practices. Nevertheless and broadly higher SEBS-DSI values correspond to drier soil moisture conditions and the global patterns are consistent. Future analysis should focus on stratification of SEBS-DSI and in situ soil moisture observation per land use (e.g. using NDVI information) and soil textures.

8.5 Summary and Conclusions

Drought phenomenon is very complex as it is caused by a multitude of factors starting from the deficit of precipitation when compared to climatic mean. Such a deficit, when combined with sustained dry weather conditions entertain high evaporation and transpiration and thus result in soil moisture drought if there is no adequate irrigation. Finally, hydrological drought emerges by a combination of lack of precipitation and excessive use of available water resources for a sustained period. To assess historical droughts and to provide drought monitoring, various indices have been proposed for different purposes and using different datasets. Despite the panoply of these indices, when one focuses on the processes that caused droughts, a set of consistent indices may adequately describe the different aspects of the droughts. We propose the use of standardised index, such as SPI for precipitation, SPEI for precipitation and potential evaporation and SDSI-ETDI for evapotranspiration, SVCI for vegetation condition and STWSI for the terrestrial water storage in the catchment, all calculated on weekly or monthly time interval from 1 to N time intervals (e.g. for 48 months) (i.e. as standardised anomalies and cumulative anomalies). As remote sensing and hydrometeorological data assimilation systems routinely provide data to derive the different indices, use of them can provide a uniform framework for drought assessment, monitoring and analysis as well as predictions.

Acknowledgements This work is supported in part by the ESA-MOST Dragon programme. We thank them for providing continuous supports in the project in terms of data, financial and logistic supports. We are grateful to colleagues for their supports to the various project activities. This project is conducted in collaboration with the EU FP7 CEOP-AEGIS and CORE-CLIMAX projects.

References

Adler RF, Huffman GJ, Chang A, Ferraro R, Xie P, Janowiak J, Arkin P (2003) The version 2 Global Precipitation Climatology Project (GPCP) monthly precipitation analysis. J Hydrometeorol 4:1147–1167

Arnold JG, Srinivasan R, Muttiah RS, Williams JR (1998) Large area hydrologic modelling and assessment. Part 1. Model development. J Am Soc Water Res Assoc 34(1):73–89

Aydin M (1994) Hydraulic properties and water balance of a clay soil cropped with cotton. Irrig Sci 15:17–23

Chen X, Su Z, Ma Y, Liu S, Yu Q, Xu Z (2014) Development of a 10-year (2001–2010) 0.1-degree dataset of land-surface energy balance for mainland China. Atmos Chem Phys 14 (23):13097–13117

Eden U (2012) Drought assessment by evapotranspiration mapping in Twente, the Netherlands. M. Sc. thesis, Faculty of Geo-Information Science and Earth Observation, (ITC), University of Twente, p 79

Guttman NB (1999) Accepting the standardized precipitation index: a calculation algorithm. J Am Water Resour Assoc 35(2):311–322

Heffernan O (2013) The dry facts. Nature 501(7468):S2–S3

Jia L, Hu G, Zhou J, Menenti M (2012) Assessing the sensitivity of two new indicators of vegetation response to water availability for drought monitoring. Paper presented at the SPIE Asia-Pacific Remote Sens. doi:10.1117/12.977416

Kerr RA (1998) Sea-floor dust shows drought felled Akkadian empire. Science 279 (5349):325–326

Kogan FN (1990) Remote sensing of weather impacts on vegetation in non-homogeneous areas. Int J Remote Sens 11(8):1405–1419

Kogan FN (1995) Application of vegetation index and brightness temperature for drought detection. Adv Space Res 15(11):91–100

Kogan FN (2001) Operational space technology for global vegetation assessment. Bull Am Meteorol Soc 82(9):1949–1964

Kriegler FJ, Malila WA, Nalepka RF, Richardson W (1969) Preprocessing transformations and their effects on multispectral recognition. In: Proceedings of the sixth international symposium on remote sensing of environment, pp 97–131

Li X (2001) Estimation of evaporation with remote sensing in the Urumqi River Basin. M.Sc. thesis, IHE, Delft, 84pp

McKee TB, Doesken NJ, Kleist J (1993) The relationship of drought frequency and duration to time scales. In: Proceedings of the 8th conference on applied climatology, American Meteorological Society, Anaheim, 17–22 January 1993

McKee TB, Doesken J, Kleist J (1995) Drought monitoring with multiple time scales. Paper presented at ninth conference on applied climatology, American Meteorological Society, Dallas

McVicar TR, Jupp DLB (1998) The current and potential operational uses of remote sensing to aid decisions on drought exceptional circumstances in Australia: a review. Agric Syst 57 (3):399–468

Narasimhan B, Srinivasan R (2005) Development and evaluation of Soil Moisture Deficit Index (SMDI) and Evapotranspiration Deficit Index (ETDI) for agricultural drought monitoring. Agric For Meteorol 133(1–4):69–88

Neitsch SL, Arnold JG, Kiniry JR, Williams JR, King KW (2002) Soil and water assessment tool. Theoretical documentation: Version 200. TWRI TR-191, Texas Water Resources Institute, College Station

Palmer WC (1965) Meteorological drought. Weather Bureau Research Paper No. 45, US Dept. Comm., Washington, DC, 58pp

Palmer WC (1968) Keeping track of crop moisture conditions, nationwide: the new crop moisture index. Weatherwise 21(4):156–161

Roerink GJ, Menenti M, Verhoef W (2000) Reconstructing cloud-free NDVI composites using Fourier analysis of time series. Int J Remote Sens 21:1911–1917

Rouse JW, Haas RH, Schell JA, Deering DW (1973) Monitoring vegetation systems in the Great Plains with ERTS. In: Third ERTS symposium, NASA SP-351 I, pp 309–317

Su Z (2001) A Surface Energy Balance System (SEBS) for estimation of turbulent heat fluxes from point to continental scale. In: Su Z, Jacobs C (eds) Advanced earth observation—land surface climate. Publications of the National Remote Sensing Board (BCRS), USP-2, 01-02, p 183

Su Z (2002) The Surface Energy Balance System (SEBS) for estimation of turbulent heat fluxes. Hydrol Earth Syst Sci 6(1):85–99

Su Z, Troch PA, De Troch FP (1997) Remote sensing of bare soil moisture using EMAC/ESAR data. Int J Remote Sensing 18(10):2105–2124

Su Z, Yacob A, He Y, Boogaard H, Wen J, Gao B, Roerink G, van Diepen K (2003a) Assessing relative soil moisture with remote sensing data: theory and experimental validation. Phys Chem Earth 28(1–3):89–101

Su Z, Wen J, Wan L (2003) A methodology for the retrieval of land physical parameter and actual evaporation using NOAA/AVHRR data. J Jilin Univ (Earth Sci Ed) 33(sup.):106–118

Su Z, Fernández-Prieto D, Timmermans J, Chen X, Hungershoefer K, Roebeling R, Schröder M, Schulz J, Stammes P, Wang P, Wolters E (2003a) First results of the earth observation Water Cycle Multi-mission Observation Strategy (WACMOS). Int J Appl Earth Obs Geoinf 26:270–285

Thornthwaite CW (1948) An approach toward a rational classification of climate. Geogr Rev 38:55–94

Tucker CJ (1979) Red and photographic infrared linear combinations for monitoring vegetation. Remote Sens Environ 8(2):127–150

van Hoek M (2016) Drought monitoring from space: a focus on indicators, early detection and development of a web-based integrated portal. Ph.D. thesis, Chinese Academy of Sciences, 168pp

Verhoef W (2000) Theory and software. In: Roerink GJ, Menenti M (eds) Time series of satellite data: development of new products. Publications of the National Remote Sensing Board (BCRS), NRSP-2, pp 99–33

Vicente-Serrano SM, Beguería S, López-Moreno JI (2010a) A multiscalar drought index sensitive to global warming: the standardized precipitation evapotranspiration index. J Clim 23 (7):1696–1718

Vicente-Serrano SM et al (2010b) A new global 0.5 gridded dataset (1901–2006) of a multiscalar drought index: comparison with current drought index datasets based on the Palmer Drought Severity Index. J Hydrometeorol 11(4):1033–1043

Zargar Z, Sadiq R, Naser B, Khan FI (2011) A review of drought indices. Environ Rev 19:333–349

Chapter 9
A Framework for Assessing Soil Moisture Deficit and Crop Water Stress at Multiple Space and Time Scales Under Climate Change Scenarios Using Model Platform, Satellite Remote Sensing, and Decision Support System

Binayak P. Mohanty, Amor V.M. Ines, Yongchul Shin, Nandita Gaur, Narendra Das, and Raghavendra Jana

9.1 Introduction

NASA has fostered, in response to its contribution to the U.S. Climate Assessment, the development of several enabling tools that incorporate satellite era (1979–present) observations into land surface models. The most relevant to land surface hydrology is the integrated terrestrial water analysis system, known as NCA-LDAS (National Climate Assessment – Land Data Assimilation System). NCA-LDAS is an end-to-end enabling tool for the development, evaluation, and dissemination of terrestrial hydrologic time series and indicators in support of the NCA (Jasinski et al. 2014; Kumar et al. 2014; Rui et al. 2014) using the Land Information System (LIS) modeling framework (Kumar et al. 2006; Peters-Lidard et al. 2007). The NCA-LDAS archive contains, for the satellite era period 1979 and 2013, over

B.P. Mohanty, Ph.D. (✉) • N. Gaur
Department of Biological and Agricultural Engineering, Texas A&M University, 2117 TAMU, College Station, TX, USA
e-mail: bmohanty@tamu.edu; nandita.gaur@tamu.edu

A.V.M. Ines
Michigan State University, East Lansing, MI, USA

Y. Shin
School of Agricultural Civil & Bio-Industry Engineering, College of Agriculture and Life Science, Kyungpook National University, Daegu, South Korea
e-mail: ycshin@knu.ac.kr

N. Das
Jet Propulsion Laboratory, NASA, Pasadena, CA, USA

R. Jana
King Abdullah University of Science and Technology, Thuwal, Saudi Arabia

V. Lakshmi (ed.), *Remote Sensing of Hydrological Extremes*, Springer Remote Sensing/Photogrammetry, DOI 10.1007/978-3-319-43744-6_9

60 data products associated with land surface fluxes and stores over the continental U.S. These products were developed though multivariate data assimilation of satellite-based Environmental Data Records (EDRs) of soil moisture (SMMR/SSMI/AMSR-E), snow depth and cover (SMMR/SSMI/AMSR-E/ASCAT), terrestrial storage anomalies (GRACE), and irrigation intensity (MODIS). For sustained assessment NCA-LDAS will expand its capability by ingesting data from more recently launched and near future sensors (e.g., SMOS, VIIRS, SMAP, AMSR-2) as well as by developing new tools to facilitate output dissemination (using NASA GES-DISC) and analyses/decision making (using NASA Giovanni system) by stakeholders. Leveraging this ongoing project we are putting a complementary effort for multiscale analysis of lagging indices such as growing aridity index (ratio between precipitation and potential evapotranspiration), available soil moisture deficit in the root zone (weekly deficit of soil water during different phases of crop growth), and crop water stress (based on available soil moisture deficit and evapotranspiration deficit). Aridity index, soil moisture deficit index, crop water stress index ranging from LDAS model grid / satellite footprint (encompassing multiple land covers and soil types) to individual farm (encompassing uniform land cover and soil type) will be developed for assessing impact of future climate change scenarios and better adaption of agricultural water/crop management practices. Although the framework later will be focused on agricultural water management issues, these indices at multiple scales and associated display, dissemination, and analyses framework can further be adapted for more holistic analysis and decision making for water sustainability at county, state, regional, and national level with increasing demand of water from competing sectors such as agriculture, municipality, energy, and industry under future climate scenarios.

9.2 Rational and Problem Statement

Using Aridity Index (AI) based on temperature and precipitation (e.g., precipitation over potential evapotranspiration, P/PET) is the simplest method to account for crop yield in a field, but it could be insufficient to explain spatial and temporal crop yield variability, eventually leading to annual agricultural production, economics, security, and sustainability in a broader sense. Soil water availability is significantly correlated with crop yield because of soil water uptake by plant roots and relations with crop water stress. Spatial yield variability can be obtained by examining the effect of landscape features (e.g., Normalized Difference Vegetation Index, NDVI, and Leaf Area Index, LAI) or site-specific soil properties (e.g., soil types, soil water retention, and hydraulic properties) that affect soil moisture patterns. Soil water availability in a field can be included in the Aridity Index or Soil Moisture Deficit Index (SMDI) based on weekly soil moisture deficit for any particular crop. Further, timing of water stress with crop growth stage is also important to understand crop yield variability. Thus, smaller time scale (e.g., weekly) and spatial scale (e.g., at the farm/field, 1-km or less) should be considered in developing the lagging indicators for improving the agricultural drought outlook, particularly under the evolving climate change scenarios.

Root zone soil moisture dynamics is a key factor in water resources/agricultural water management, rainfall-runoff processes, and ecosystem dynamics and alteration. Its spatiotemporal variability in a field influences runoff at the soil surface and subsurface, evapotranspiration, and groundwater recharge. Root zone soil moisture content also plays a pivotal role in ecological processes at individual plant to system scale. Importance of root zone soil moisture prediction has increased in agricultural water management because of unbalanced water supply and demand due to agricultural drought, differences in irrigation and drainage practices, farm level water distribution and scheduling, etc. Our effort aims at developing algorithms, tools for improved prediction of Soil Moisture Deficit Index (SMDI), and Crop Water Stress Index (CWSI) at multiple space–time scales under NCA themes leading to adaptive water/crop management schemes and decision support systems (DSS).

9.3 State of Land Information System (LIS)

LIS is a framework for high resolution distributed land surface modeling. LIS is the software framework for the Land Data Assimilation System (LDAS) projects of Global Land Data Assimilation System (GLDAS) and North America Land Data Assimilation System (NLDAS) developed within the Hydrological Sciences Laboratory at NASA Goddard. The LIS software takes into account most of the physical aspects and governing equations of hydrology and boundary layer physics in modeling and therefore expected to provide reasonable simulation of soil moisture profile (Arsenault et al. 2014; Kumar et al. 2013, 2014; Santanello et al. 2013; Zaitchik et al. 2013; and Liu et al. 2013). The LIS is capable to use ensemble of Land Surface Models (LSMs). The LSMs in LIS are Community Land Model (CLM) (Dai et al. 2003), Noah (Ek et al. 2003), Variable Infiltration Capacity (VIC) (Liang et al. 1996), Mosaic (Koster and Suarez 1996), and Hydrology with Simple SIB (HySSIB) (Sud and Mocko 1999). The LSMs require three types of inputs: (a) initial conditions; (b) boundary conditions (fluxes and forcings); and (c) parameters that are functions of vegetation, soil, topography, and other surface properties. Using these, the LSMs solve the governing equations. The input datasets to LIS include satellite-based remote sensing products such as precipitation, temperature, normalized difference vegetation index (NDVI), Leaf Area Index (LAI), surface albedo, etc. The soil moisture and fluxes fields from LIS are the resultant of physical modeling of hydrology within the CONUS region.

9.4 State of Remotely Sensed Soil Moisture Products and Disaggregation to Local Scale

A remote sensing (RS) scheme has advantages for mapping soil moisture measurements near the land surface. Many space-borne (Advanced Microwave Scanning Radiometer-Earth Observing System, AMSR-E, Soil Moisture and Ocean Salinity, SMOS satellites) have been used to observe pixel-based (tens of km scale) soil moisture products (Jackson et al. 1999, 2005). Recently launched Soil Moisture Active and Passive (SMAP) satellite provide soil moisture at 36 km (passive) resolutions in the recent years (Entekhabi et al. 2010, 2014). However, most of these available RS datasets suffer from its poor spatial resolution for agricultural/water management at local/field scales. Engman (1991) and Entekhabi et al. (1999) reported that the satellite-based passive microwave radiometers have disadvantage for predicting hydrologic processes due to the scaling discrepancy between the pixel resolution (~tens of km) and the local scale (1-km or less) at which the hydrological processes occur. Thus, the RS soil moisture product needs to be downscaled for improving its spatial resolution through a disaggregation approach. Kim and Barros (2002) reported that fine-scale surface information can be extracted with fractal interpolation methods. Bindlish and Barros (2002) studied a scaling and fractal interpolation approach for downscaling passive microwave data to the spatial resolution of active radar data. Recently, Ines and Mohanty (2008a, b; 2009), Shin and Mohanty (2013), and Ines et al. (2013) have developed disaggregation methods to obtain various land surface information (representing soil properties and subarea fractions, leading to soil moisture) within a RS footprint using optimization algorithm and uncertainty analyses techniques.

9.5 Limitation of Soil Moisture Prediction

Historically, very few studies have been conducted to forecast soil moisture changes based on climate forecast models coupled with a local land surface parameterization approach. The scale incompatibility between hydrologic and the climate models leads to errors at the catchment-, watershed-, and basin-scales for predicting root zone soil moisture dynamics. Root zone soil moisture content at the local scale is affected by precipitation, soil texture, which determines the water holding capacity, vegetation/land cover, which affects evapotranspiration and percolation, topography, which impacts runoff and infiltration. To address proper water balance in the root zone at the local/field scale, we need to develop better long-term root zone soil moisture prediction schemes at localized regions by scaling down the coarse-scale LDAS simulated and RS assimilated data.

9.6 Motivation for National Climate Assessment Indicators at Field Scale

Most of the processes in hydrology that influence drought monitoring, agricultural productivity, and watershed/basin management occur at much finer spatial resolution. Especially, agricultural and hydrological drought monitoring/delineation require high spatial and temporal resolution soil moisture data. As discussed earlier, there have been no dedicated studies: (1) to understand the relation of the soil moisture dynamics between the LDAS grid/RS pixel resolution and local/field scale (1 km or less) and its implications on a disaggregation method, or (2) to predict the distributed root zone soil moisture dynamics with climatic models at local regions for addressing potential agricultural drought risk assessment and efficient water/crop management. Specific goals and outcomes of our current agricultural drought assessment platform effort include:

1. *develop* lagging climate indicators as Aridity Index (AI), Soil Moisture Deficit Index (SMDI), and Crop Water Stress Index (CWSI) at spatial (farm/field, county, state, regional, and national) and temporal (weekly, biweekly, monthly, and seasonal) scale using past (1995–2015) and future (GCM predicted 2015–2045) climatic data and LDAS outputs for the conterminous USA,
2. *calibration/validation* of the projected climate indicators in several hydroclimatic regions (e.g., Great Plains, Midwest, West, Northeast, Southeast, and Southwest) during the course of the project using soil moisture networks (USCRN, SCAN) and past/upcoming field campaigns data,
3. *disseminate* the lagging climate indicators as AI, SMDI, and CWSI using interactive/searchable graphical (zoomable) and tabular displays over the web,
4. *develop* decision support tools/scenarios for application in planning and adaptation/mitigation during crop growing season by farmers, water resource managers, and decision makers at local, state, regional, and national level.

9.7 Soil Moisture Physical Controls and Multiscale Physics and Relevant Activities

At a particular point in time soil moisture content is influenced by: (1) the precipitation history; (2) the texture of the soil, which determines the water-holding capacity; (3) the slope of the land surface, which affects runoff and infiltration; and (4) the vegetation and land cover, which influences evapotranspiration and deep percolation.

Soil: Soil heterogeneity (Gaur and Mohanty 2013, 2015; Joshi and Mohanty 2010; Joshi et al. 2011; Das et al. 2008a, b, 2010a, b; Mohanty and Skaggs 2001) affects the distribution of soil moisture through variations in texture, organic matter content, porosity, structure, and macroporosity. Each of these factors affects the fluid transmission and retention properties. Variability in soil hydraulic conductivity and soil water retention greatly influences the vertical and lateral transmission properties (Mohanty and Zhu 2007). Significant soil moisture variation may exist over very small distances due to variation in soil particle/pore sizes which in turn determines transient upward and downward soil water fluxes.

Topography: Topography (macro or micro) (Gaur and Mohanty 2013, 2015; Joshi and Mohanty 2010; Joshi et al. 2011; Mohanty et al. 2000a; Mohanty and Skaggs 2001; Jacobs et al. 2004) also plays an important role in the spatial organization of soil moisture at different scales. Variations in slope, aspect, curvature, upslope contributing area, and relative elevation all affect the distribution of soil moisture near the land surface. At the small catchment and hill-slope scales, soil moisture varies as a result of water-routing processes, radiative (aspect) effects, and heterogeneity in vegetation and soil textural deposition/characteristics.

Vegetation: Land cover is also critical for understanding the soil moisture regimes as it affects infiltration, runoff, and evapotranspiration through root water uptake. Vegetation type, density, and uniformity are some of the associated features that contribute to soil moisture variation at different space and time scales. Furthermore, the influence of vegetation on soil moisture is more dynamic (Mohanty et al. 2000b; Das and Mohanty 2008; Gaur and Mohanty 2013) as compared to soil and topographic factors. Literature also shows that the variability of soil moisture is lowest with full canopy cover and highest with partial coverage.

Climate: Precipitation, solar radiation, temperature, wind, and humidity are some of the important climatic factors that contribute to the space–time dynamics of soil moisture and resultant fluxes. Precipitation is the single most important climatic forcing for soil moisture content and its distribution. As shown by Sivapalan et al. (1987), the dominant runoff producing mechanism may vary with storm characteristics and antecedent soil moisture conditions resulting in the spatiotemporal variability in soil moisture. During the SGP97 hydrology campaign, Famiglietti et al. (1999) found a distinct trend in mean soil moisture for SGP97 locations with a south-to-north precipitation gradient.

Mohanty and Skaggs (2001), Jacobs et al. (2004), Cosh et al. (2004), Joshi and Mohanty (2010), Joshi et al. (2011), and Gaur and Mohanty (2013, 2016) conducted analyses of ground and remote sensing soil moisture data collected during the SGP97, SGP99, SMEX02, SMEX03, SMEX04, CLASIC 2007, and SMAPVEX12 and concluded that characteristic differences were observed in the space–time dynamics of soil moisture within selected remote sensing footprints with various combinations of soil texture, slope, vegetation, and precipitation. *To better understand the dynamics of soil moisture in different hydroclimatic regions scaling factors relating to dominant physical controls at different spatial scales are being developed in a current effort*. We (Jana 2010; Jana and Mohanty 2012a, b, c; Crow et al. 2012; Gaur and Mohanty 2015) hypothesized that different landscape features (soil, topography, vegetation, and precipitation) are dominant /controlling factors governing soil moisture variability at different spatial scales (Fig. 9.1). In other

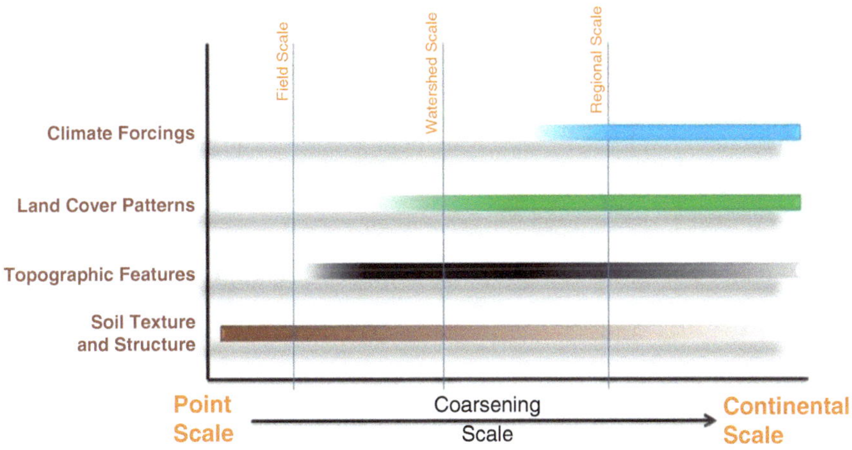

Fig. 9.1 Multiple/dominant physical controls for soil moisture at different spatial scales ranging from local, field, catchment, watershed, regional, to continental scale by Jana (2010), Jana and Mohanty (2012a, b, c) as cited in Crow et al. (2012)

words, soil moisture spatiotemporal pattern reflects a conjoint variability of soil properties, topography, vegetation, and precipitation attributes. Soil moisture variability is *dominated* by soil properties at the field/remote sensor footprint (e.g., of the order of 10–100 m) scale, topographic features at the catchment scale, vegetation characteristics at watershed scale, and precipitation patterns at the regional scale and beyond. Ensemble hydrologic fluxes (including evapotranspiration, infiltration, shallow ground water recharge) within and across the root zone reflect the evolution of soil moisture at a particular spatial scale (field, catchment, watershed, or region) and can be "effectively" represented by one or more linear/nonlinear hydrologic (soil/field, topography/catchment, vegetation/watershed, or precipitation/region) scale parameters reflecting dominant heterogeneity of the landscape.

Our ongoing/previous studies also revealed that the soil moisture statistics change whenever heterogeneity in any of these physical controls is encountered. This happens across all spatial scales. Heterogeneity evolves differently for various physical controls across observation scales. Soil texture can change across a scale of a few meters and as we go further in scale the variability in soil texture increases. However, beyond a certain scale, similar soil textures may be observed resulting in variability in soil texture at larger scale being averaged out in comparison to other physical controls. Similarly, topographic heterogeneity becomes important at scales larger than the plot/field scale, whereas vegetation heterogeneity becomes important at the scale of biomes, and beyond that we notice heterogeneity in precipitation, which changes with hydro-climate. However, this is a very generic assumption for the land surface without accounting for agricultural management practices, crop root growth across season, or land use change over time. Based on the dominant physical control concept a comprehensive multiscale soil moisture data analyses and assimilation scheme (Das and Mohanty 2008) using data from various in situ and remote sensing platforms has been developed. This long-term data, analyses, and process understanding with dominant physical controls provide us the basis for our proposed downscaling scheme at various scales efficiently (e.g., LDAS grid/ satellite footprint, watershed, catchment, field) described later.

9.8 Disaggregation of Soil Moisture from Larger Scale Soil Moisture Imagery/LIS Output

Primarily, the use of large-scale RS/LIS/GCM products at the local scale is the major challenge and question is how to extract valuable information from satellite/ model-based products at the local scale. In the proposed effort, we develop several tools/methods to disaggregate large-scale remote sensing/LDAS soil moisture products, and on the predictions of effective soil moisture in the rooting zone (top ~1 m), and hydrologic fluxes as surface runoff, evapotranspiration, and groundwater recharge. The downscaling approach includes statistical and numerical methods, e.g., mapping probability distributions between satellite measurements and observations in the field, nonhomogenous hidden Markov modeling (Robertson et al. 2004), mixture modeling (Ines and Honda 2005; Ines et al. 2013), multiscale artificial/Bayesian neural network with bias correction (Jana et al. 2007, 2008), adaptive Markov Chain Monte Carlo (Das et al. 2008b), and data-driven (Das et al. 2016) schemes. For accomplishing downscaling of satellite-based soil moisture to subgrid scales, few (selected) *alternative approaches* (based on types of available data in a particular region) are elaborated as follows:

Method 1—Mapping Probability Distributions: Pixel-based soil moisture data will result from its individual components based on soil types (or similarly land cover types or topographic features) based on our recent findings on dominant soil moisture physical controls at different scales. Assuming that the individual components can be predicted by the pixel-based soil moisture data, a simple approach to downscale the synoptic data into its possible components can be developed by mapping the probability distribution of the synoptic data with the respective probability distributions of the measured components.

Mapping probabilities from one scale to another would perhaps account for the nonlinearity involved on the physical processes (controls) occurring between scales (Ines and Hansen 2006). Assuming further that a time series of pixel-based soil moisture and its respective components (i.e., subpixel data) are available, the distribution functions $F_p(x_p;\theta_p)$ and $F_{sub}(x_{sub\text{-}p};\theta_{sub\text{-}p})$ can be determined, where x is a vector of soil moisture data and θ is a vector of parameters (i.e., the scaling and shape parameters of the distribution functions), the subscripts p and sub-p denote pixel-based and subpixel data, respectively. The downscaling procedure is schematized in Fig. 9.2 where it is simply a mapping of distributions between the scales as a function of time. The downscaled soil moisture data can be determined by,

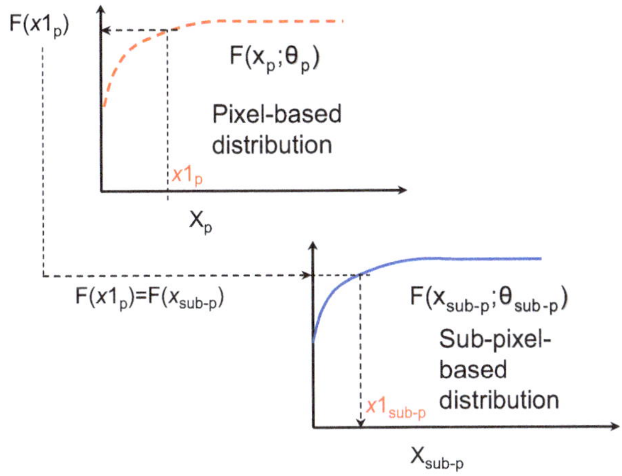

Fig. 9.2 Mapping of cumulative distribution functions of pixel-based soil moisture and subpixel data at time $= t$

$$x_{\text{sub-p}} = F^{-1}_{t,\text{sub-p}}\left(F_{t,\text{p}}\left(x_{\text{p}}\right)\right) \tag{9.1}$$

where t is an index for day of the year. We investigate the appropriate probability distributions that would best describe the soil moisture data at different scales matching various ground-based, air-borne, and space-borne RS footprints. We also examine how the distribution functions would be constructed, either by a general distribution for each scale or by calendar month basis for each scale (synoptic vs. components). The problem that may arise when using only one distribution function to describe the data series is the seasonality on the downscaled soil moisture. Prescribing the probability distributions on a weekly/monthly basis might address this problem (e.g., Ines 2004; Wood et al. 2002). Once developed these hydroclimate and scale-specific distribution functions will help downscale the satellite and LDAS (grid-based) soil moisture forecasts and national climate indicators (2015–2045) to finer scales. Data from soil moisture networks (USCRN, SCAN), available field campaigns, and other physical control-based estimations at various scales are being used for this "pdf"-based downscaling method.

Method 2—Nonhomogeneous Hidden Markov Model: This model aims to relate regional or local hydrologic process (discrete or continuous), e.g., soil moisture dynamics to large-scale measurements from remote sensing or simulated variables from SVAT models. The unifying feature of the model is the hypothesis of an unobserved "soil moisture state" so-called here "hidden state" which transcends the differences in scales between the two observed processes (synoptic vs. regional/local). The hidden state is regarded as an automatic classifier of large-scale observational/simulated patterns into classes that are associated with the local process patterns (Hughes and Guttorp 1994).

In NHMM, there are two underlying assumptions given as follows (Robertson et al. 2004):

$$P\left(\mathbf{R}_t \middle| S_1^T, \mathbf{R}_1^{t\text{-}1}, \mathbf{X}_1^T\right) = P(\mathbf{R}_t | S_t) \tag{9.2}$$

$$P\left(S_t \middle| S_1^{t\text{-}1}, \mathbf{X}_1^T\right) = P\left(S_t \middle| S_{t\text{-}1}, \mathbf{X}_t\right) \tag{9.3}$$

where $\mathbf{R}_t$ is the measurement of the regional (multivariate from network of points) or local hydrologic process at time t, S_t is the hidden state at time t (unobserved), and $\mathbf{X}_t$ is the measurement (summary or ensemble average if multivariate) of remote sensing data at time t for $1 \leq t \leq T$.

Equation (9.2) assumes that the local process $\mathbf{R}_t$ is conditionally independent given the hidden state S_t. Meaning, the temporal persistence in the local process is captured by the persistence in the hidden state derived from Eq. (9.3). Equation (9.3) assumes that the hidden state process is first-order Markov, i.e., given the history of the hidden state up to $t - 1$ and the entire record of the remote sensing data (past and future), the hidden state at time t depends only on the previous hidden state and the current remote sensing data (Robertson et al. 2004; Hughes and Guttorp 1994; Hughes et al. 1999). From these underlying principles, large-scale near-surface soil moisture data could be possibly downscaled into smaller continuous soil moisture fields. $P(\mathbf{R_t}|S_t)$ can be modeled as a mixture of 2 or 3 normal distributions, representing a wetting, drying, and/or an extended drying phase. $P\left(S_t \middle| S_{t\text{-}1}, \mathbf{X}_t\right)$ is modeled in NHMM as nonstationary transition probability matrix. Hughes and Guttorp (1994) defined this function by multiplying the baseline transition probabilities γji (when S is not dependent on X), with a function of the process predictors. Robertson et al. (2004) defined the hidden state transition by a polytomous (multinomial) logistic regression.

The NHMM is a general spatial disaggregator and can be used in downscaling of large-scale soil moisture and its predictions into network of points/subgrids, perhaps, also at various soil depths given that there are available data at these depths for model learning. This hypothesis will be also examined in the study. Moreover, the conditional independence assumption for $\mathbf{R}_t$ in Eq. (9.2) may not well represent the **spatial structure** of the network (of points/subgrids). An extension of NHMM to capture both dependencies across time and network variables can be implemented by embedding Chow-Liu tree within the HMM framework (see Kirshner et al. 2004). This may account for the effect of overland/subsurface flow from higher to lower elevation (e.g., pattern, connectedness) or any mutual relationships among network variables existing in the landscape, e.g., the effects of vegetation activities, soil management, crop management, irrigation, planning and harvest timing, tillage, etc.

Method 3—Mixture Modeling: Mixture modeling using satellite remote sensing data, LDAS outputs, and genetic algorithms (see Ines et al. 2013; Shin and Mohanty 2013) could be potentially used to downscale pixel-based soil moisture data at least to area fractions of land covers and/or soil types of the study pixel/region.

For example, let S_i be the soil types in the pixel, and V_i represents their vegetation/crop area fractions (Fig. 9.3). The mixture problem for one pixel can be formulated as follows,

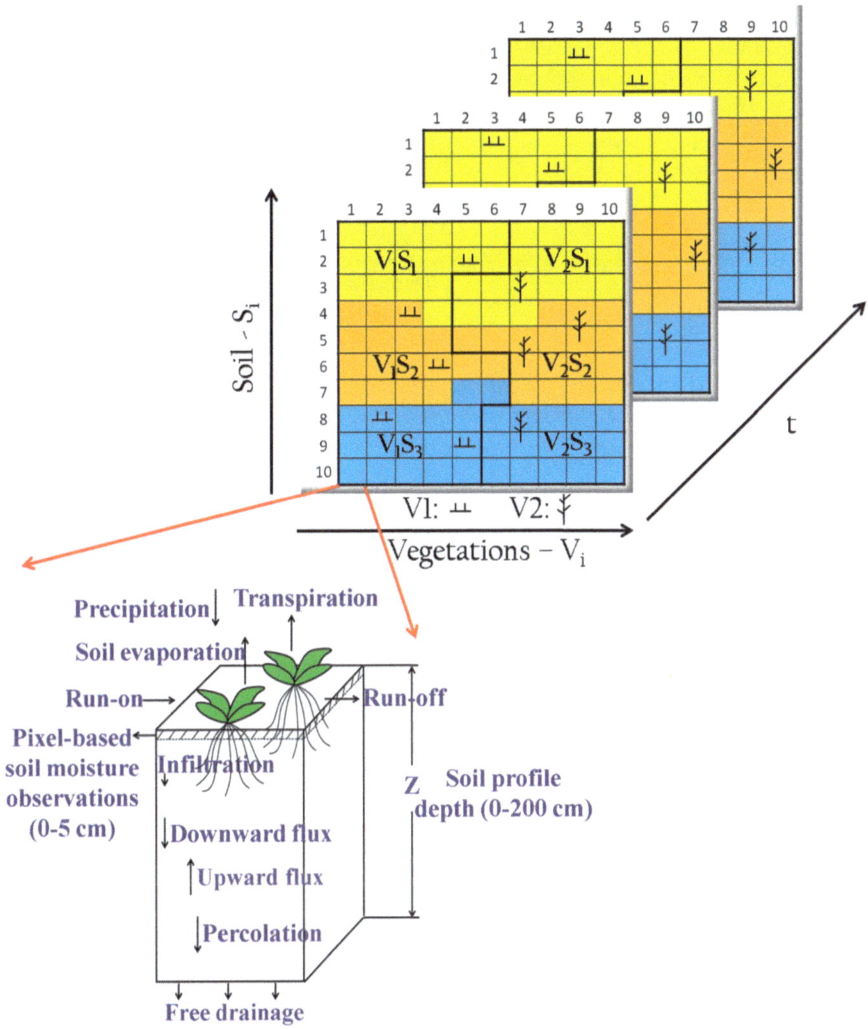

Fig. 9.3 Sample grid (10 × 10) domain for the complex soil-vegetation pixel experiments (e.g., two vegetation and three soil types) for downscaling using a mixture model

$$\theta(k)_t = \sum_{i=1}^{m} V_i \theta_{ti} \quad \forall t \tag{9.4}$$

$$\theta_{ti} = \theta(S_i) \quad \forall t \forall i \tag{9.5}$$

$$k = \{V_i, S_i\} \quad \forall i \tag{9.6}$$

Equation (9.4) defines the mixed signature of soil moisture at any time t (i.e., the remote sensing pixel data or model grid data), Eq. (9.5) defines the possible components of this signature, and Eq. (9.6) defines the mixing parameters, which are the vegetation/crop area fractions (V_i) and soil types (S_i); i and t are the indices for the number of soil types and time domain, respectively, m denotes the total number of soil types in the pixel. $\theta(S_i)$ can be determined using a simulation model, e.g., Noah (Ek et al. 2003) or SMAT (Das and Mohanty 2006). To determine $\mathbf{k}$, we must apply an un-mixing algorithm; objective function (simulated $\theta(\mathbf{k})$ vs. observed, θ°) can be defined as

$$\text{Obj}(k)_{\text{Min}} = \text{Min}\left\{\frac{1}{n}\sum_{t=1}^{n}\left|\theta(k)_t - \hat{\theta}_t^{o}\right|\right\} \tag{9.7}$$

subject to the constraints.

The presented mixture model can be extended to account for the variability of other ***dominant physical controls*** *within the pixel at different scales by factoring the possible combinations of soils and vegetation types, etc., and then applying the same principle.*

Figure 9.4 shows a sample result from the proposed grid-based disaggregation approach reflecting the proof of the concept under the hypothetical conditions at a site in Little Washita, Oklahoma. It is apparent that the mixed soil moisture estimated by the grid-based disaggregation method was identified well with the observations. The soil ID (soil properties) values and subgrid fractions derived by Ensemble Multiple Operators Genetic Algorithm (EMOGA) (Shin and Mohanty 2013) respond well to the actual soil moisture values (Fig. 9.4), although there are

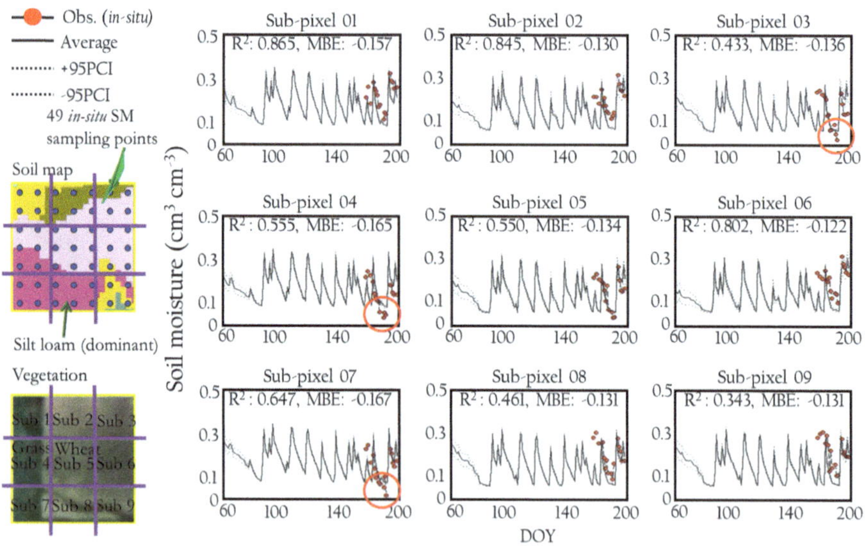

Fig. 9.4 Sample results from the grid-based disaggregation algorithm; (*Left*) Little Washita (LW 21) field site with mixed vegetation and soil type including the in situ distributed soil moisture sampling points, and (*Right*) Downscaled soil moisture dynamics at subpixel scale time series (*bullets*: target values, *lines*: estimated) (after Shin and Mohanty 2013)

uncertainties due to the inherent weakness of RS products, modeling and measurement errors, and optimization errors. It indicates that this approach is applicable for disaggregating a large-scale RS/LDAS model products and climate indicators to field scale.

9.9 Prediction of Future Climate Change Scenarios Using General Circulation Models (GCMs)

In general, the purpose of General Circulation Models (GCMs) is to numerically simulate atmospheric processes based on physical principles. GCMs reproduce observed measurements of recent/past climate changes and provide credible estimates of future climate change (e.g., precipitation, temperature, humidity, wind speed, solar radiation, etc.) on various large scales. The Intergovernmental Panel on Climate Change (IPCC) provides the spatiotemporal distributed scenarios of climate change and related environmental and socioeconomic factors for use in climate impact and adaptation assessment. The IPCC reported Scientific Assessment scenario in 1990 (SA90), which assumes the future population, CO_2 emissions, circulation, etc., and provides scenarios in 2000, published in the Third Assessment Report (Special Report on Emissions Scenarios—SRES). In 2013, the IPCC published Fifth Assessment Report (AR5) and Representative Concentration Pathways (RCPs) scenarios. The SRES/RCP scenarios explore future developments in the global environment due to the production of greenhouse gases and aerosol precursor emissions.

As shown by Nakicenovic et al. (2000), the SRES provides the four storyline (labeled A1, A2, B1, and B2) describing the relationships between the forces driving greenhouse gas and aerosol emissions and their evolution during the twenty-first century world. Each scenario has different demographic, social, economic, technological, and environmental developments that diverge in increasingly irreversible ways. In simple terms, the four storylines combine two sets of divergent tendencies: one set varying between strong economic values and strong environmental values, the other set between increasing globalization and increasing regionalization. The storylines are summarized as follows:

A1 storyline and scenario family: *based on scenarios with respect to very rapid economic growth, global population that peaks in mid-century and declines thereafter, and rapid introduction of new and more efficient technologies.*

A2 storyline and scenario family: *based on scenarios with regionally oriented economic growth which is fragmented and slower than in other storylines and continuously increasing global population.*

B1 storyline and scenario family: based on the same global population with the A1 storyline, but B1 is rapidly changed in economic structures toward a service and information economy, with reductions in material intensity, and the introduction of clean and resource-efficient technologies.

B2 storyline and scenario family: B1 is assumed under the emphasis is on local solutions to economic, social, and environmental sustainability, with continuously increasing population (lower than A2) and intermediate economic development.

Many GCMs have been developed/ improved to simulate earth's climate changes (Ines and Hansen 2006; Chervin 1981; among others) based on the above storylines. In Goddard Institute for Space Studies (GISS, http://www.giss.nasa.gov/tools/), NASA provides various GCMs versions (e.g., Model E, Model AOM-GR, and Model II). Using ECHAM v4.5 GCM model (Roeckner et al. 1996), Gong et al. (2003) simulated global climate that is available at IRI/LDEO Climate Data Library (http://iri.columbia.edu). Canadian Centre for Climate Modeling and Analysis (CCCma: http://www.cccma.ec.gc.ca/) also provides a number of climate models (AGCM1, AGCM2, AGCM3, AGCM4/CanAM4, CGCM1, CGCM2, CGCM3, and CGCM/CanCM4) for understanding climate changes and variability under the various processes.

Despite better understanding in climatic forecasts, hydrologists still struggle to improve a bias of GCM scenarios in applications (e.g., water resource planning, management, etc.). One of bias factors of forecast information is the scale discrepancy between the climate model outputs and local scale. Various studies have conducted for downscaling algorithm with how best to reproduce realistic forecasted outputs. The widely used approaches are dynamical modeling such as a regional climate model (RCM—see Leung et al. 2004) within GCM scenarios, statistical or empirical transfer functions (Hewitson and Crane 1996; Wilby and Wigley 1997; Wilby et al. 1998), etc. Wood et al. (2004) has used a quantile-based mapping for a bias correction of GCM/RCM scenarios. Ines and Hansen (2006) developed the bias correction method for the frequency and the intensity distribution of daily GCM rainfall. The output of GCM models is extremely dynamic based on the modeling types, assumptions, and initial and bottom boundary conditions, etc. Thus, we consider the uncertainty range of GCM scenarios with agricultural drought index representing the average, maximum, and minimum GCM (GCM_{Ave}, GCM_{Max}, and GCM_{Min}) scenarios.

9.10 LIS-Based Modeling/Bayesian Multimodel Averaging (BMA) for Root Zone Soil Moisture

Once regional features and GCM/RCM scenarios were obtained, surface soil moisture estimated and downscaled to local scale, the local-scale daily root zone (top 1 m) soil moisture dynamics can be forecasted using a suite of SVAT (SWAP, CLM, VIC, Noah from LIS, e.g., Fig. 9.5) models. A multimodel combination using the Bayesian Model Averaging (BMA) scheme will help exploit the diversity of skillful predictions made by different hydrologic models. BMA is a probabilistic scheme for model combination that infers more reliable and skillful predictions

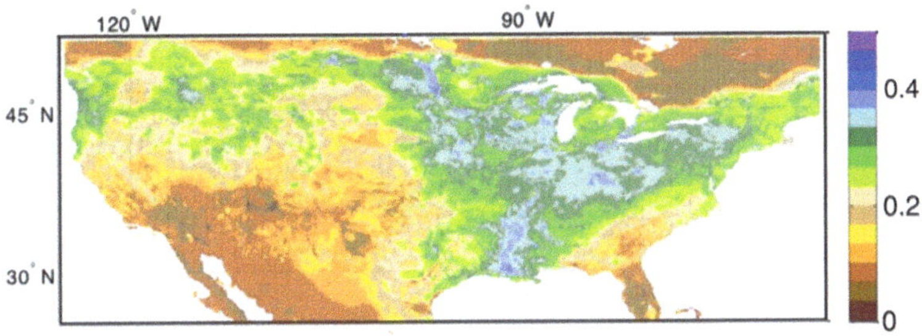

Fig. 9.5 A sample surface soil moisture snapshot map across USA at 9 km from **LIS** platform

from several competing models, by weighing individual predictions based on their probabilistic likelihood measures, with the better performing predictions receiving higher weights than the worse performing ones (Madigan et al. 1996; Duan et al. 2007; Kim et al. 2015). The BMA prediction is the average of predictions weighted by the likelihood that an individual model is correct (Ajami et al. 2007). With the soil ID values and subgrid fractions extracted from a RS soil moisture product and bias corrected GCM scenarios during the past two decades (1995–2015), we forecast the field-scale long-term root zone (0–5, 0–30, 0–60, 60–90 cm) soil moisture changes using GCM scenarios across various hydroclimatic regions of USA for the next three decades (2015–2045). Although inherent uncertainties of land–atmosphere interaction models remain, the proposed approach provides a process-based unified platform for better understanding of soil moisture dynamics with climate variability and changes across the USA.

9.11 Estimating Climate Indicators at Different Space–Time Scales

Root zone soil moisture and fluxes (e.g., ET, groundwater recharge) are predicted using the integrated LIS framework (LDAS-NCA) in a distributed manner. Run-on and run-off routing are included to account for the estimation of runoff at the watershed(s) outlet. Using the derived effective soil hydraulic parameters, we can simulate the water balance within a specified period, giving opportunity to evaluate the effective soil moisture in the rooting zone, surface run-off, evapotranspiration, and groundwater recharge. Auxiliary remote sensing data such as land use-land cover (LANDSAT), vegetation (MODIS), DEM (SRTM), etc., are being used to characterize our modeling/study domain. The grid-based disaggregation algorithm is tested with pixel-based soil moisture (e.g., PSR, ESTAR, AMSR-E, SMOS, SMAP) and network (NOAA-USCRN and USDA-SCAN) products at selected hydroclimatic regional sites (e.g., *Oklahoma (Southern Great Plains), Iowa (Mid-west), Arizona (Semiarid Southwest), Idaho (Arid Northwest), Georgia (Humid Southeast), Pennsylvania (Northeast), and California (Mediterranean West)*) for multiple years from the National Snow and Ice Data Center (NSIDC: http://nsidc.org/). We extracted field-scale (~100 m to 1 km) land surface information (soil ID representing soil properties and subgrid fractions) from a RS product using the downscaling-simulation–optimization scheme. With the soil ID values derived by the simulation–optimization scheme and various GCM/RCM scenarios from climatic models (described earlier), the long-term daily root zone (0–5, 0–30, 0–60, 0–90 cm) soil moisture dynamics are forecasted using SVAT modeling and Bayesian multimodel averaging (BMA) at the subgrid/field scales. *These root zone soil moisture and fluxes are used to develop various agricultural drought indicators (AI, SMDI, and CWSI) and aggregate them at multiple spatiotemporal scales ranging spatial (farm/field, county, catchment, state, region, to national) and temporal (weekly, biweekly, monthly, and seasonal) scales.* Once validated at selected locations, approach will be replicated to populate the findings across similar regions for the conterminous USA.

Aridity Index (AI): Using GCM/RCM estimation of precipitation (P) and temperature (T) and land-cover and crop-specific potential evapotranspiration (PET) based on SVAT models in LIS, aridity index (P/PET) is calculated at field scale for study regions and aggregated to higher space and time scales (Products 1–4) in incremental fashion (as shown later).

Soil Moisture Deficit Index (SMDI): The weekly SMDI is calculated using the daily available soil water content in the root zone as shown in Eqs. (9.8), (9.9a), and (9.9b) (defined in Narasimhan and Srinivasan 2005). The soil water contents of 7 days are averaged for each of 52 weeks in a year at the field scale and the median value is obtained during the simulation period (52 weeks) (Fig. 9.6). Then the maximum and minimum soil water contents for each week are also taken for the simulation period,

$$\mathrm{SMDI}_j = \frac{\sum_{t=1}^{j} \mathrm{SD}_t}{25t + 25} \tag{9.8}$$

$$\mathrm{SD}_{i,j} = \frac{\mathrm{SW}_{i,j} - \mathrm{MSW}_j}{\mathrm{MSW}_j - \mathrm{minSW}_j} \times 100, \quad \text{if} \quad \mathrm{SW}_{i,j} = \mathrm{MSW}_j \tag{9.9a}$$

$$\mathrm{SD}_{i,j} = \frac{\mathrm{SW}_{i,j} - \mathrm{MSW}_j}{\mathrm{maxSW}_j - \mathrm{MSW}_j} \times 100, \quad \text{if} \quad \mathrm{SW}_{i,j} > \mathrm{MSW}_j \tag{9.9b}$$

where $\mathrm{SD}_{i,j}$ is the soil water deficit in the root zone (%), $\mathrm{SW}_{i,j}$ is the mean weekly soil water available in the soil profile (mm), MSW_j is the long-term median available soil water in the soil profile (mm), maxSW_j the long-term maximum available soil water in the soil profile (mm), minSW_j is the long-term minimum available soil water in the soil profile (mm), i is the simulation (hindcast: 1995–2015, forecast: 2015–2045) period, and j is the (1–52) weeks in a year, respectively. According to this definition, SD value during any week gives dryness

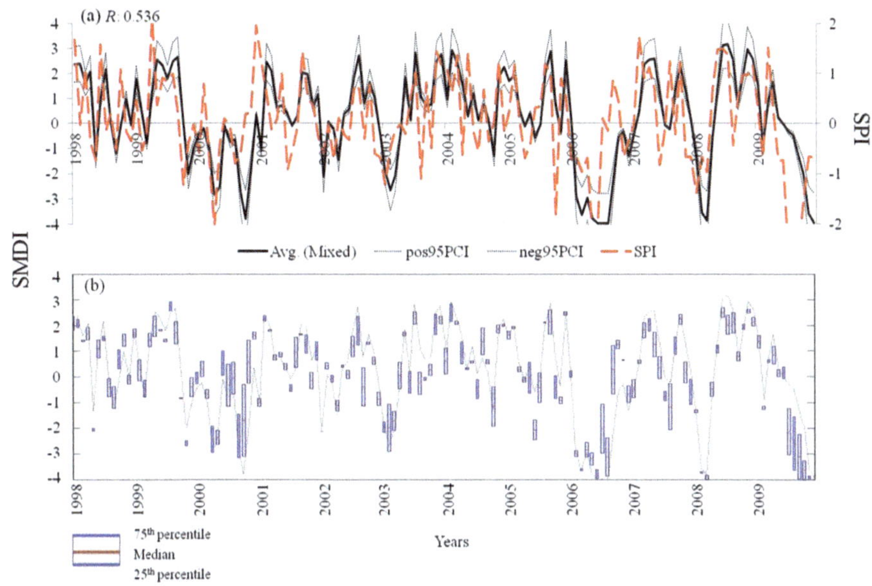

Fig. 9.6 Comparison of the monthly soil moisture deficit index (SMDI) and standardized precipitation index (SPI) at the field sites using the un-filtered solutions derived by EMOGA; (**a**) mixed-scale (s1v1 + s2v1 + s3v1) for the airborne sensing (AS)-scale Walnut Creek (WCAS 1) site. Note that the average root zone soil moisture dynamics in the depth of 0–30 cm were used. (**b**) Diaggregated SMDI for s1v1 (Shin and Mohanty, WRR, under review, 2016)

during that week when compared to long-term historical data. Agricultural drought occurs when dryness prolongs that can affect crop growth.

Crop Water Stress Index (CWSI): Based on Narashiman and Srinivasan (2005) crop water stress index on a weekly basis can be calculated as

$$\mathrm{CWSI} = \frac{\mathrm{PET} - \mathrm{AET}}{\mathrm{PET}} \tag{9.10}$$

where CWSI is the weekly water stress ratio, PET the weekly reference crop evapotranspiration, and AET the actual weekly evapotranspiration. From the long-term median, maximum, and minimum crop water stress, percentage water stress anomaly during any week (j) in year (i) can be calculated as.

$$\mathrm{CWSA}_{i,j} = \frac{\mathrm{MCWSI}_j - \mathrm{CWSI}_{i,j}}{\mathrm{MCWSI}_j - \mathrm{minCWSI}_j} \times 100, \quad \text{if} \quad \mathrm{CWSI}_{i,j} = \mathrm{MCWSI}_j \tag{9.11a}$$

$$\mathrm{CWSA}_{i,j} = \frac{\mathrm{MCWSI}_j - \mathrm{CWSI}_{i,j}}{\mathrm{maxCWSI}_j - \mathrm{CWSI}_j} \times 100, \quad \text{if} \quad \mathrm{CWSI}_{i,j} > \mathrm{MCWSI}_j \tag{9.11b}$$

where CWSA is the weekly water stress anomaly, $MCWSI_j$ is the long-term median water stress of week j, $maxCWSI_j$ is the long-term maximum water stress of week j, and $minCWSI_j$ is the long-term minimum water stress of week j. Most widely used drought indices as the Palmer Drought Severity Index (PDSI) or Standardized Precipitation Index (SPI) are too slow to evaluate short-term dry conditions, which could influence severe impacts on crop growth. On the other hand, proposed SMDI and CWSI evaluate drought severity based on weekly soil moisture and ET deficit.

How to read Fig. 9.7: Water resources allocation for agriculture requires intensive planning at different levels of the governance hierarchy. The decision-making process can initiate at the national scale when droughts are predicted for certain regions and subsequently decisions can be made given the present and projected water resources availability status for different states. The final allocation of water resources from a governance point of view is logistically feasible only at the county scale. However, due to the nonlinear scaling of moisture and water fluxes, these estimates made at coarser time–space scales may not reflect the actual need of the farmers and as such lead to wastage and mismanagement of water resources. In order to optimize the agricultural water use, we propose to develop

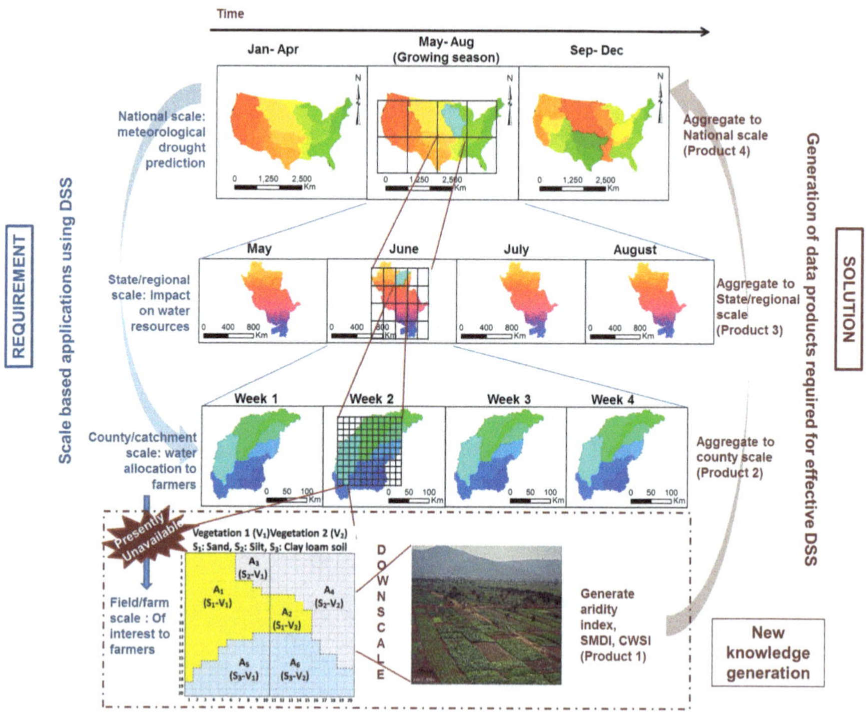

Fig. 9.7 Proposed Agricultural Drought Decision Support System (DSS) based on **NCA Indicators** at Different Space (field/farm, county/catchment, and state/regional/national) and Time (weekly, monthly, to seasonal) **Scales**

indices (aridity index, soil moisture deficit index, crop water stress index) that are crucial for a farmer to make decisions at a temporal and spatial scale suitable to him. These indices will then be ***aggregated*** *at subsequently coarser space–time scales for the decision-making authorities. This study bridge the scale gap and enables the decisions made by the governing bodies with limited manageable data to still positively impact the farmers at their scale of interest.*

9.12 Risk Assessment for Agricultural Drought and Adaptive Water Management Strategies

In addition to understand the impact of climate change on root zone soil moisture reflected by AI, SMDI, and CWSI at multiple space and time scales, we conduct scenario analyses with various management strategies within the soil hydrology and crop growth model (SWAP) including water management practices and cropping rotations (Fig. 9.8). We also evaluate the agricultural drought severity under various conditions (e.g., vegetation covers, soil types, precipitation/temperature timings, cropping/ soil management practices, soil organic matter status, irrigation and drainage, the presence of shallow water table, etc.). For example, implementing drip irrigation instead of furrow or flood irrigation, alternate cropping with drought resistant characteristics, soil amendment in the root zone for manipulating water holding capacity, adjusting sowing dates in the irrigated area, and recycling drainage water across neighboring fields will be investigated. Figure 9.8 shows the schematic outline of the data flow, modeling, and analyses leading to agricultural drought risk assessment and possible management strategies. This task is very useful for potential risk assessment during agricultural drought and enables greater understanding of agricultural and hydrological water management at the selected watershed/regions.

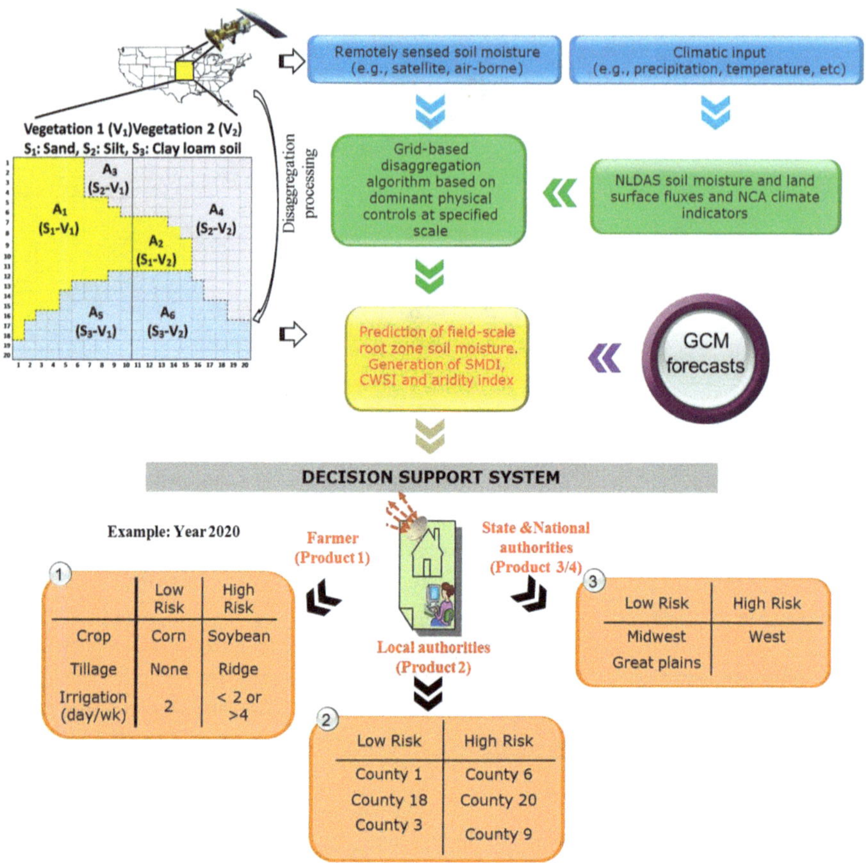

Fig. 9.8 Information flow schematics to conduct agricultural drought severity assessment and water/crop management / adaption and resources allocation strategies under climate change scenarios

References

Ajami NK, Duan Q, Sorooshian S (2007) An integrated hydrologic Bayesian multimodel combination framework: confronting input, parameter, and model structural uncertainty in hydrologic prediction. Water Resour Res 43(W01403):1–19

Arsenault KR, Houser PR, De Lannoy GJM (2014) Evaluation of the MODIS snow cover fraction product. Hydrol Process 28(3):980–998. doi:10.1002/hyp.9636

Bindlish R, Barros AP (2002) Subpixel variability of remotely sensed soil moisture: an inter-comparison study of SAR and ESTAR. IEEE Trans Geosci Remote Sens 40:326–337

Chervin RM (1981) On the comparison of observed GCM simulated climate ensembles. J Atmos Sci 38:885–901

Cosh MH, Jackson TJ, Bindlish R, Prueger JH (2004) Watershed scale temporal and spatial stability of soil moisture and its role in validating satellite estimates. Remote Sens Environ 92:427–435

Crow WT, Berg A, Cosh MH, Loew A, Mohanty BP, Panciera R, De Rosnay P, Ryu D, Walker J (2012) Upscaling sparse ground-based soil moisture observations for the validation of satellite surface soil moisture products. Rev Geophys 50:RG2002. doi:10.1029/2011RG000372

Dai Y, Zeng X, Dickinson R, Baker I, Bonan G, Bosilovich M, Denning S, Dirmeyer P, Houser P, Niu G, Oleson K, Schlosser A, Yang Z-L (2003) The common land model (CLM). Bull Am Meteorol Soc 84(4):1013e1023. doi:10.1175/BAMS-84-8-1013

Das NN, Mohanty BP (2006) Root zone soil moisture assessment using remote sensing and vadose zone modeling. Vadose Zone J 5:296–307

Das NN, Mohanty BP (2008) Temporal dynamics of PSR-based soil moisture across spatial scales in an agricultural landscape during SMEX02: a wavelet approach. Remote Sens Environ 112 (2):522–534

Das NN, Mohanty BP, Cosh MH, Jackson TJ (2008a) Modeling and assimilation of root zone soil moisture using remote sensing observations in walnut gulch watershed during SMEX04. Remote Sens Environ 112(2):415–429

Das NN, Mohanty BP, Njoku EG (2008b) A Markov chain Monte Carlo algorithm for upscaled soil-vegetation-atmosphere-transfer modeling to evaluate satellite-based soil moisture measurements. Water Resour Res 44. doi:10.1029/2008WR006472

Das NN, Mohanty BP, Njoku EG (2010a) Profile soil moisture across spatial scales under different hydroclimatic conditions. Soil Sci 175(7):315–319

Das NN, Mohanty BP, Efendiev Y (2010a) Characterization of saturated hydraulic conductivity in agricultural field using Karhunen-Loève expansion with the Markov chain Monte Carlo technique. Water Resour Res 46. doi:10.1029/2007WR007100

Das NN, Mohanty BP, Seo D, Efendiev Y (2016) Data-driven downscaling of satellite-based surface soil moisture using high resolution physical controls information. *Remote Sensing of Environment*, in Review.

Duan Q, Ajami NK, Gao X, Sorooshian S (2007) Multi-model ensemble hydrologic prediction using Bayesian model averaging. Adv Water Resour 30:1371–1386

Ek M, Mitchell K, Yin L, Rogers P, Grunmann P, Koren V, Gayno G, Tarpley J (2003) Implementation of Noah land surface model advances in the NCEP operational mesoscale Eta model. J Geophys Res 108(D22):8851. doi:10.1029/2002JD003296

Engman T (1991) Application of remote sensing of soil moisture for water resources and agriculture. Remote Sens Environ 35:213–226

Entekhabi D, Asrar GR, Betts AK, Beven KJ, Bras RL, Duffy CJ (1999) An agenda for land surface hydrology research and a call for the second international hydrological decade. Bull Am Meteorol Soc 80(10):2043–2058

Entekhabi D, Njoku EG, O'Neill PE, Kellogg KH, Crow WT, Edelstein WN, Entin JK, Goodman SD, Jackson TJ, Johnson J, Kimball J, Piepmeier JR, Koster RD, Martin N, McDonald KC, Moghaddam M, Moran S, Reichle R, Shi JC, Spencer MW, Thurman SW, Tsang L, Zyl JV (2010) The soil moisture active passive (SMAP) mission. Proc IEEE 98(5):704–716

Entekhabi et al. 2014, SMAP Handbook, Jet Propulsion Lab, NASA, p. 180

Famiglietti JS, Devereaux JA, Laymon CA, Tsegaye T, Houser PR, Jackson TJ, Graham ST, Rodell M, van Oevelen P (1999) Ground-based investigation of soil moisture variability within remote sensing footprints during the Southern Great Plains (1997) hydrology experiment. Water Resour Res 35:1839–1851

Gaur N, Mohanty BP (2013) Evolution of physical controls for soil moisture in humid and sub-humid watersheds. Water Resour Res 49:1–15. doi:10.1002/wrcr.20069

Gaur N, Mohanty BP (2015) Land-surface controls on near-surface soil moisture dynamics: 1 traversing remote sensing footprints. Water Resour Res. Revised

Gaur N, Mohanty BP (2016) Land-surface controls on near-surface soil moisture dynamics: Traversing remote sensing footprints. Water Resour Res, 52, doi: 10.1002/2015WR018085

Gong X, Barnston AG, Ward MN (2003) The effect of spatial aggregation on the skill of seasonal rainfall forecasts. J Clim 16:3059–3071

Hewitson BC, Crane RG (1996) Climate downscaling: techniques and application. Clim Res 7:85–95

Hughes JP, Guttorp P (1994) A class of stochastic models for relating synoptic atmospheric patterns to regional hydrologic phenomena. Water Resour Res 30:1535–1546

Hughes JP, Guttorp P, Charles SP (1999) A non-homogenous hidden Markov model for precipitation occurrence. Appl Stat 48:15–30

Ines AVM (2004) GCM bias correction tool. Version 0.3a. IRI-Columbia University, New York

Ines AVM, Hansen JW (2006) Bias correction of daily GCM rainfall for crop simulation studies. Agric For Meteorol 138:44–53

Ines AVM, Honda K (2005) On quantifying agricultural and water management practices from Low spatial resolution RS data using genetic algorithms: a numerical study for mixed pixel environment. Adv Water Resour 28:856–870

Ines AVM, Mohanty BP (2008a) Near-surface soil moisture assimilation to quantify effective soil hydraulic properties using genetic algorithm. I. Conceptual modeling. Water Resour Res 44. doi10.1029/2007WR005990

Ines AVM, Mohanty BP (2008b) Near-surface soil moisture assimilation for quantifying effective soil hydraulic properties using genetic algorithms: II. Using airborne remote sensing drying SGP97 and SMEX02. Water Resour Res 45. doi10.1029/2008WR007022

Ines AVM, Mohanty BP (2009) Parameter conditioning with a noisy Monte Carlo genetic algorithm for estimating effective soil hydraulic properties from space. Water Resour Res 44:W08441. doi:10.1029/2007WR006125

Ines AVM, Mohanty BP, Shin Y (2013) An unmixing algorithm for remotely sensed soil moisture. Water Resour Res 49:408–425. doi:10.1029/2012WR012379

Jackson TJ, Le Vine DM, Hsu AY, Oldak A, Starks PJ (1999) Soil moisture mapping at regional scales using microwave radiometry: the Southern Great Plains hydrology experiment. IEEE Trans Geosci Remote Sens 37:2136–2151

Jackson TJ, Bindlish R, Gasiewski AJ, Stankov B, Klein M, Njoku EG, Bosch D, Coleman TL, Laymon C, Starks PJ (2005) Polarimetric scanning radiometer C and X band microwave observations during SMEX03. IEEE Trans Geosci Remote Sens 43(11):2418–2429

Jacobs J, Mohanty BP, Hsu E-C, Miller D (2004) Field scale variability and similarity of soil moisture during SMEX02. Remote Sens Environ 92:436–446

Jana R (2010) Scaling characteristics of soil hydraulic parameters at varying spatial resolutions. Ph.D. Dissertation, Texas A&M University, p 264

Jana R, Mohanty BP (2012a) On topographic controls of soil hydraulic parameter scaling at hillslope scales. Water Resour Res 48:W02518. doi:10.1029/2011WR011204

Jana R, Mohanty BP (2012b) A topography-based scaling algorithm for soil hydraulic parameters at hillslope scales: field testing. Water Resour Res 48:W02519. doi:10.1029/2011WR011205

Jana R, Mohanty BP (2012c) A comparative study of multiple approaches to soil hydraulic parameter scaling applied at the hillslope scale. Water Resour Res 48:W02520. doi:10.1029/2010WR010185

Jana R, Mohanty BP, Springer EP (2007) Multi-scale pedo-transfer functions for soil water retention. Vadose Zone J 6(4):868–878

Jana R, Mohanty BP, Springer E (2008) Bayesian implementation of multi-scale pedo-transfer functions with Non-linear bias corrections. Water Resour Res 44:W08408. doi:10.1029/2008WR006879

Jasinski M, Arsenault K, Beaudoing H, Bolten J, Borak J, Kempler S, Kumar S, Peters-Lidard C, Li B, Liu Y, Mocko D, Rodell M, Vollmer B (2014) NCA-LDAS: an integrated terrestrial water analysis system for development, evaluation, and dissemination of climate indicators. In: American geophysical union fall meeting, San Francisco, p GC51B–0405

Joshi C, Mohanty BP (2010) Physical controls of near-surface soil moisture across varying spatial scales in an agricultural landscape during SMEX02. Water Resour Res 46:12503. doi:10.1029/2010WR009152

Joshi C, Mohanty BP, Jacobs J, Ines AVM (2011) Spatiotemporal analyses of soil moisture from point to footprint scale in two different hydroclimatic regions. Water Resour Res 47. doi:10.1029/2009WR009002
Kim G, Barros AP (2002) Downscaling of remotely sensed soil moisture with a modified fractal interpolation method using contraction mapping and ancillary data. Remote Sens Environ 83:400–413
Kim J, Mohanty BP, Shin Y (2015) Effective soil moisture estimates and its uncertainty using multi-model simulation based on Bayesian model averaging. J Geophys Res Atmos 120:8023–8042. doi:10.1002/2014JD022905
Kirshner S, Smyth P, Robertson AW (2004) Conditional Chow-Liu tree structures for modeling discrete-valued vector time series. Technical Report UCI-ICS 04-04. Information and Computer Science. University of California, Irvine
Koster R, Suarez M (1996) Energy and water balance calculations in the Mosaic LSM. Technical memorandum 104606, NASA Goddard Space Flight Center
Kumar SV, Peters-Lidard CD, Tian Y, Houser PR, Geiger J, Olden S, Lighty L, Eastman JL, Doty B, Dirmeyer P, Adams J, Mitchell K, Wood EF, Sheffield J (2006) Land information system: an interoperable framework for high resolution land surface modeling. Environmental Modeling & Software 21:1402–1415
Kumar SV, Peters-Lidard CD, Mocko D, Tian Y (2013) Multiscale evaluation of the improvements in surface snow simulation through terrain adjustments to radiation. J Hydrometeorol 14:220–232. http://dx.doi.org/10.1175/JHM-D-12-046.1
Kumar SV, Peters-Lidard CD, Mocko D, Reichle R, Liu Y, Arsenault K, Xia Y, Ek M, Riggs G, Livneh B, Cosh M (2014) Assimilation of remotely sensed soil moisture and snow depth retrievals for drought estimation. J Hydrometeorol 15(6):2446–2469. doi:10.1175/JHM-D-13-0132.1
Leung LR, Qian Y, Bian X, Washington WM, Han J, Roads JO (2004) Mid-century ensemble regional climate change scenarios for the Western United States. Climate Change 62:75–113
Liang X, Lettenmaier D, Wood E (1996) One-dimensional statistical dynamic representation of subgrid spatial variability of precipitation in the two-layer variable infiltration capacity model. J Geophys Res 101(D16):21403e21422
Liu Y, Peters-Lidard CD, Kumar SV, Foster JL, Shaw M, Tian Y, Fall GM (2013) Assimilating satellite-based snow depth and snow cover products for improving snow predictions in Alaska. Adv Water Resour 54:208–227. http://dx.doi.org/10.1016/j.advwatres.2013.02.005
Madigan D, Raftery AE, Volinsky C, Hoeting J (1996) Bayesian model averaging. In: Proceedings of the AAAI workshop on integrating multiple learned models, AAAI Press, Portland, p 77–83
Mohanty BP, Skaggs TH (2001) Spatio-temporal evolution and time stable characteristics of soil moisture within remote sensing footprints with varying soils, slopes, and vegetation. Adv Water Resour 24:1051–1067
Mohanty BP, Zhu J (2007) Effective hydraulic parameters in horizontally and vertically heterogeneous soils for steady-state land–atmosphere interaction. J Hydrometeorol 8(4):715–729
Mohanty BP, Skaggs TH, Famiglietti JS (2000a) Analysis and mapping of field-scale soil moisture variability using high-resolution ground based data during the Southern Great Plains 1997 (SGP97) hydrology experiment. Water Resour Res 36:1023–1032
Mohanty BP, Famiglietti JS, Skaggs TH (2000b) Evolution of soil moisture spatial structure in a mixed-vegetation pixel during the SGP97 hydrology experiment. Water Resour Res 36 (12):3675–3686
Nakicenovic N, Alcamo J, Davis G, de Vries B, Fenhann J, Gaffin S, Gregory K, Grübler A, Jung TY, Kram T, La Rovere EL, Michaelis L, Mori S, Morita T, Pepper W, Pitcher H, Price L, Raihi K, Roehrl A, Rogner H-H, Sankovski A, Schlesinger M, Shukla P, Smith S, Swart R, van Rooijen S, Victor N, Dadi Z (2000) Emissions scenarios. A Special Report of Working Group III of the Intergovernmental Panel on Climate Change. Cambridge University Press, Cambridge/New York

Narasimhan B, Srinivasan R (2005) Development and evaluation of soil moisture deficit index (SMDI) and evapotranspiration deficit index (ETDI) for agricultural drought monitoring. Agric For Meteorol 133:69–88

Peters-Lidard CD, Houser PR, Tian Y, Kumar SV, Geiger J, Olden S, Lighty L, Doty B, Dirmeyer P, Adams J, Mitchell K, Wood EF, Sheffield J (2007) High-performance earth system modeling with NASA/GSFC's land information system. Innov Syst Softw Eng 3 (3):157–165

Robertson AW, Kirshner S, Smyth P (2004) Downscaling of daily rainfall occurrence over Northeast Brazil using a hidden Markov model. J Clim 17:4407–4424

Roeckner E, Arpe K, Bengtsson L, Claussen CM, Dümenil L, Esch M, Giorgetta M, Schiese U, Schulzweida U (1996) The atmospheric general circulation model ECHAM-4: model description and simulation of present-day climate. Max-Planck-Institute for Meteorology. Report No. 218, Hamburg

Rui H, Teng W, Vollmer B, Jasinski M, Mocko D, Kempler S (2014) National Climate Assessment-Land Data Assimilation System (NCA-LDAS) Data at NASA GES DISC. In: American geophysical union fall meeting, San Francisco, p GC51B–0405

Santanello JA, Kumar SV, Peters-Lidard CD, Harrison KW, Zhou S (2013) Impact of land model calibration on coupled land-atmosphere prediction. J Hydrometeorol 14:1373–1400. http://dx.doi.org/10.1175/JHM-D-12-0127.1

Shin Y, Mohanty BP (2013) Development of a deterministic downscaling algorithm for remote sensing soil moisture footprint using soil and vegetation classifications. Water Resour Res 49:10. doi:10.1002/wrcr.20495

Shin Y, Mohanty BP, Ines AVM (2012) Soil hydraulic properties in one-dimensional layered soil profile using layer-specific soil moisture assimilation scheme. Water Resour Res 48:W06529. doi:10.1029/2010WR009581

Sivapalan M, Beven K, Wood EF (1987) On hydrologic similarity 2. A scaled model of storm runoff production. Water Resour Res 23:2266–2278

Sud Y, Mocko D (1999) New snow-physics to complement Ssib Part I: design and evaluation with ISLSCP initiative I datasets. J Meteorol Soc Jpn 77(1B):335–348

Wilby RL, Wigley TML (1997) Downscaling general circulation model output: a review of methods and limitations. Prog Phys Geogr 21:530–548

Wilby RL, Wigley TML, Conway D, Jones PD, Hewistson BC, Main J, Wilks DS (1998) Statistical downscaling of general circulation model output: a comparison of methods. Water Resour Res 34:2995–3008

Wood AW, Maurer EP, Kumar A, Lettenmaier DP (2002) Long-range experimental hydrologic forecasting for the Eastern United States. J Geophys Res 107(D20):4429. doi:10.1029/2001JD000659

Wood AW, Leung LR, Sridhar V, Lettenmaier DP (2004) Hydrologic implications of dynamical and statistical approaches to downscaling climate model outputs. Clim Chang 62:189–216

Zaitchik BF, Santanello JA, Kumar SV, Peters-Lidard CD (2013) Representation of soil moisture feedbacks during drought in NASA Unified WRF (NU-WRF). J Hydrometeorol 14:360–367. http://dx.doi.org/10.1175/JHM-D-12-069.1

Chapter 10
Monitoring Drought in Brazil by Remote Sensing

Vitor Paiva Alcoforado Rebello, Augusto Getirana, Venkat Lakshmi, and Otto Corrêa Rotunno Filho

10.1 Introduction

Drought is considered a natural hazard that affects most people around the globe. It is a normal and recurring climate extreme event in most parts of the word affecting both wet and dry regions. Especially in previous but relatively recent drought studies, it has been pointed out that frequency has increased mainly as a function of the variability of hydro-meteorological variables due to climate change (Mishra and Singh 2011).

Drought is in fact a complex and multicomponent phenomena, and in the absence of a precise and objective definition, it turns out to be a difficult task to quantify their social impact and corresponding mitigation (Wilhite 1993). However, in a broad sense, drought is associated with water shortage regarding several water needs. Consequently in a supply and demand context, drought can be seen as

---

V. Paiva Alcoforado Rebello (✉) • O. Corrêa Rotunno Filho
Laboratório de Recursos Hídricos e Meio Ambiente (LABH2O), Programa de Engenharia Civil, Instituto Alberto Luiz Coimbra de Pós-Graduação e Pesquisa em Engenharia—COPPE, Universidade Federal do Rio de Janeiro, Rio de Janeiro, RJ, Brazil
e-mail: vitoralcoforado@gmail.com; otto@coc.ufrj.br

A. Getirana
Hydrological Sciences Laboratory, NASA Goddard Space Flight Center, Greenbelt, MD, USA

Earth System Science Interdisciplinary Center, University of Maryland, College Park, MD, USA
e-mail: augusto.getirana@nasa.gov

V. Lakshmi
Department of Geophysics, Stanford University, Stanford, CA, USA

School of Earth, Ocean and Environment, University of South Carolina, Columbia, SC, USA
e-mail: vlakshmi@geol.sc.edu

V. Lakshmi (ed.), *Remote Sensing of Hydrological Extremes*, Springer Remote Sensing/Photogrammetry, DOI 10.1007/978-3-319-43744-6_10

a water deficit, which can be traditionally classified in meteorological, hydrological, or agricultural drought (Dracup et al. 1980).

Traditionally, drought is associated with a long period of deficit in precipitation with respect to climatological normal values. As stated above, droughts can be classified into three categories. Meteorological drought is associated with precipitation levels lower than the climatological normal values. For this type of drought, we have to define a threshold value to assume a drought occurrence when lower levels are recorded. Agricultural drought corresponds to soil moisture content that is below the minimum required to fully support crop development for a certain month. Hydrological drought is associated with negative anomalies of surface water, notably surface storage and flowing water in rivers and streams. However, a fourth type of drought has emerged known as the sociological drought which focuses on the impacts of the meteorological, agricultural, and hydrological droughts on human and ecological water demands.

From a hydrological cycle point of view, drought can be assessed by examining the cycle's three main components: precipitation, soil moisture, and streamflow, all of which are strongly correlated with each other (Dracup et al. 1980). First, precipitation is the driving force triggering the onset of drought. Changes in precipitation can reduce, delay, or shift the beginning of the wet period and are related to atmospheric phenomena such as El Nino-Southern Oscillation (ENSO) and Sea Surface Temperature (SST) climate conditions which causes drought in tropical regions (Lyon 2004). Soil moisture of the vadoze zone is the intermediate system component, acting as the water reservoir and accounts for the interchange of water between the surface and atmosphere through evapotranspiration. Water retention capacity is strongly influenced by antecedent moisture conditions, so water depletion in soil will strongly affect water availability for plants. As dry periods are frequently associated with high temperatures, potential evapotranspiration also increases in these conditions, which in turn will provoke higher evapotranspiration from the land surface. Lastly, streamflow is the water system output reflecting an integrated effect of drought via hydrological component interactions all over the catchment.

Compared to other natural hazards, drought effects accumulate at a lower time step, leading to an extension of the hydrological cycle. Consequently, the time series for drought are usually at monthly or yearly temporal scales. The temporal scale will also depend on different variables. Thus, for example, precipitation requires usually at least a monthly temporal approach since rainfall varies seasonally. On the other side, a reliable analysis for streamflow requires a time series that is longer than 2 years to discriminate low streamflow from normal streamflows. The spatial scale is also an important issue to be considered as we extend the study area from regional to global where longer time period analyses are more representative.

To quantify drought severity and corresponding extent, it is necessary to convert the value of the studied variable into an index for further interpretation. A single index cannot be applied to all situations, leading to a large variety of proposed indices to address each type of drought for which they have been developed. The calculation can be done for different temporal scales, ranging from yearly, especially for global analysis or large areas, to monthly which are more suitable for regional scales (Mishra and Singh 2010). The formulated index can make use of any variable of interest, depending on the intended application.

The widely used Palmer Drought Severity Index (PDSI) is based on anomalies in the water balance equation. PDSI uses a two-layer soil model to quantify water balance by monthly water input (rain) and it measures outputs (runoff and evaporation) by considering the previous soil moisture status. The data is standardized so that different location values can be compared. For agriculture/crop monitoring, the use of Crop Moisture Index (CMI) is more suitable since it works in a shorter time scale than a weekly scale to access plant drought stress. The CMI is calculated from precipitation and surface moisture deficits instead of the ratio between actual and potential evapotranspiration (Palmer 1968). Although suitable for agriculture, this index cannot be used to monitor long-term changes.

The Standardized Precipitation Index (SPI) is a simple index that utilizes precipitation as the only input data. An interesting feature of the index is that it can be calculated on different time scales. Initially, the precipitation time series is fitted to a gamma distribution and then transformed into a normal distribution having the mean value of zero. Values below zero are indicative of negative anomalies of precipitation values (Hayes et al. 1999).

In drought studies, some commonly used variables derived from the drought index are: (a) duration, (b) intensity, and (c) severity. The starting point is to establish a threshold level, which is a constant value over time, so that the portion of time series below this truncation level is the duration of the drought. The threshold can be defined arbitrarily or by using a stochastic approach such as deciles. Drought intensity, often called magnitude, is the mean value below the critical level. The severity is the cumulative deficit of critical level, or simply the multiplication between duration and intensity. Once this information is gathered, it is possible to explore data mining techniques, multivariate distribution, and spatial–temporal analysis of indexes for forecasting and impact assessment (Mishra and Singh 2011). The impact of droughts on water resources management goes beyond its frequency, severity, and duration analysis and should also comprise drought spatial patterns. For this purpose, the severity-area-frequency curve (SAF) depicts drought severity and drought area with respect to drought return periods so as to describe and characterize the spatial and recurrence patterns of drought (Mishra and Desai 2005). González and Valdés (2004) created a model for regional drought distribution analysis using a synthetic precipitation time series to design a SAF and used the same return period of the historical time series. The same SAF approach was applied by Kim et al. (2002) in his study for kriging-interpolated PSDI.

Their Principal Component Analysis (PCA) and Empirical Orthogonal Functions (EOF) represent a second method capable of revealing spatial patterns of the drought as the variable of interest. These multivariate techniques are often applied to dimensionality reduction and also emphasize data patterns and reveal relationships among variables themselves and between variables and observations (Santos et al. 2010). van der Schrier et al. (2013) have applied the EOF method in a self-calibrated version of PSDI through a global analysis. The first mode was able to show different regions with wet and dry patterns, although this tendency changes over different decades. In a regional assessment for Turkey, Tatli and Türkeş (2011) compared several drought indexes, including PSDI, and they were able to group several of them that were driven by the same physically driven forces. Santos et al. (2010) also showed that, in addition to PCA, a nonhierarchical cluster analysis (KMC) was able to separate regions with distinct behaviors in a 94-year rainfall dataset with SPI ranging along three time scales.

10.2 Monitoring Drought Using Remote Sensing Data

To better monitor and minimize drought impact, weather data availability is necessary. However, ground source information is costly and scarce in many regions of the world. Also, to use and assess the spatial extent of this information, interpolation techniques with various limitations are required. From this point of view, remote sensing can play an essential role and can contribute to significant advances in drought phenomena research. After more than four decades of several spatial missions, we have currently a broad data set of observations from the space continuously providing us global and reliable information from a wide range of sensors, each with different physical principles.

Remote sensing is the technique of acquiring information from an object area of phenomenon through a sensor from a remote location. Several satellites orbit the Earth that record information through different physical approaches such as electric-magnetic energy, including reflectance, emission and backscattering, acoustic waves, or gravimetric field, which can be used for remote sensing. One major remote sensing classification is regarding the sensor energy source which can be a passive system dependent on solar radiation or an active system if it emits energy and registers its own energy back. Passive systems can be subdivided into three more groups. The first detects sun-reflected radiation on the visible and near-infrared. The following two groups register the Earth's thermal emission, one in the so-called infrared part of the spectrum and the other at longer wavelengths of the microwave part of spectrum (Rees 2012).

From a historical perspective, the Advanced Very High Resolution Radiometer (AVHRR) on the board of the NOAA series of satellites since 1978 provided a major contribution to global drought monitoring. This sensor operates from the visible to thermal-infrared regions of the spectrum, highlighting low to moderate spatial resolution and a relatively high temporal frequency pattern with a great potential for large-scale monitoring of vegetation dynamics. More recently, the Moderate Resolution Imaging Spectroradiometer (MODIS) started to operate on 2002 until today and followed AVHRR characteristics with improvements mainly in spatial resolution, reaching 1000–250 m, and also higher spectral resolution.

Vegetation indexes (VI) are spectral transformations of two or more bands designed to enhance the contribution of vegetation properties. Such indexes allow us to monitor seasonal, annual, and long-term variations of structural, phenological, and biophysical vegetation parameters (Huete et al. 2002). For instance, the

Normalized Differential Vegetation Index (NDVI) is a widely used index defined by the difference between the reflectance in the red part of the spectrum, which interacts with the chlorophyll content, and near-infrared spectra, which reflects changes in spongy mesophyll (Gu et al. 2007). Nevertheless, there are still limitations in the performance of this index due to a certain degree of saturation in closed canopies, sensitivity to atmospheric aerosols, and soil background (Anderson et al. 2010). The Enhanced Vegetation Index (EVI) is a trial to improve NVDI by including the blue band reflectance to minimize canopy background and aerosol influence (Huete et al. 2002). There is also the Soil Adjust Vegetation Index (SAVI) and the Normalized Difference Water Index (NDWI). The NDWI was proposed after NDVI, in such a way that this new index uses short wave infrared and near-infrared channels, with the former channel having a higher liquid absorption compared to the latest one, which then makes this index more sensitive to water content (Gao 1996).

On the other side, passive microwave radiometers measure the natural emission of microwave energy and the brightness temperature from land surfaces. Active sensors, on the other hand, including Synthetic Aperture Radar (SAR) and scatterometers, transmit signals to a target surface area and measure the scattering return, called the backscattering. These sensors are characterized by their frequency. Some examples with their respective frequency/band are: (radiometers) the Special Sensor Microwave/Imager (SSM/I) ranging from 19.35 to 85.5 GHz bands, Tropical Rainfall Measuring Mission (TRMM), Microwave Imager (TMI), Advanced Microwave Scanning Radiometer (AMSR-E) ranging from 6.9 to 89 GHz bands and Soil Moisture and Ocean Salinity (SMOS) 1.4 GHz band (L-band) along with (active systems) scatterometers, European Remote Sensing (ERS) with a 5.3 GHz band (C-band), and QuikSCAT (QSCAT) 5.255 GHz band (C-band). The microwave system has an advantage because of its relative transparency with respect to the atmosphere and sensitivity to soil moisture variation. At the range of microwave frequencies, the soil dielectric constant has a remarkable dependence on the soil water content (Huffman et al. 2007), making these systems more prone to potentially monitor soil moisture globally. Some current limitations of microwave systems include soil moisture measurements that are limited to the interaction with very top soil moisture layer (0–10 cm) and constrained to the high noise-to-signal ratio in areas fully vegetated. Another constraint for passive systems is related to its coarse image resolution.

The global remote sensing provided a broader view and more suitable conditions for exploring the formulation of drought indexes previously limited to weather station data sources. Fundamental hydroclimate variables such as surface temperature, precipitation, and soil moisture that jointly provide reliable information of land parameters status are available, thanks to the launch of various recent space missions. Moreover, remote sensing information nowadays provide appropriate spatial coverage information and convenient time resolution for addressing different physical processes varying from the local scale to the global perspective.

10.2.1 MODIS

The Moderate Resolution Imaging Spectroradiometer (MODIS) is an instrument onboard the Earth Observation System (EOS) Terra satellite launched on 1999 and also onboard the NASA Aqua satellite launched on 4 May 2002. This mission was designed to provide reliable information for atmosphere, ocean, and land disciplines through different high-quality spectral products. MODIS is composed of 36 spectral bands of 12-bit radiometric resolution and acquires data at three spatial resolutions: 250 m, 500 m, and 1000 m (Justice et al. 2002). The first seven bands are designed for land surface studies, noting that the band 1 (red, 620–670 nm) and band 2 (near infrared, 841–876 nm) have spatial resolutions of 250 m and band 3 (blue, 459–479 nm), band 4 (green, 545–565 nm), band 5, 6, 7 (1230–1250, 1628–1652, 2105–2155 nm) have spatial resolutions of 500 m. Also crucial for land surface studies are the MODIS emissive bands in the infrared spectrum (3000–15,000 nm) with seven of them used as input for surface temperature and emissivity land products.

The data is distributed across different levels. Level-2 is based on geophysical-derived variables at the same resolution and location as Level-1 source data. Level-3 includes gridded variables in derived spatial and/or temporal resolutions. Level-4 is a model output or results from analyses of lower-level data. Our focus will be on MODLAND, which represents a suite of global land products of special interest for drought studies, including spectral vegetation indices, surface temperature, and other biophysical variables associated to global carbon and to hydrologic cycles.

The MODIS standard surface reflectance product is MOD09, which is distributed as daily and 8-day composite images involving from band 1 to 7. The algorithm corrects for aerosol and atmospheric effects and Bidirectional Reflectance Distribution Function (BRDF) (Justice et al. 2002; Vermote and Vermeulen 1999). The MOD 13 provides two gridded 16-day products of vegetation indices products, namely NDVI and EVI, and a quality analysis image with flags indicating pixel information. This is a Level-3 product generated by MOD09, with spatial resolution of 250 m, 500 m, 1 km, and a 25 km for a monthly coarser climate modeling grid product version (Justice et al. 2002). The MOD11 is a product for LST and emissivity collected during daytime and nighttime for a 1 km spatial resolution and for daily and 8-day temporal resolution. A view-angle-dependent split-window algorithm has been developed for retrieval of LST, taking into account the incorporation of atmospheric correction and emissivity effects for the broad range of land cover types. It is also available for a monthly quarter degree latitude/longitude global climate modeling grid (Wan 1999). Other MODLAND products of importance for drought studies are: MOD16 for actual and potential evapotranspiration (EVP and PET) (Mu et al. 2011), MOD17 for gross and net primary production (GPP and NPP) (Running et al. 2004), and MOD 15 for Leaf Area Index (LAI) and Fractional Photosynthetically Active Radiation (FPAR) (Privette et al. 2002).

Over non-homogeneous landscapes, a wide range of vegetation indices (VI) might be observed mainly due to different vegetation categories. However, a seasonal pattern is expected over this entire region induced by a local weather (Singh et al. 2003). Therefore, the easiest method to detect drought periods using VI is by calculating NDVI anomalies for each region. Anomalies are the deviation between the NDVI value and its long-term average value, which can be calculated for the month or annual time period. Similarly, a standardized anomaly is the deviation divided by a standard deviation. Despite vegetation is a good indicator for drought, the time lag between rainfall and vegetation change in addition to variability of vegetation poses still some limitations to be overcome. LST is highly related to canopy water content (Yao et al. 2010) and it is more sensible to moisture condition especially after long drought periods (Singh et al. 2003). For this reason, coupling between NDVI and LST might provide a better characterization of drought.

The Vegetation Condition Index (VCI) and the Temperature Condition Index (TCI) are estimated relative to the maximum and minimum pixel values with respect to NDVI and LST, respectively. In sequence, both are combined additively resulting in the Vegetation Health Index (VHI) (Kogan 1995). Moreover, the negative relationship observed between NDVI and LST could be explored as a proxy of surface soil moisture condition. A strong relationship between surface moisture condition and vegetation indices has been already observed and verified through remotely sensed data (Gu et al. 2007; Sun and Kafatos 2007; Brown et al. 2008; Wu et al. 2015). Thus, NDVI will present a relatively high value for a pixel in wet regions when considering a large enough mapped region, whereas LST will be expected to be relatively low, since the surface radiant temperatures are dependent on the surface soil water content (Carlson 2007 and Karnieli et al. 2010). On the other hand, the opposite relationship will be observed in dry regions. The scatterplot of NDVI and LST images form a triangular shape, and by its simple interpretation, one can distinguish between dry and wet pixels (Carlson et al. 1995 and Sandholt et al. 2002). A detailed and complete review of this approach is provided by Petropoulos et al. (2009). The triangle's upper and lowest boundary correspond, respectively, to the dry and wet edge and both regions can be delineated using a linear regression curve (Sandholt et al. 2002).

Drought affects vegetation structure and plant physiology and also impacts entire ecosystems by reducing CO_2 uptake and increasing CO_2 concentration in the atmosphere (AghaKouchak et al. 2015). The NPP is the quantification of net CO_2 exchange between atmosphere and plant biomass, while GPP is the amount of carbon uptake by daily photosynthesis. Remote sensing techniques enable us to estimate both variables over an entire region, noticeably at ecosystem scales. The incident solar radiation and air temperature are limiting factors for NPP. However, water availability seems to be the factor that most limits NPP around the globe (Running et al. 2004), although one should consider lag responses between water stress and ecosystem response.

The adoption of rain use efficiency metric (RUE), which is the rate of annual NPP by precipitation, can better explain CO_2 exchanges in different biomes between dry and wet years (Zhang et al. 2012; Zhang et al. 2014). Another direct result of precipitation reduction is the decrease of evapotranspiration, which is also linked to carbon uptake by plants. Water use efficiency (WUE), which corresponds to NEE divided by evapotranspiration, measures the amount of carbon fixed per water unit transpirated and it is more sensible to RUE because each vegetation type has different evapotranspiration patterns as stated by Liu et al. (2015). Their findings have shown that different regions in China respond positively or negatively to either slight or severe droughts.

Evapotranspiration is one of the main components of the hydrological cycle and it is the connecting link between the water budget and the energy budget in the soil-water-vegetation system. In past years, several attempts have focused on estimating distributed evapotranspiration from remote sensing data. Some methods are based on physical models which estimate evapotranspiration from the energy balance residual using thermal infra-red band, VI, and auxiliary meteorological data (Bastiaanssen et al. 1998; Allen et al. 2007; Anderson 1997; Su 2002). Potential evapotranspiration is defined as the amount of water that would evapotranspirate if water were plentiful for a set of standard meteorological conditions, such as temperature, radiation, wind, and relativity humidity. Once PET could be seen as an atmospheric demand to extract water from the surface, one can consider a balance between precipitation as an input and potential evapotranspiration as an output. This is the basic idea of the Reconnaissance Drought Index (RDI) proposed by Tsakiris and Vangelis (2005), which is the ratio between two aggregated quantities of precipitation and potential evapotranspiration. RDI is a more physically based index and can be applied to hydrological and agricultural drought and it is more sensible when compared to SPI (Tsakiris et al. 2006).

Drought severity can be intensified when below-normal rainfall combines with heat waves, strong winds, and sunny skies that enhance vegetation evaporative stress (Otkin et al. 2014). The evaporative stress index (ESI) quantifies anomalies in the actual and potential evapotranspiration ratio using thermal band remotely sensed or modeled data (Yao et al. 2011 and Mu et al. 2013). Using orbital data, ESI can provide finer spatial resolution compared to precipitation-derived indices. Anderson et al. (2011) demonstrated that ESI has a fast-drought response and agrees well with United States Drought Monitor (USDM) compared to other remotely sensed drought indexes. A recent study assessed ESI as an indicator of agricultural drought in Brazil for the 2003–2013 period (Anderson et al. 2016). ESI, precipitation from Tropical Rainfall Mapping Mission (TRMM), and leaf area index (LAI) from MODIS were compared with state and municipality level crop productivity. In general, all metrics provided similar spatial and temporal correlation with agricultural statistics at southeast and northeast regions of Brazil. Although, at a finer municipality level, ESI had higher annual yield correlation and strength of correlative signal. Specifically in the northeastern dry region, ESI better emphasizes flash drought events compared to the other satellite-derived information.

10.2.2 AMSR

The Advanced Microwave Scanning Radiometer (AMSR) is a multi-channel passive microwave-radiometer system on board NASA Aqua satellite that was launched in 2002 and provides terrestrial and oceanic global coverage brightness temperature values at five frequencies in the range of 6.9–89 GHz (Bolten et al. 2010; Zhang et al. 2013). Since 2001, AMSR has provided soil moisture values derived from global surface temperature coverage at 3–5 daily intervals of $0.25° \times 0.25°$ spatial resolution.

Many research centers have established and validated algorithms to retrieve soil moisture from AMSR microwave brightness temperature. This can be achieved through land–surface radiative transfer models which accounts for soil moisture, soil temperature, and vegetation, factors that contribute to evaluate passive microwave emissions. Some common algorithms, as listed by Draper et al. (2009), are from the following institutes: NASA, Japanese Aerospace Exploration Agency (JAXA), United States Department of Agriculture (USDA), and Vrije Universiteit Amsterdam (VUA).

Soil moisture is a simple and effective indicator of drought and weather extremes (Hirschi et al. 2014; Lakshmi et al. 2004; Cai et al. 2009; Tang and Piechota 2009). To evaluate and monitor drought onset and progress, the use of an index is necessary. A soil moisture index aims to describe the water content deficit at the root zone profile. One of the first indices used as a proxy to estimate soil moisture at the watershed level in the beginning of a rainfall event was the Antecedent Precipitation Index (API) (Kohler and Linsley 2008). The index is a weighted sum of daily precipitations values previous to the precipitation event of interest, as can be seen in Hong et al. (2007) for further details.

Other indices have been developed applying statistical methods, such as parametric (Sheffield 2004; Rahmani et al. 2016) and non-parametric probability distribution (Carrão et al. 2016). Additional indices measure deviation from historical soil moisture values (Narasimhan and Srinivasan 2005). Soil moisture can also be combined with other hydrological variables to produce a multivariable drought index.

Integrating soil moisture, precipitation, and surface temperature data in such a unique procedure from microwave remote sensing data sets has been shown to be more suitable tracking with respect to a short-term drought over semi-arid environment (Zhang et al. 2013). In East Africa during 2010–2011 drought period, the Multivariate Standardized Drought Index (MSDI), which integrates soil moisture and precipitation, was able to detect the drought onset and persistence with observations consistent with previous studies (AghaKouchak 2015).

At this point, we should mention that agricultural drought is potentially the most complex drought category. While meteorological and hydrological categories can be more easily assessed by direct precipitation measurements, stream flow records, and reservoir levels, soil moisture has high spatial-temporal variability and heterogeneous distribution over soil layers Once made this remark, it can be emphasized that soil moisture index, as derived from models or remote sensing data, is the most straightforward tool for agricultural monitoring at local, regional, and global scale. Different crop types production well correlate with soil moisture index at different time scales and locations (Carrão et al. 2016; Narasimhan and Srinivasan 2005).

Narasimhan and Srinivasan (2005) were able to identify critical time periods of growth stages affecting crops production through a soil moisture index. Assimilation of remote sensing data into models by means of Kalman filter is another way to improve model results during drought periods. For instance, AMSR-E soil moisture and MODIS LAI improved the estimation of aggregated yields when assimilated into crop model framework (Mishra et al. 2015). The same sensor was assimilated on PDSI-based model to better predict root-zone soil moisture (Bolten et al. 2010).

Soil moisture from microwave sensors is constrained to the first few centimeters, which is considered a shallow depth compared to the root zones. Wagner et al. (1999) developed an exponential filter to extrapolate moisture into the root zone. This simple model takes into account water propagation through the profile considering a defined time lag and a homogeneous and isotropic porous media with a constant soil hydraulic conductivity.

Microwave soil moisture measurements or modeling estimations are subject to errors or driven forces' uncertainties (Gruber et al. 2016). One way to overcome this problem is to use the Triple Collocation Analysis (TCA), which is a statistical method to cross compare geophysical variables (Stoffelen 1998). This method has been suggested to spatially evaluate different soil moisture data sets (de Jeu and Dorigo 2016). Anderson et al. (2012) assessed the TCA with three different soil moisture products using three soil moisture data sets, more specifically AMSR-E, thermal remote sensing using ALEXI, and modeled soil moisture simulations in Africa during a drought period. The final results showed the corresponding data limitation.

Beyond the limitation pointed out above, the coarse resolution typical of microwave sensor can narrow some of the potential use or application of soil moisture content information. To overcome this, many efforts have been made to disaggregate spatial data into a finer resolution. In this sense, most common methods employ NDVI, LST, and other surface variables for disaggregate pixel value. A method based on the linear thermal inertial relationship under different vegetation conditions was proposed and validated using AMSR-E at Oklahoma Mesonet, USA (Fang et al. 2013). Using the same sensor, Zhao and Li (2013) fitted a second-order multivariable polynomial for vegetation, diurnal surface temperature variation, and albedo at REMEDHUS soil moisture network, Spain. Other algorithms are based on the so-called "universal triangle" concept (Piles et al. 2011) and the evaporative fraction (Merlin et al. 2008).

10.2.3 NDVI and Soil Moisture Products for Drought Monitoring

The southeast region of Brazil (SE) suffered a great drought event observed in the austral summer of 2013/2014 and 2014/2015 and early 2015 caused by a significant precipitation deficit (Coelho et al. 2015a). The SE region is the most populated and industrial area in Brazil, corresponding to the most important socioeconomic region of the country with the highest Gross National Product (GNP) and featuring a significant set of reservoirs for hydroelectricity power generation. This recent drought event caused a huge water resources shortage, although it has not caused an energy crisis as observed in the drought of 1999/2001, because at that time the electric system was also facing simultaneously important changes in policy making procedures for integrated power generation. These two recent years of drought (2013/2015) have led to a water consumption crisis in the state of São Paulo, where the precipitation deficit was extreme in such a way to compromise the operation of one of the main systems of reservoirs responsible by supplying water for the city of São Paulo, namely the Cantareira system, with a volume level that reached 5–15 % of its capacity including the dead volume of the reservoir (Coelho et al. 2015b; Seth et al. 2015).

The Northeast region of Brazil (NE) contrasts with SE, noting that the former is less populated and has a relatively poor economical development. Regarding the climate, NE region climate is mainly drier and its central part encompasses a large semi-arid landscape, covered by a xeromorphic vegetation (Moura et al. 2013). The NE region is dominated by two different biomes, the Atlantic tropical rain forest, located in the east coast and in the western part, and the Caatinga (desert shrubs) located in the central part (Liu and Juárez 2001). NE is a drought-prone region due to high rainfall interannual variability and many of these past drought events were responsible for several social impacts, mainly poor local farms, over the last years (Silva et al. 2015). Drought occurrence in NE is in general associated with sea surface temperature variations in the tropical Pacific ocean expressed as the extremes of El Niño Southern Oscillation (ENSO), such as the droughts of 1998 and 2002. However, the 2012–2013 drought occurred during La Niña event, when more precipitation was expected (Marengo et al. 2013). This most recent drought has been considered the worst in the last 60 years (Marengo and Bernasconi 2014) and some authors have reported its persistence until 2015 (Silva et al. 2015).

Figure 10.1a, b display monthly average of AMSR-E (2002–2012) and AMSR-2 (2012–2014) soil moisture and MODIS MYD13C2 NDVI product for the two studied regions, southeast and northeast of Brazil. The central NE region is less vegetated compared to the whole western Brazil. Higher NDVI values are located close to the coast, in the northeastern and eastern NE and SE regions, respectively. Soil moisture spatial distribution has a great correspondence with NDVI. A long-term soil moisture and NDVI time series (Fig. 10.1c–f) show considerable drought impact in the NE region especially from 2012 to 2013, when a great decay in average-area had been observed. The NDVI was slightly lower in SE region during

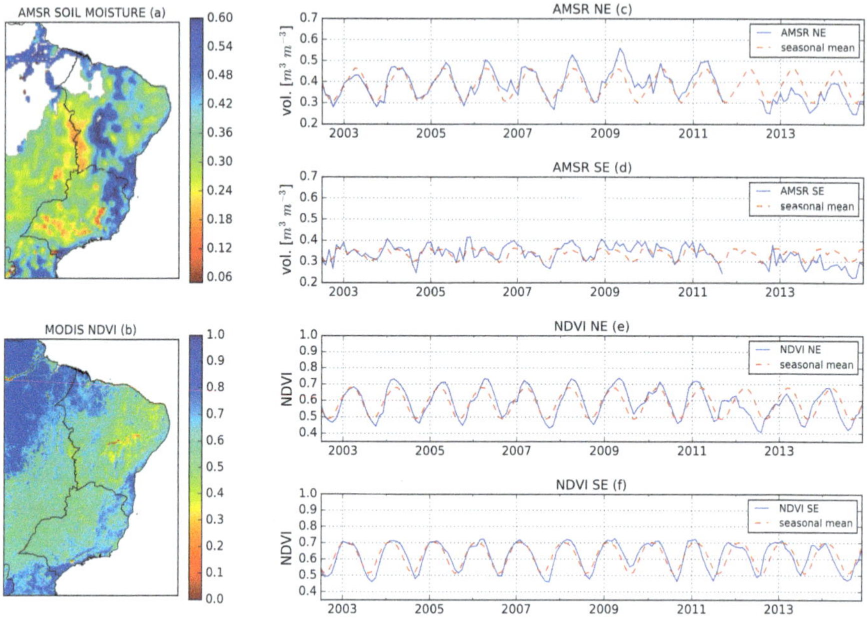

Fig. 10.1 AMSR (**a**) and NDVI (**b**) average images and time series of soil moisture (**c**, **d**) and NDVI (**e**, **f**) for, respectively, Northeast (NE) and Southeast (SE) Brazilian regions from 2002 to 2014

2014, when compared with other years, although this region had a strong reduction on soil moisture values during the last years. However, AMSR soil moisture in the SE region seems noisy and less representative of real soil moisture condition as observed in the NE region, where soil moisture has a typical seasonal annual cycle. Northeast region landscape is dominated by flat and bare soils, which better contributes to the remote sensing soil moisture retrievals algorithm.

EOF analysis was conducted for monthly anomalies of AMSR soil moisture and MODIS NDVI in order to better understand the spatio-temporal variations in both regions. The EOF approach allows to determine the modes that govern the variability of a climatic phenomenon (Björnsson and Venegas 1997). The corresponding time series of each EOF, named expanded coefficients, elucidates the temporal evolution of the variance in the areas of maximum variability (Syed et al. 2004). In general, the first mode usually corresponds to the seasonal oscillation of the studied variable, while subsequent modes can display annual perturbations, such as the ones produced by El Niño events (Toumazou and Cretaux 2001). In remote sensing multi-temporal images applications, the most part of variance is concentrated until the fourth to fifth component. Thus, we have selected the first two components for conducting the analysis.

Figure 10.2a, b show EOF 1 and 2 for AMSR in the NE region, depicting that the first two components explain 38 % and 12 %, respectively, of the corresponding

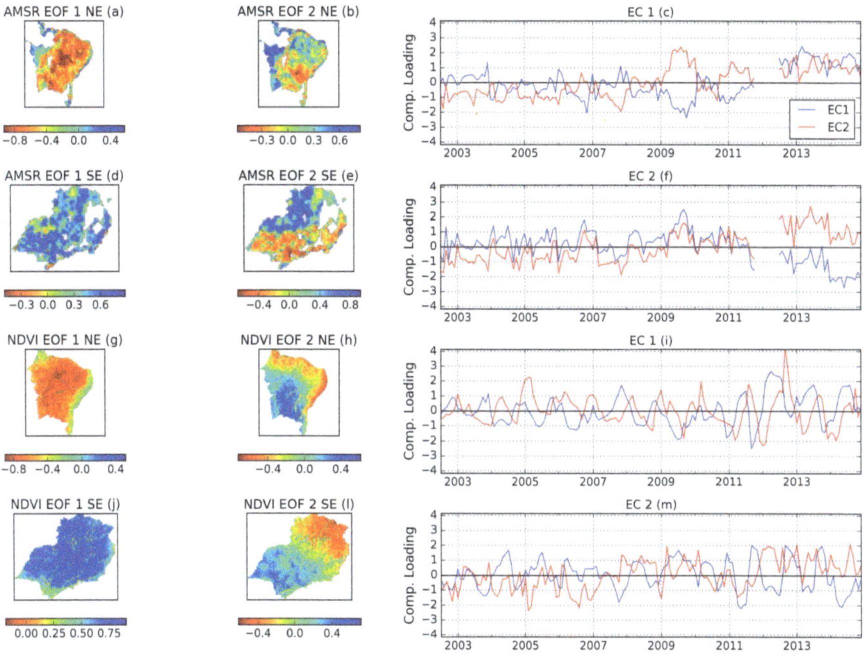

Fig. 10.2 AMSR EOF loadings (**a**, **b**, **d**, **e**) and expansion coefficients (**c**, **f**) and NDVI (**g**–**j**, **l**, **m**) for, respectively, Northeast (NE) and Southeast (SE) regions

variability for AMSR. The spatial pattern of EOF 1 is characterized by a strong negative value in the central part and positive value in each margin. The second EOF has a negative anomaly centered over the south-central region and a positive anomaly located in the northwest board of the NE region. The associated time series does not reveal a seasonal pattern, but it shows an increasing trend after 2012.

The first EOF for SE region (30 % of explained variability), Fig. 10.2d, is dominated by a positive homogeneous signal over the entire region, while the second EOF mode (21 % of explained variability), Fig. 10.2e, seems to indicate a subregional dipole like between the north and the southwest part of SE region. This is in accordance with the observed northwest–southeast diagonally oriented spatial pattern of precipitation over this region (Coelho et al. 2015b). The temporal series, Fig. 10.2c, f, depict a similar temporal pattern until 2011, followed by an increasing trend after 2012, EOF 1, and decreasing trend, EOF 2.

EOF analysis results of NDVI developed for the NE represent 41 % of total variance as given in Fig. 10.2g, h. The EOF 1 shows similar results with AMSR EOF 1, Fig. 10.2a, with a major negative signal otherwise, while EOF 2, which explains about 9 % of variance , shows positive values in southwestern NE contrasting with negatives anomalies over the NE shore. The rainfall distribution is a dominant factor in the dynamics of vegetation in the NE region (Barbosa and Lakshmi Kumar 2016). The EOF 1 might be strongly connected to vegetation cover-averaged yearly rainfall. EOF 2 probably captures different local wet regimes as the NE is subjected to irregular spatio-temporal distribution of precipitation over the year, which occurs from February to May in the semiarid zone and from April to August over the east coast (Cunha et al. 2015; Marengo and Bernasconi 2014). The time series show a typical seasonal oscillation with an up to 3 months lag between the first and second expanded coefficient. One might observe that both time series have their lowest values during 2012–2013 drought, followed by an intense vegetation vigor recovery.

The remaining EOF 1 and 2 for NDVI in SE region explains 48 and 8 % of total variance, Fig. 10.2j, l, with the first EOF been more significant and with a homogeneous positive value, while the second EOF seems to separate the signal in a north–south pattern in the second mode and an east–west pattern in the third mode. The temporal plot, Fig. 10.2m, shows two minimum peaks in 2011 and 2014 and an increase of variability after 2011.

In summary, the EOF analyses of the soil moisture and NDVI indicate that there was a sudden decrease over both studied regions during the drought periods reported. For NDVI in the NE region, we were able to describe the time series of two distinct regions, semi-arid and the east-coast, and how they reacted during 2012–2013 drought event. Further studies could describe the link between NDVI and soil moisture, and possibly other variables like rainfall through statistical multivariate analyses, such as the single value decomposition method.

10.3 GRACE

The Gravity Recovery and Climate Experiment (GRACE) mission is a joint US-German satellite mission launched in March 2002, designed to detect the spatiotemporal variation of the Earth's gravity field. GRACE is composed of two orbiting tandem satellites approximately 250 km apart. Water mass variation in large areas has a gravitational potential strong enough to alter the orbit of satellites, which, for instance, can alter the distance between GRACE in both aircrafts. The distance is measured using onboard microwave tracking systems, and in this way, can be used to quantify terrestrial water storage. Consequently, mass variations in land will comprise of changes in total column-water storage due to changes in groundwater , soil moisture due to snow, ice, and lakes, rivers, and other bodies of water-containing biomass.

Time series of spherical harmonic coefficients of Earth's Gravity field named GRACE Level-2 RL05 are available in three data processing centers: the Center of Spatial Research of the University of Texas (CSR), the NASA's Jet Propulsion Laboratory (JPL), and the German GeoForschungsZentrum Information System and Data Center (GFZISDC).

Previous studies have taken advantage of terrestrial water storage anomalies (TWSA) derived from the Gravity Recovery and Climate Experiment (GRACE) mission (Tapley 2004) in order to identify droughts and quantify their severity worldwide (Rodell et al. 2009; Famiglietti and Rodell 2013; Thomas et al. 2014). The GRACE mission, designed to measure changes in Earth's gravimetric field, enables estimation of TWSA over large areas. TWSA is an aggregate estimate, including surface water (rivers, wetlands, and lakes) and sub-surface (soil moisture and groundwater) reservoirs, along with water on the leaves and snow. GRACE data have been used to determine the evapotranspiration over land (Ramillien et al. 2006; Rodell et al. 2011), groundwater storage change (Frappart et al. 2011) and depletion (Rodell et al. 2009), impacts of severe floods on water storage (Espinoza et al. 2013), and drought detection (Famiglietti and Rodell 2013) and its characterization (Thomas et al. 2014). GRACE estimates of TWSA have also been used in data assimilation schemes (Zaitchik et al. 2008; Li et al. 2012; Kumar 2014) and in the calibration (Werth et al. 2009) and evaluation (Getirana et al. 2014) of hydrological models.

10.3.1 Recent Drought Studies with Grace in Brazil

A recent study presented by Getirana (2016) has characterized and quantified an extended drought occurring in Southeast Brazil using GRACE TWSA, GLDAS vertical flux data, and ground-based reservoir water storage change (RWSC). That region has an approximate area of 930,000 km^2 and accommodates major hydropower plants in the country, numerous industrial centers, and irrigation fields, which together require a continuous and large amount of water.

In order to identify major breaks in GRACE TWSA time series, the author used two break tests, Pettitt's (1979) method and Hubert's segmentation procedure (Hubert et al. 1989). The application of the Hubert's segmentation points out to a few breaks in the 13-year time series, but two dates seem to be especially significant: the transition January–February 2004 and February–March 2012, both occurring during wet seasons (austral summer). The first one defines the end of an extended and severe dry season that began in the early 2000s, causing extreme reservoir depletions and subsequent nationwide electricity crisis and blackouts in 2001. The second one is identified in the time series as the transition between an unusually wet season (with a few dry months) and an extended more-than usual dry season. The Pettitt test indicates that a breakpoint occurred in February 2012. This result corroborates with those obtained with Hubert's segmentation, evidencing that the latter was the most significant break occurring in the 13-year monthly time series.

Comparisons between GRACE TWSA and RWSC resulted in high correlations, demonstrating the ability of GRACE in detecting water volume change in reservoirs covering small fractions of TWSA product's spatial resolution. Despite the small reservoir surface area relative to the spatial resolution of GRACE, statistically significant correlations were found between RWSC and TWSA averaged over 3×3 grid cell boxes. These results could be explained by RWSC amplitude, reservoir size, and their proximity to the drought nucleus. However, the actual impact of reservoirs on GRACE TWSA is inconclusive since other factors can influence the observed correlations. These factors include the natural hydrological seasonality observed in the surrounding area (other water bodies, such as rivers and other lakes), soil moisture, and aquifers. Additionally, GRACE TWSA amplitude errors over those boxes should be in the order of 2 km^3, indicating that the RWSC of small reservoirs might not significantly impact GRACE TWSA anomalies. An effective way to demonstrate such dependencies would be to perform model runs (LSMs coupled with river routing schemes capable of simulating reservoir operation rules), considering multiple scenarios, and quantitatively determine a correlation between variables. However, such analysis is beyond the scope of this paper.

Based on complementary data from model outputs derived from GLDAS, it was demonstrated that the recent water deficit observed over Eastern Brazil is mostly due to lower-than-usual precipitation rates. Though the reasons for the recent low precipitation are still unknown and are under research investigation, there is an increasing speculation that man-induced climate change (Escobar 2015) and deforestation in Amazonia (Nazareno and Laurance 2015) may be altering the moisture transport from Amazonia to Southeast South America (SESA). Indeed, recent studies have demonstrated the existence of a dipole-like structure between SESA and the South American Convective Zone (SACZ) (e.g., Junquas et al. 2011) and moisture transport across the continent through low-level jets and aerial rivers from tropical to subtropical regions of South America (e.g., Marengo et al. 2004; Poveda et al. 2014). Future studies investigating possible causes for the reduced precipitation are recommended. These studies should consider, among other factors, the simultaneous drought over Eastern Brazil and floods over the Amazon (Marengo et al. 2013), potential relationships with low-level jets and aerial rivers carrying moisture from tropical to subtropical regions in South America (Marengo et al. 2004; Poveda et al. 2014), and impacts of deforestation on these fluxes.

Lastly, the use of remote sensing data in environmental studies can empower policy and decision makers to develop social, economic, and environmental policies better adapted to extreme events and capable of preventing major socioeconomic losses due to extreme droughts. Due to the short time series and a non-negligible time shift between data acquisition, processing, and availability to the scientific community, GRACE is not yet ready for use in seasonal predictions, nor near real-time drought forecasts (Thomas et al. 2014). However, such data, combined with additional hydrological information from current and next-generation satellite missions (Alsdorf et al. 2007; Getirana and Peters-Lidard 2013; Papa et al. 2013; Kumar 2014), as well as model outputs, can be used to diagnose recent extreme water-related events and can contribute to water scarcity predictions for upcoming dry seasons (Landerer and Swenson 2012) and be integrated into model calibration frameworks and decision support systems. Combining such techniques with GRACE data assimilation framework (Li and Rodell 2015) and model forecasts could produce useful decision-making tools for minimizing the impacts of future extreme and prolonged drought events on water supply and energy generation.

References

AghaKouchak A (2015) A multivariate approach for persistence-based drought prediction: application to the 2010–2011 east Africa drought. J Hydrol 526:127–135. doi:10.1016/j.jhydrol.2014.09.063

AghaKouchak A, Farahmand A, Melton FS, Teixeira J, Anderson MC, Wardlow BD, Hain CR (2015) Remote sensing of drought: progress, challenges and opportunities. Rev Geophys 53 (2):452–480. doi:10.1002/2014rg000456

Allen RG, Tasumi M, Morse A, Trezza R, Wright JL, Bastiaanssen W, Kramber W, Lorite I, Robison CW (2007) Satellite-based energy balance for mapping Evapotranspiration with internalized calibration (METRIC)—applications. J Irri Drain Eng 133(4):395–406. doi:10.1061/(asce)0733-9437(2007)133:4(395)

Alsdorf D, Bates P, Melack J, Wilson M, Dunne T (2007) Spatial and temporal complexity of the Amazon flood measured from space. Geophys Res Lett 34(8), L08402. doi:10.1029/2007gl029447

Anderson LO, Malhi Y, Aragão LEOC, Ladle R, Arai E, Barbier N, Phillips O (2010) Remote sensing detection of droughts in Amazonian forest canopies. New Phytol 187(3):733–750. doi:10.1111/j.1469-8137.2010.03355.x

Anderson M (1997) A Two-Source time-integrated model for estimating surface fluxes using thermal infrared remote sensing. Remote Sens Environ 60(2):195–216. doi:10.1016/s0034-4257(96)00215-5

Anderson MC, Hain C, Wardlow B, Pimstein A, Mecikalski JR, Kustas WP (2011) Evaluation of drought indices based on thermal remote sensing of Evapotranspiration over the continental United States. J Climate 24(8):2025–2044. doi:10.1175/2010jcli3812.1

Anderson MC, Zolin CA, Sentelhas PC, Hain CR, Semmens K, Tugrul Yilmaz M, Gao F, Otkin JA, Tetrault R (2016) The evaporative stress index as an indicator of agricultural drought in brazil: an assessment based on crop yield impacts. Remote Sens Environ 174:82–99. doi:10.1016/j.rse.2015.11.034

Anderson WB, Zaitchik BF, Hain CR, Anderson MC, Yilmaz MT, Mecikalski J, Schultz L (2012) Towards an integrated soil moisture drought monitor for east Africa. Hydrol Earth Syst Sci 16 (8):2893–2913. doi:10.5194/hess-16-2893-2012

Barbosa HA, Lakshmi Kumar TV (2016) Influence of rainfall variability on the vegetation dynamics over northeastern brazil. J Arid Environ 124:377–387. doi:10.1016/j.jaridenv.2015.08.015

Bastiaanssen WGM, Pelgrum H, Wang J, Ma Y, Moreno JF, Roerink GJ, van der Wal T (1998) A remote sensing surface energy balance algorithm for land (SEBAL). J Hydrol 212–213:213–229. doi:10.1016/s0022-1694(98)00254-6

Björnsson H, Venegas SA (1997) A manual for EOF and SVD analyses of climatic data

Bolten JD, Crow WT, Zhan X, Jackson TJ, Reynolds CA (2010) Evaluating the utility of remotely sensed soil moisture Retrievals for operational agricultural drought monitoring. IEEE J Sel Top Appl Earth Obs Remote Sens 3(1):57–66. doi:10.1109/jstars.2009.2037163

Brown JF, Wardlow BD, Tadesse T, Hayes MJ, Reed BC (2008) The vegetation drought response index (VegDRI): a new integrated approach for monitoring drought stress in vegetation. GIScі Remote Sens 45(1):16–46. doi:10.2747/1548-1603.45.1.16

Cai W, Cowan T, Briggs P, Raupach M (2009) Rising temperature depletes soil moisture and exacerbates severe drought conditions across southeast Australia. Geophys Res Lett 36(21), L21709. doi:10.1029/2009gl040334

Carlson T (2007) An overview of the "triangle method" for estimating surface Evapotranspiration and soil moisture from satellite imagery. Sensors 7(8):1612–1629. doi:10.3390/s7081612

Carlson TN, Gillies RR, Schmugge TJ (1995) An interpretation of methodologies for indirect measurement of soil water content. Agric For Meteorol 77(3–4):191–205. doi:10.1016/0168-1923(95)02261-u

Carrão H, Russo S, Sepulcre-Canto G, Barbosa P (2016) An empirical standardized soil moisture index for agricultural drought assessment from remotely sensed data. Int J Appl Earth Obs Geoinfo 48:74–84. doi:10.1016/j.jag.2015.06.011

Coelho CAS, Cardoso DHF, Firpo MAF (2015a) Precipitation diagnostics of an exceptionally dry event in São Paulo, brazil. Theor Appl Climatol. doi:10.1007/s00704-015-1540-9

Coelho CAS, de Oliveira CP, Ambrizzi T, Reboita MS, Carpenedo CB, Campos JLPS, Tomaziello ACN, Pampuch LA, Custódio M de S, Dutra LMM, Da Rocha RP, Rehbein A (2015b) The 2014 southeast brazil austral summer drought: regional scale mechanisms and teleconnections. Climate Dynam 46(11–12):3737–3752. doi:10.1007/s00382-015-2800-1

Cunha APM, Alvalá RC, Nobre CA, Carvalho MA (2015) Monitoring vegetative drought dynamics in the Brazilian semiarid region. Agric For Meteorol 214–215:494–505. doi:10.1016/j.agrformet.2015.09.010

Dracup JA, Lee KS, Paulson EG (1980) On the definition of droughts. Water Resour Res 16 (2):297–302. doi:10.1029/wr016i002p00297

Draper CS, Walker JP, Steinle PJ, de Jeu RAM, Holmes TRH (2009) An evaluation of AMSR–E derived soil moisture over Australia. Remote Sens Environ 113(4):703–710. doi:10.1016/j.rse.2008.11.011

Escobar H (2015) Drought triggers alarms in brazil's biggest metropolis. Science 347(6224):812. doi:10.1126/science.347.6224.812

Espinoza JC, Ronchail J, Frappart F, Lavado W, Santini W, Guyot JL (2013) The major floods in the Amazonas river and Tributaries (western Amazon basin) during the 1970–2012 period: a focus on the 2012 flood. J Hydrometeorol 14(3):1000–1008. doi:10.1175/jhm-d-12-0100.1

Famiglietti JS, Rodell M (2013) Water in the balance. Science 340(6138):1300–1301. doi:10.1126/science.1236460

Fang B, Lakshmi V, Bindlish R et al (2013) Passive microwave soil moisture downscaling using vegetation index and skin surface temperature. Vadose Zone J 12(3). doi:10.2136/vzj2013.05.0089

Frappart F, Papa F, Güntner A, Werth S, Santos da Silva J, Tomasella J, Seyler F, Prigent C, Rossow WB, Calmant S, Bonnet M-P (2011) Satellite-based estimates of groundwater storage variations in large drainage basins with extensive floodplains. Remote Sens Environ 115 (6):1588–1594. doi:10.1016/j.rse.2011.02.003

Gao B (1996) NDWI—a normalized difference water index for remote sensing of vegetation liquid water from space. Remote Sens Environ 58(3):257–266. doi:10.1016/s0034-4257(96)00067-3

Getirana A (2016) Extreme water deficit in brazil detected from space. J Hydrometeorol 17 (2):591–599. doi:10.1175/jhm-d-15-0096.1

Getirana ACV, Dutra E, Guimberteau M, Kam J, Li H-Y, Decharme B, Zhang Z, Ducharne A, Boone A, Balsamo G, Rodell M, Toure AM, Xue Y, Peters-Lidard CD, Kumar SV, Arsenault K, Drapeau G, Ruby Leung L, Ronchail J, Sheffield J (2014) Water balance in the Amazon basin from a land surface model ensemble. J Hydrometeorol 15(6):2586–2614. doi:10.1175/jhm-d-14-0068.1

Getirana ACV, Peters-Lidard C (2013) Estimating water discharge from large radar altimetry datasets. Hydrol Earth Syst Sci 17(3):923–933. doi:10.5194/hess-17-923-2013

González J, Valdés JB (2004) The mean frequency of recurrence of in-time-multidimensional events for drought analyses. Nat Hazards Earth Syst Sci 4(1):17–28. doi:10.5194/nhess-4-17-2004

Gruber A, Su C, Zwieback S, Crow W, Dorigo W, Wagner W (2016) Recent advances in (soil moisture) triple collocation analysis. Int J Appl Earth Obs Geoinfo 45:200–211. doi:10.1016/j.jag.2015.09.002

Gu Y, Brown JF, Verdin JP, Wardlow B (2007) A five-year analysis of MODIS NDVI and NDWI for grassland drought assessment over the central great plains of the United States. Geophys Res Lett 34(6), L06407. doi:10.1029/2006gl029127

Hayes MJ, Svoboda MD, Wilhite DA, Vanyarkho OV (1999) Monitoring the 1996 drought using the standardized precipitation index. Bull Am Meteorol Soc 80(3):429–438. doi:10.1175/1520-0477(1999)080<0429:mtduts>2.0.co;2

Hirschi M, Mueller B, Dorigo W, Seneviratne SI (2014) Using remotely sensed soil moisture for land–atmosphere coupling diagnostics: the role of surface vs. Root-zone soil moisture variability. Remote Sens Environ 154:246–252. doi:10.1016/j.rse.2014.08.030

Hong Y, Adler RF, Hossain F et al (2007) A first approach to global runoff simulation using satellite rainfall estimation. Water Resour Res 43(8). doi:10.1029/2006wr005739

Hubert P, Carbonnel JP, Chaouche A (1989) Segmentation des séries hydrométéorologiques—application à des séries de précipitations et de débits de l'afrique de l'ouest. J Hydrol 110 (3–4):349–367. doi:10.1016/0022-1694(89)90197-2

Huete A, Didan K, Miura T, Rodriguez EP, Gao X, Ferreira LG (2002) Overview of the radiometric and biophysical performance of the MODIS vegetation indices. Remote Sens Environ 83(1–2):195–213. doi:10.1016/s0034-4257(02)00096-2

Huffman GJ, Bolvin DT, Nelkin EJ, Wolff DB, Adler RF, Gu G, Hong Y, Bowman KP, Stocker EF (2007) The TRMM Multisatellite Precipitation analysis (TMPA): Quasi-Global, multiyear, combined-sensor Precipitation estimates at fine scales. J Hydrometeorol 8(1):38–55. doi:10.1175/jhm560.1

de Jeu R, Dorigo W (2016) On the importance of satellite observed soil moisture. Int J Appl Earth Obs Geoinfo 45:107–109. doi:10.1016/j.jag.2015.10.007

Junquas C, Vera C, Li L, Le Treut H (2011) Summer precipitation variability over southeastern South America in a global warming scenario. Climate Dynam 38(9–10):1867–1883. doi:10.1007/s00382-011-1141-y

Justice CO, Townshend JRG, Vermote EF, Masuoka E, Wolfe RE, Saleous N, Roy DP, Morisette JT (2002) An overview of MODIS land data processing and product status. Remote Sens Environ 83(1–2):3–15. doi:10.1016/s0034-4257(02)00084-6

Karnieli A, Agam N, Pinker RT, Anderson M, Imhoff ML, Gutman GG, Panov N, Goldberg A (2010) Use of NDVI and land surface temperature for drought assessment: merits and limitations. J Climate 23(3):618–633. doi:10.1175/2009jcli2900.1

Kim T-W, Valdés JB, Aparicio J (2002) Frequency and spatial characteristics of droughts in the Conchos river basin, Mexico. Water Int 27(3):420–430. doi:10.1080/02508060208687021

Kogan FN (1995) Application of vegetation index and brightness temperature for drought detection. Adv Space Res 15(11):91–100. doi:10.1016/0273-1177(95)00079-t

Kohler MA, Linsley RK (2008) Predicting the runoff from storm rainfall. http://www.nrc.gov/docs/ML0819/ML081900279.pdf. Accessed 18 June 2016

Kumar 2014 Multivariate satellite data assimilation in NLDAS. AMS Annual meeting, 95th AMS annual meeting, Phoenix

Lakshmi V, Piechota T, Narayan U, Tang C (2004) Soil moisture as an indicator of weather extremes. Geophys Res Lett 31(11), L11401. doi:10.1029/2004gl019930

Landerer FW, Swenson SC (2012) Accuracy of scaled GRACE terrestrial water storage estimates. Water Resour Res 48(4), W04531. doi:10.1029/2011wr011453

Li B, Rodell M (2015) Evaluation of a model-based groundwater drought indicator in the conterminous U.S. J Hydrol 526:78–88. doi:10.1016/j.jhydrol.2014.09.027

Li B, Rodell M, Zaitchik BF, Reichle RH, Koster RD, van Dam TM (2012) Assimilation of GRACE terrestrial water storage into a land surface model: evaluation and potential value for drought monitoring in western and central Europe. Journal of Hydrology 446–447:103–115. doi:10.1016/j.jhydrol.2012.04.035

Liu J, Rambal S, Mouillot F (2015) Soil drought anomalies in MODIS GPP of a Mediterranean Broadleaved evergreen forest. Remote Sens 7(1):1154–1180. doi:10.3390/rs70101154

Liu WT, Juárez RIN (2001) ENSO drought onset prediction in northeast brazil using NDVI. Int J Remote Sens 22(17):3483–3501. doi:10.1080/01431160010006430

Lyon B (2004) The strength of El Niño and the spatial extent of tropical drought. Geophys Res Lett 31(21), L21204. doi:10.1029/2004gl020901

Marengo JA, Alves LM, Soares WR, Rodriguez DA, Camargo H, Riveros MP, Pabló AD (2013) Two contrasting severe seasonal extremes in tropical south America in 2012: flood in Amazonia and drought in northeast brazil. J Climate 26(22):9137–9154. doi:10.1175/jcli-d-12-00642.1

Marengo JA, Bernasconi M (2014) Regional differences in aridity/drought conditions over northeast brazil: present state and future projections. Clim Change 129(1–2):103–115. doi:10.1007/s10584-014-1310-1

Marengo JA, Soares WR, Saulo C, Nicolini M (2004) Climatology of the low-level jet east of the Andes as derived from the NCEP–NCAR Reanalyses: characteristics and temporal variability. J Climate 17(12):2261–2280. doi:10.1175/1520-0442(2004)017<2261:cotlje>2.0.co;2

Merlin O, Walker J, Chehnbouni A, Kerr Y (2008) Towards deterministic downscaling of SMOS soil moisture using MODIS derived soil evaporative efficiency. Remote Sens Environ 112 (10):3935–3946. doi:10.1016/j.rse.2008.06.012

Mishra AK, Desai VR (2005) Spatial and temporal drought analysis in the Kansabati river basin, India. Int J River Basin Manage 3(1):31–41. doi:10.1080/15715124.2005.9635243

Mishra AK, Ines AVM, Das NN, Prakash Khedun C, Singh VP, Sivakumar B, Hansen JW (2015) Anatomy of a local-scale drought: application of assimilated remote sensing products, crop model, and statistical methods to an agricultural drought study. J Hydrol 526:15–29. doi:10.1016/j.jhydrol.2014.10.038

Mishra AK, Singh VP (2010) A review of drought concepts. J Hydrol 391(1–2):202–216. doi:10.1016/j.jhydrol.2010.07.012

Mishra AK, Singh VP (2011) Drought modeling—a review. J Hydrol 403(1–2):157–175. doi:10.1016/j.jhydrol.2011.03.049

Moura F de BP, Mendes Malhado AC, Ladle RJ (2013) Nursing the caatinga back to health. J Arid Environ 90:67–68. doi:10.1016/j.jaridenv.2012.10.010

Mu Q, Zhao M, Kimball JS, McDowell NG, Running SW (2013) A remotely sensed global terrestrial drought severity index. Bull Am Meteorol Soc 94(1):83–98. doi:10.1175/bams-d-11-00213.1

Mu Q, Zhao M, Running SW (2011) Improvements to a MODIS global terrestrial evapotranspiration algorithm. Remote Sens Environ 115(8):1781–1800. doi:10.1016/j.rse.2011.02.019

Narasimhan B, Srinivasan R (2005) Development and evaluation of soil moisture deficit index (SMDI) and Evapotranspiration deficit index (ETDI) for agricultural drought monitoring. Agric For Meteorol 133(1–4):69–88. doi:10.1016/j.agrformet.2005.07.012

Nazareno AG, Laurance WF (2015) Brazil's drought: beware deforestation. Science 347 (6229):1427–1427. doi:10.1126/science.347.6229.1427-a

Otkin JA, Anderson MC, Hain C, Svoboda M (2014) Examining the relationship between drought development and rapid changes in the evaporative stress index. J Hydrometeorol 15 (3):938–956. doi:10.1175/jhm-d-13-0110.1

Palmer WC (1968) Keeping track of crop moisture conditions, nationwide: the new crop moisture index. Weatherwise 21(4):156–161. doi:10.1080/00431672.1968.9932814

Papa F, Frappart F, Güntner A, Prigent C, Aires F, Getirana ACV, Maurer R (2013) Surface freshwater storage and variability in the Amazon basin from multi-satellite observations, 1993-2007. J Geophys Res Atmos 118(21):11951–11965. doi:10.1002/2013jd020500

Petropoulos G, Carlson TN, Wooster MJ, Islam S (2009) A review of Ts/VI remote sensing based methods for the retrieval of land surface energy fluxes and soil surface moisture. Prog Phys Geo 33(2):224–250. doi:10.1177/0309133309338997

Pettitt AN (1979) A non-parametric approach to the change-point problem. Appl Stat 28(2):126. doi:10.2307/2346729

Piles M, Camps A, Vall-llossera M, Corbella I, Panciera R, Rudiger C, Kerr YH, Walker J (2011) Downscaling SMOS-Derived soil moisture using MODIS visible/infrared data. IEEE Trans Geosci Remote Sens 49(9):3156–3166. doi:10.1109/tgrs.2011.2120615

Poveda G, Jaramillo L, Vallejo LF (2014) Seasonal precipitation patterns along pathways of south American low-level jets and aerial rivers. Water Resour Res 50(1):98–118. doi:10.1002/2013wr014087

Privette JL, Myneni RB, Knyazikhin Y, Mukelabai M, Roberts G, Tian Y, Wang Y, Leblanc SG (2002) Early spatial and temporal validation of MODIS LAI product in the southern Africa Kalahari. Remote Sens Environ 83(1–2):232–243. doi:10.1016/s0034-4257(02)00075-5

Rahmani A, Golian S, Brocca L (2016) Multiyear monitoring of soil moisture over Iran through satellite and reanalysis soil moisture products. Int J Appl Earth Obs Geoinf 48:85–95. doi:10.1016/j.jag.2015.06.009

Ramillien G, Frappart F, Güntner A, Ngo-Duc T, Cazenave A, Laval K (2006) Time variations of the regional evapotranspiration rate from gravity recovery and climate experiment (GRACE) satellite gravimetry. Water Resour Res 42(10), W10403. doi:10.1029/2005wr004331

Rees G (2012) Electromagnetic waves in free space. In: Physical principles of remote sensing. Cambridge University Press, Cambridge

Rodell M, McWilliams EB, Famiglietti JS, Beaudoing HK, Nigro J (2011) Estimating evapotranspiration using an observation based terrestrial water budget. Hydrol Process 25 (26):4082–4092. doi:10.1002/hyp.8369

Rodell M, Velicogna I, Famiglietti JS (2009) Satellite-based estimates of groundwater depletion in India. Nature 460(7258):999–1002. doi:10.1038/nature08238

Running SW, Nemani RR, Heinsch FA, Zhao M, Reeves M, Hashimoto H (2004) A continuous satellite-derived measure of global terrestrial primary production. BioScience 54(6):547. doi:10.1641/0006-3568(2004)054[0547:acsmog]2.0.co;2

Sandholt I, Rasmussen K, Andersen J (2002) A simple interpretation of the surface temperature/vegetation index space for assessment of surface moisture status. Remote Sens Environ 79 (2–3):213–224. doi:10.1016/s0034-4257(01)00274-7

Santos JF, Pulido-Calvo I, Portela MM (2010) Spatial and temporal variability of droughts in Portugal. Water Resour Res 46(3), W03503. doi:10.1029/2009wr008071

van der Schrier G, Barichivich J, Briffa KR, Jones PD (2013) A scPDSI-based global data set of dry and wet spells for 1901–2009. J Geophys Res Atmos 118(10):4025–4048. doi:10.1002/jgrd.50355

Seth A, Fernandes K, Camargo SJ (2015) Two summers of São Paulo drought: origins in the western tropical pacific. Geophys Res Lett 42(24):10816–10823. doi:10.1002/2015gl066314

Sheffield J (2004) A simulated soil moisture based drought analysis for the United States. J Geophys Res 109(D24), D24108. doi:10.1029/2004jd005182

Silva ACS, Galvão CO, Silva GNS (2015) Droughts and governance impacts on water scarcity: an analysis in the Brazilian semi-arid. Proc Int Assoc Hydrol Sci 369:129–134. doi:10.5194/piahs-369-129-2015

Singh RP, Roy S, Kogan F (2003) Vegetation and temperature condition indices from NOAA AVHRR data for drought monitoring over India. Int J Remote Sens 24(22):4393–4402. doi:10.1080/0143116031000084323

Stoffelen A (1998) Toward the true near-surface wind speed: Error modeling and calibration using triple collocation. J Geophy Res Oceans 103(C4):7755–7766. doi:10.1029/97jc03180

Su Z (2002) The surface energy balance system (SEBS) for estimation of turbulent heat fluxes. Hydrol Earth Syst Sci 6(1):85–100. doi:10.5194/hess-6-85-2002

Sun D, Kafatos M (2007) Note on the NDVI-LST relationship and the use of temperature-related drought indices over north America. Geophys Res Lett 34(24), L24406. doi:10.1029/2007gl031485

Syed TH, Lakshmi V, Paleologos E, Lohmann D, Mitchell K, Famiglietti JS (2004) Analysis of process controls in land surface hydrological cycle over the continental United States. J Geophys Res Atmos 109(D22), D22105. doi:10.1029/2004jd004640

Tang C, Piechota TC (2009) Spatial and temporal soil moisture and drought variability in the upper Colorado river basin. J Hydrol 379(1–2):122–135. doi:10.1016/j.jhydrol.2009.09.052

Tapley BD (2004) GRACE measurements of mass variability in the earth system. Science 305 (5683):503–505. doi:10.1126/science.1099192

Tatli H, Türkeş M (2011) Empirical Orthogonal function analysis of the palmer drought indices. Agric For Meteorol 151(7):981–991. doi:10.1016/j.agrformet.2011.03.004

Thomas AC, Reager JT, Famiglietti JS, Rodell M (2014) A gRACE-based water storage deficit approach for hydrological drought characterization. Geophys Res Lett 41(5):1537–1545. doi:10.1002/2014gl059323

Toumazou V, Cretaux J-F (2001) Using a Lanczos Eigensolver in the computation of empirical Orthogonal functions. Mon Weather Rev 129(5):1243–1250. doi:10.1175/1520-0493(2001) 129<1243:ualeit>2.0.co;2

Tsakiris G, Pangalou D, Vangelis H (2006) Regional drought assessment based on the reconnaissance drought index (RDI). Water Resour Manag 21(5):821–833. doi:10.1007/s11269-006-9105-4

Tsakiris G, Vangelis H (2005) Establishing a drought index incorporating evapotranspiration. Eur Water 9:3–11

Vermote E, Vermeulen A (1999) Atmospheric correction algorithm: spectral reflectances (MOD09), Algorithm Theoretical Basis Documents (ATBD) Version 4.0. Tech Report April, EOS Project Science Office. NASA Goddard Space Flight Center, Greenbelt

Wagner W, Lemoine G, Rott H (1999) A method for estimating soil moisture from ERS Scatterometer and soil data. Remote Sens Environ 70(2):191–207. doi:10.1016/s0034-4257 (99)00036-x

Wan, Z. (1999) Land-surface temperature algorithm theoretical basis document (LST ATBD). Technical Report April. Institute for Computational Earth System Science, University of California

Werth S, Güntner A, Petrovic S, Schmidt R (2009) Integration of GRACE mass variations into a global hydrological model. Earth Planet Sci Lett 277(1–2):166–173. doi:10.1016/j.epsl.2008. 10.021

Wilhite DA (1993) The enigma of drought. In: Wilhite DA (ed) Drought assessment, management, and planning: theory and case studies. Kluwer Academic, Berlin

Wu D, Qu JJ, Hao X (2015) Agricultural drought monitoring using MODIS-based drought indices over the USA corn belt. Int J Remote Sens 36(21):5403–5425. doi:10.1080/01431161.2015. 1093190

Yao Y, Liang S, Qin Q, Wang K (2010) Monitoring drought over the Conterminous United States using MODIS and NCEP Reanalysis-2 data. J Appl Meteorol Climatol 49(8):1665–1680. doi:10.1175/2010jamc2328.1

Yao Y, Liang S, Qin Q, Wang K, Zhao S (2011) Monitoring global land surface drought based on a hybrid evapotranspiration model. Int J Appl Earth Obs Geoinf 13(3):447–457. doi:10.1016/j. jag.2010.09.009

Zaitchik BF, Rodell M, Reichle RH (2008) Assimilation of GRACE terrestrial water storage data into a land surface model: results for the Mississippi river basin. J Hydrometeorol 9 (3):535–548. doi:10.1175/2007jhm951.1

Zhang F, Zhang L, Wang X, Hung J (2013) Detecting agro-droughts in southwest of china using MODIS satellite data. J Integr Agric 12(1):159–168. doi:10.1016/s2095-3119(13)60216-6

Zhang L, Xiao J, Li J, Wang K, Lei L, Guo H (2012) The 2010 spring drought reduced primary productivity in southwestern china. Environ Res Lett 7(4):045706. doi:10.1088/1748-9326/7/ 4/045706

Zhang X, Susan Moran M, Zhao X, Liu S, Zhou T, Ponce-Campos GE, Liu F (2014) Impact of prolonged drought on rainfall use efficiency using MODIS data across china in the early 21st century. Remote Sens Environ 150:188–197. doi:10.1016/j.rse.2014.05.003

Zhao W, Li A (2013) A downscaling method for improving the spatial resolution of AMSR-E derived soil moisture product based on MSG-SEVIRI data. Remote Sens 5(12):6790–6811. doi:10.3390/rs5126790

Chapter 11
Multi-Sensor Remote Sensing of Drought from Space

M. Sadegh, C. Love, A. Farahmand, A. Mehran, M.J. Tourian, and A. AghaKouchak

Abstract Drought monitoring is vital considering the immense costs of this natural hazard. The root cause for all types of drought (meteorological, agricultural, hydrological, and socio-economic) is sustained below average precipitation. Since regional precipitation variability depends on large-scale climatic and oceanic circulation patterns, it is necessary to study droughts from a global perspective which requires satellite observations. Satellite data allow comprehensive assessment of drought onset, development, and recovery through a multi-sensor multivariate monitoring of hydrological variables. However, there are major challenges in using satellite data, including consistency, reliability, uncertainty, and length of record that merit more in-depth research.

11.1 Introduction

Droughts are among the costliest natural disasters that pose food and water security concerns around the world (Wilhite 2005; Hoerling 2013; Godfray et al. 2010; AghaKouchak et al. 2015a). While droughts are typically caused by sustained below average precipitation, they are classified into four types based on the different hydrological variables they impact. Deficit in precipitation is typically referred to as meteorological drought (Hayes et al. 1999). Deficit in soil moisture (i.e., below average moisture in the soil) is known as agricultural drought

M. Sadegh (✉) • C. Love • A. Farahmand • A. AghaKouchak
University of California, Irvine, CA 92697, USA
e-mail: msadegh@uci.edu

A. Mehran
University of California, Irvine, CA 92697, USA

University of California, Los Angeles, CA 90095, USA

M.J. Tourian
University of Stuttgart, Stuttgart, Germany

V. Lakshmi (ed.), *Remote Sensing of Hydrological Extremes*, Springer Remote Sensing/Photogrammetry, DOI 10.1007/978-3-319-43744-6_11

(AghaKouchak 2014). Hydrological drought corresponds to a deficit in runoff or groundwater resources (Wood and Lettenmaier 2008; Van Loon 2015). Socio-economic drought (Mehran et al. 2015), also termed anthropogenic drought (AghaKouchak et al. 2015b), refers to water stress intensified by human activities and increasing water demands (see also Van Loon et al. 2016). Typically, sustained deficit in precipitation leads to deficit in soil moisture, runoff, and groundwater resources with a certain time lag. For this reason, a comprehensive assessment of drought onset, development, and recovery requires a multi-sensor approach in which multiple hydrological variables are monitored.

Drought monitoring and prediction have long relied on ground-based observations (Shen and Tabios 1996; Sheffield et al. 2012; Aghakouchak et al. 2014; Gallagher et al. 1976; Palmer 1965). However, understanding the causes of drought and explaining the underlying processes requires a global (or at least, continental) perspective, since regional precipitation variability has been linked to large-scale climatic and oceanic circulation patterns (Hoerling and Kumar 2003; Rasmusson et al. 1983; Keyantash and Dracup 2004; Golian et al. 2015; Hidalgo 2004; Wegren 2011). This highlights the need for monitoring droughts from a global perspective to better understand regional drought trends and patterns (Grasso and Singh 2011). However, many areas of the world are not sufficiently covered with ground-based observations, making large-scale monitoring and assessment challenging (Easterling 2013). Moreover, ground-based observations are not necessarily available in a timely manner for drought monitoring and assessment. This underscores the need for global satellite observations which can provide improved initial and boundary condition information for drought monitoring and prediction (AghaKouchak and Nakhjiri 2012).

In the past four decades, satellite-based observations have provided valuable information to the hydrology and climate communities with respect to precipitation, temperature, evapotranspiration, soil moisture, vegetation greenness, land cover condition, and total water storage at the global scale (NASA 2010; AghaKouchak et al. 2015a; Wardlow et al. 2012; Krajewski et al. 2006; Whitcraft et al. 2015). Satellite observations not only provide near real-time quasi-global to global information, but also allow multi-sensor drought monitoring from different viewpoints, such as precipitation, evapotranspiration, soil moisture, and vegetation health (Barrett and Herschy 1989; Morgan 1989; Heumann 2011; Sorooshian et al. 2011; Entekhabi et al. 2004; Anderson and Kustas 2008; Running et al. 1989; Karnieli et al. 2010; McVicar and Jupp 1998; Price 1982; Fensholt et al. 2006; Wang and Qu 2007; Anderson et al. 2011b; Allen et al. 2007; Cashion et al. 2005).

Remotely sensed observations have also been widely used to assess droughts based on corresponding impacts on the ecosystem, such as photosynthetic capacity, vegetation growth rate, and vegetation greenness (Tucker and Choudhury 1987; Asrar et al. 1984, 1989; Asner and Alencar 2010; Wardlow et al. 2012; Hatfield et al. 1984). While the abundance of remotely sensed observations provides opportunities for advancing drought monitoring and prediction, there are major challenges which include data consistency, reliability, underlying uncertainties, length of record, and interoperability of data sources.

In a recent study, AghaKouchak et al. (2015a) reviewed opportunities and challenges in remote sensing of drought from meteorological and ecological perspectives. In this chapter, however, we focus solely on drought monitoring from a climatic perspective. After a brief review of the available satellite-based data products, we concentrate further on opportunities for multi-sensor multivariate drought monitoring and assessment using multiple satellite data products.

11.2 Remotely Sensed Precipitation Information

Precipitation information is necessary for the computation of widely used drought indicators, such as Standardized Precipitation Index (SPI) (McKee et al. 1993) and Percent of Normal Precipitation (PNP) (Werick et al. 1994). This often requires the use of Geostationary (GEO) and Low Earth Orbit (LEO) satellites, which retrieve precipitation information using infrared (IR), typically from GEOs, and passive microwave (PMW) from LEOs. GEOs, positioned at altitudes of 35,880 km, offer higher temporal resolution since they hover over the same location on the Earth, whereas LEOs orbit the Earth at much lower altitudes visiting any location on Earth typically twice each day (Levizzani 2008).

IR-based precipitation estimates typically rely on cloud-top temperature and albedo information (Joyce and Arkin 1997; Arkin et al. 1994; Turk et al. 1999). Passive microwave sensors offer more accurate, but infrequent, instantaneous precipitation estimates (Kummerow et al. 1996, 2001). The infrequency of PMW information can be compensated for by merging both IR and PMW information for improving precipitation retrieval (Sorooshian et al. 2011; Joyce et al. 2004). Several satellite-based precipitation products are now available for research and operational applications including (Levizzani et al. 2007; Kidd 2001):

- Precipitation Estimation from Remotely Sensed Information using Artificial Neural Networks (PERSIANN) (Sorooshian et al. 2000; Hsu et al. 1997; Hong et al. 2004; Sorooshian et al. 2014).
- Tropical Rainfall Measuring Mission (TRMM) Multi-satellite Precipitation Analysis (TMPA) (Huffman et al. 2007).
- CPC Morphing Technique (CMORPH) (Joyce et al. 2004).
- Precipitation Estimation from Remotely Sensed Information using Artificial Neural Networks—Climate Data Record (PERSIANN-CDR) (Ashouri et al. 2015).
- Global Precipitation Climatology Project (GPCP) (Adler et al. 2003).
- Global Satellite Mapping of Precipitation (GSMaP) (Kubota et al. 2007).
- Integrated Multi-satellitE Retrievals for GPM (IMERG) (Huffman et al. 2013; Prakash et al. 2016a).

These satellite precipitation products have been used in a wide variety of studies including drought monitoring and assessment (Damberg and AghaKouchak 2014; Paridal et al. 2008; Anderson et al. 2008; Sheffield et al. 2006). These products all

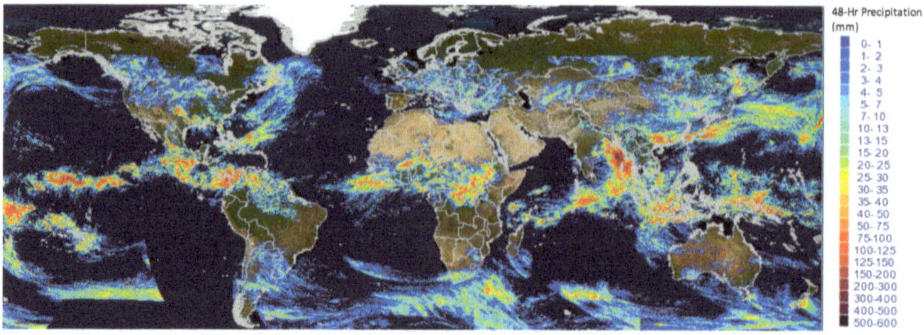

Fig. 11.1 UCI's GWADI map server (http://hydis.eng.uci.edu/gwadi/) offers near real-time 0.04 ° precipitation information based on PERSIANN-Cloud Classification System (PERSIANN-CCS) (Hong et al. 2004)

use similar inputs but provide information at different spatial and temporal scales. For a detailed discussion, interested readers are referred to AghaKouchak et al. (2012), Tian et al. (2009), Ebert et al. (2007), Chappell et al. (2013), Prakash et al. (2016b), Prakash et al. (2015), Anagnostou et al. (2010), Maggioni et al. (2016), AghaKouchak et al. (2011), Hirpa et al. (2010), Gebremichael (2010), and Hossain and Anagnostou (2005).

One limitation of the available satellite precipitation products for drought monitoring is their relative short length of record. Among the ones listed above, GPCP and PERSIANN-CDR offer over 30 years of data, making them more desirable for drought monitoring, although these two products are not currently available in real-time. The diversity of satellite precipitation products, however, does allow for the merging of long-term records (e.g., PERSIANN-CDR, GPCP) with near real-time satellite products (e.g., TMPA, PERSIANN, and CMORPH) for drought monitoring purposes (see, for example, AghaKouchak and Nakhjiri 2012). There are multiple satellite data visualization systems that can be used for viewing and extracting precipitation information, such as UCI's GWADI System (see Fig. 11.1), which offers near real-time high resolution (0.04 °) precipitation data based on PERSIANN-Cloud Classification System (PERSIANN-CCS) (Hong et al. 2004). This data product has been used in a variety of applications, an example of which is provided by Nguyen et al. (2015).

11.3 Satellite Soil Moisture Estimates

Soil moisture deficit is a key variable in defining and monitoring agricultural drought, plant growth, and vegetation stress (Liu et al. 2016; D'Odorico et al. 2007; Boken et al. 2005). Soil moisture information is important for assessing the persistence of droughts which is closely related to drought duration (AghaKouchak 2014). Several soil moisture-based indicators, such as soil moisture percentile (Sheffield et al. 2004; Wang et al. 2009) and Standardized Soil Moisture Index (SSI) (Hao and AghaKouchak 2013), have been used for both drought monitoring and prediction.

The basic concept for retrieving soil moisture estimates from satellite observations relies on the association between the soil's liquid water content and its permittivity. There are empirical relationships that can be used to link passive microwave information to soil's volumetric water content (Njoku et al. 2003; Jackson 1997; Njoku and Entekhabi 1996). Similar relationships exist between active microwave backscatter and soil moisture (Wagner et al. 1999; Takada et al. 2009). There exist soil moisture retrieval methods that combine information from multiple sensors to improve both the spatial coverage/resolution and the quality of estimates (Wilson et al. 2001; Gruhier et al. 2010; Entekhabi et al. 2010a; Liu et al. 2011; Kim and Hogue 2012; Wagner et al. 2012; Liu et al. 2011). Satellite-based soil moisture estimates typically provide information only about the top soil layer (around 5 cm of soil depth Entekhabi et al. 2010a; Njoku et al. 2003; Wang and Qu 2009). However, for agricultural drought monitoring, root zone soil moisture is the key variable of interest. While satellite observations do not provide root zone soil moisture, they can be assimilated into land-surface or land-atmosphere models to improve simulations of root zone soil moisture (Reichle et al. 2004; Wang and Qu 2009).

The long-term satellite-based soil moisture time series obtained from the Water Cycle Multi-Mission Observation Strategy (WACMOS) have been used for drought detection and monitoring in the Horn of Africa region (Ambaw 2013). The United States Department of Agriculture (USDA) International Production Assessment Division (IPAD) estimates surface and root zone soil moisture with a two-layer modified Palmer soil moisture model forced by global precipitation and near-surface air temperature measurements (Palmer and Havens 1958). In this approach, only near-surface air temperature is used to approximate potential evapotranspiration, which has limitations in estimating evapotranspiration (McVicar et al. 2012; Donohue et al. 2010; Hobbins et al. 2008). Soil moisture data retrieved from the Advanced Microwave Scanning Radiometer on Earth Observing System (AMSR-E) (Jackson 1993) have been integrated into the real-time USDA IPAD soil model to improve drought monitoring and prediction (Bolten et al. 2010).

Recently, NASA's Soil Moisture Active Passive (SMAP) mission (Entekhabi et al. 2010a) began providing near real-time top soil moisture estimates (Fig. 11.2). Currently, SMAP's length of record is not sufficient for analyzing droughts from a climatological perspective. However, it can be merged with other data products for improving agricultural drought monitoring. Potential applications of SMAP's soil moisture estimates for drought assessment are yet to be explored, but we anticipate broad applications of SMAP data in drought-related studies in the near future.

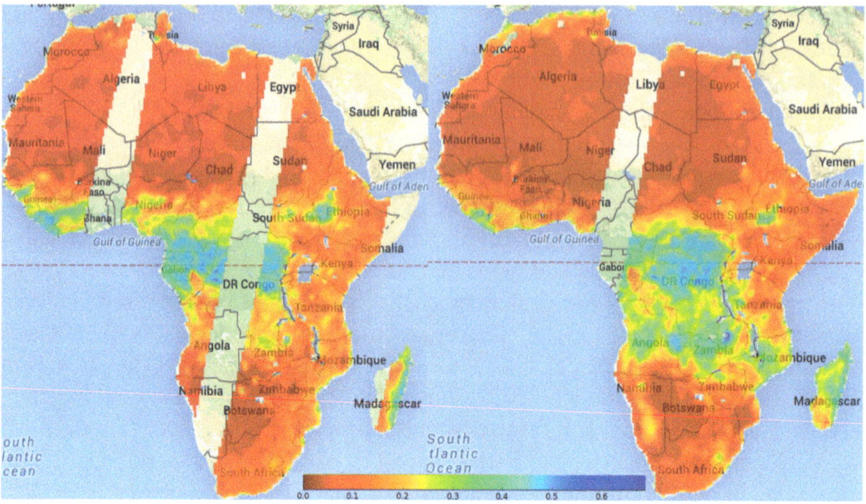

Fig. 11.2 3-day satellite-based soil moisture (m^3/m^3) information from the SMAP mission on (*left*) June 4th, 2016, and (*right*) March 4th, 2016 (Source: Princeton University's African Flood and Drought Monitor, http://stream.princeton.edu/AWCM/WEBPAGE/interface.php?locale=en)

11.4 Relative Humidity

Remotely sensed water vapor and relative humidity are available from NASA's Atmospheric Infrared Sounder (AIRS) instrument. While the AIRS instrument was not specifically designed for drought analysis, recent studies show that near-surface air relative humidity can provide valuable information on drought onset and development (Farahmand et al. 2015; AghaKouchak et al. 2015a). Precipitation is closely associated with relative humidity. Near-surface air relative humidity also affects local evaporation, which has feedback on precipitation. For these reasons, relative humidity is expected to provide relevant information for drought monitoring and assessment.

While many studies indicate that precipitation is perhaps the best indicator for early drought onset detection (Mo 2011; Hao and AghaKouchak 2013), recent studies show that integrating relative humidity from AIRS could improve the drought onset detection. Figure 11.3 displays the 3-month (top) SPI and (middle) Standardized Relative Humidity Index (SRHI; Farahmand et al. 2015). The latter is obtained by standardizing the AIRS-based relative humidity information using an empirical approach outlined in Farahmand and AghaKouchak (2015). The two panels show similarities due to the relationship between precipitation and humidity. Exploring time series of precipitation- and relative humidity-based indicators (Fig. 11.3 (bottom)) highlights the opportunity for early drought onset detection (see Farahmand et al. 2015 for more information).

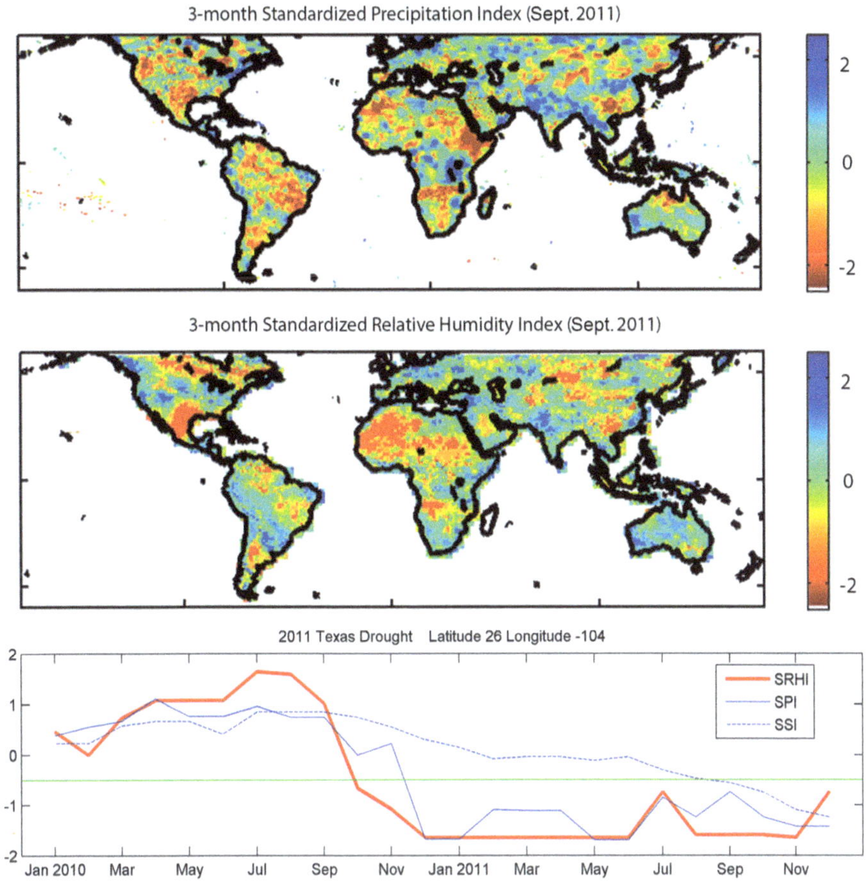

Fig. 11.3 (*top*) 3-month standardized precipitation index (SPI) and (*middle*) standardized relative humidity index (SRHI) for Sept. 2011; (*bottom*) 3-month standardized precipitation index (SPI), standardized soil moisture index (SSI), and standardized relative humidity index (SRHI) for a location affected by the 2011 Texas-Mexico drought (modified after Farahmand et al. 2015; AghaKouchak et al. 2015a)

11.5 Terrestrial Water Storage from Space

The Gravity Recovery and Climate Experiment (GRACE) mission, launched in 2002, provides an invaluable source of information to evaluate drought impacts, at regional to global scales, on terrestrial water storage and groundwater conditions (Tapley et al. 2004). GRACE employs traces of global variations in Earth's gravity field which can be used to estimate terrestrial water storage (TWS) (Rodell and Famiglietti 2002). The principle of gravimetry that governs TWS estimation posits that the gravitational potential of surface and subsurface water mass alters the

Earth's gravity field, indicators of which can be utilized to estimate the total amount of water in the vertical medium. The twin GRACE satellites, being approximately 220 km apart along-track, orbit the earth at an altitude of 500 km. The distance between the two satellites changes as they speed up, or slow down, in response to variations in Earth's gravity field. Utilizing this concept, GRACE provides a measure of temporal TWS (or Equivalent Water Height) anomalies by observing the relative motion of the center-of-mass of the two satellites with a highly precise inter-satellite K-band microwave link, in addition to high precision accelerometers and Global Positioning System (GPS) receivers (Rodell 2012).

GRACE's estimation of TWS enables quantification of groundwater changes, ΔG, as $\Delta G = \Delta TWS - \Delta SM - \Delta SWE$, in which SM and SWE signify soil moisture and snow water equivalence, respectively (Rodell et al. 2007). The TWS data from GRACE has empowered several drought monitoring and water storage assessment studies over a diverse pool of regions, such as the Canadian Prairie (Yirdaw et al. 2008), Australia (Leblanc et al. 2009; van Dijk et al. 2011, 2013), the Amazon River basin (Chen et al. 2009), the Lake Urmia in Iran (Tourian et al. 2015), and western and central Europe (Li et al. 2012), among others. This data set has also offered invaluable information to link meteorological and hydrological drought conditions during the 2011 Texas drought (Long et al. 2013).

Such great advantages of GRACE-based data, however, coincide with some limitations. GRACE provides only about 14 years of data, restricting its application for climatological drought assessment. An example of such data is presented in Fig. 11.4, which demonstrates anomalies of Equivalent Water Height in California. Moreover, a coarse spatial resolution of $> 150{,}000\,\text{km}^2$ for the GRACE TWS data (Houborg et al. 2012) limits its ability to be used for subregional drought assessments.

To diminish the limitations of a coarse spatial resolution, GRACE data has recently been assimilated into models and/or hydrological data, and down-scaled

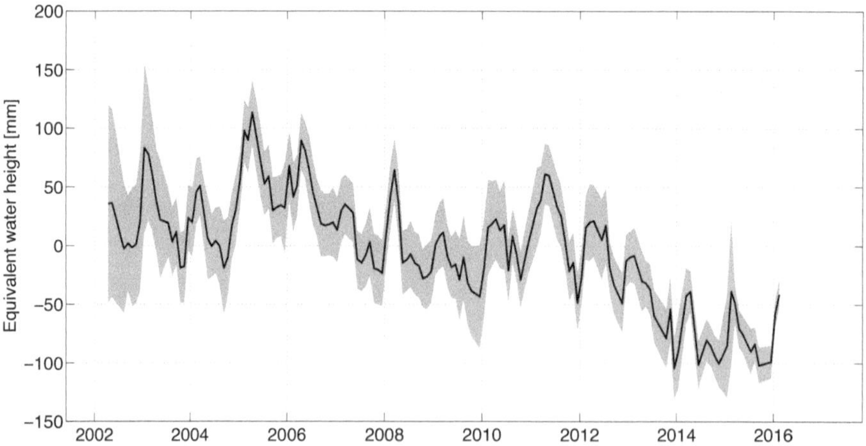

Fig. 11.4 Spaced-based equivalent water height (mm) in California from GRACE satellites

to higher resolutions (Zaitchik et al. 2008; Reager et al. 2015; Lorenz et al. 2015). The assimilated data set offers a potential improvement in drought monitoring and a more advanced evaluation of groundwater resources at finer spatial resolutions. Indeed, a GRACE-based drought indicator has been developed and integrated into the United States and North America Drought Monitor (Houborg et al. 2012), using the GRACE Data Assimilation System (GRACE-DAS) (Zaitchik et al. 2008) and the Catchment Land Surface Model (CLSM) (Koster et al. 2000). Since GRACE provides detailed information regarding the total water storage deficit, it can also be employed to quantify the amount of water input, i.e., precipitation, required to recover from a drought (Thomas et al. 2014).

11.6 Evapotranspiration

Evapotranspiration (ET) is a key constituent of the water and energy cycle (Senay et al. 2012; Wang and Dickinson 2012); manifests mass and energy exchange among terrestrial ecosystems and atmosphere; and depends on net radiation, air temperature, wind speed, and relative humidity (Donohue et al. 2010; Hobbins et al. 2008; McVicar et al. 2012; Yin et al. 2014). Obtaining ground-based measurements of ET is a challenging, if not impossible, task at regional to global scales. Therefore, scientists have resorted to remotely sensed data to estimate ET at large scales. The availability of such invaluable data and the unique characteristic of ET which simultaneously reflects the water/moisture availability and its consumption rate (Anderson et al. 2012a) offer a great opportunity for drought analysis and monitoring studies.

Indeed, several drought indicators, such as Crop Water Stress Index (CWSI) (Idso et al. 1981; Jackson et al. 1981), Water Deficit Index (WDI) (Moran et al. 1994), Evaporative Stress Index (ESI) (Anderson et al. 2011a, b, 2013a), Evaporative Drought Index (EDI) (Yao et al. 2010), Drought Severity Index (DSI) (Mu et al. 2013), and Reconnaissance Drought Index (RDI) (Tsakiris and Vangelis 2005; Tsakiris et al. 2007), which integrate ET as an input variable in the drought analysis, have been proposed in the literature. These indices are briefly explained in the following paragraphs.

CWSI is expressed on the basis of the ratio of actual to potential ET (AET to PET) as $CWSI = 1 - \frac{AET}{PET}$. WDI is developed on the same concept, but using rate, λ, instead of the original value of AET and PET: $WDI = 1 - \frac{\lambda_{AET}}{\lambda_{PET}}$. ESI, developed on the standardized anomalies in the ratio of AET to PET (Anderson et al. 2011b), has shown to be a propitious drought indicator for characterizing streamflow and soil moisture anomalies (Choi et al. 2013), and offers indispensable information for early warning of "flash" (swiftly developing) droughts (Anderson et al. 2013b; Otkin et al. 2014). EDI, similar to ESI in development, has also been employed for drought monitoring at continental and global scales (Yao et al. 2011).

DSI is computed by adding the normalized ratio of $\frac{AET}{PET}$ to the Normalized Difference Vegetation Index (NDVI) (Mu et al. 2013). Short wave satellite observations obtained from the Moderate-resolution Imaging Spectroradiometer (MODIS) are used, in this approach, within a Penman-Monteith ET formulation to quantify the ratio of $\frac{AET}{PET}$ (Mu et al. 2007, 2009, 2011). DSI is a dimensional indicator ranging between $[-\infty, \infty]$, with lower values indicating more severe drought condition. This indicator has been shown in the literature to be consistent with the drought indices based on precipitation and satellite-based measures of vegetation net primary production (NPP) (Running et al. 2004). However, DSI is distinct from other indicators in the sense that it is not a standardized measure of drought severity.

RDI, also coined as Aridity Index (UNESCO 1979), is formulated as a ratio of aggregated precipitation (P) to PET ($\frac{P}{PET}$) for drought monitoring purposes (Tsakiris and Vangelis 2005; Tsakiris et al. 2007). Being widely used in the literature, this index can be standardized for cross comparison with other drought indices. RDI is dissimilar to other indicators in not using AET in its formulation. PET estimates in RDI have been derived from satellite-based air temperature data (Dalezios et al. 2012), disregarding other dominant meteorological drivers, such as net radiation, wind speed, and relative humidity (McVicar et al. 2012; Donohue et al. 2010), that shed concern on the accuracy of such estimation.

11.7 Snow

Snow is an important natural water bank. Indeed, some regions depend to a large extent on this water reservoir, and snow melt constitutes a significant portion of their annual runoff (Kongoli et al. 2012; Bales et al. 2006). Monitoring snow pack is essential for drought assessment in such regions, as decreases in this reservoir may potentially result in a summer hydrological drought (e.g., diminished stream flows and/or groundwater levels) or agricultural drought (e.g., exhausted soil moisture reserves). In this sense, Snow Water Equivalent (SWE), Snow Covered Area (SCA), Snow Depth (SD), and Snow Albedo (SA) (Kongoli et al. 2012; Painter et al. 2013; Molotch and Bales 2006) are among the most important hydrological variables. These snow-based indices are usually derived from the remotely sensed estimation of snow pack including (a) optical; (b) microwave (MW); and (c) composite optical and MW methods. Information provided by the optical approach only yield the SCA index, whereas the other two approaches can be used to estimate SCA, SD, and SWE.

The characteristic of snow to exhibit a strong spectral gradient in reflectance constructs the basis for optical monitoring of snow (Dozier et al. 2009; Kongoli et al. 2012; Wiscombe and Warren 1980). Indeed, the ratio of visible reflectance (R_{VIS}) and middle infrared (IR) reflectance (R_{mIR}) helps monitor and record snow pack (Romanov et al. 2000). Snow can, alternatively, be detected through the Normalized Difference Snow Index (NDSI), formulated as $\frac{R_{VIS}-R_{mIR}}{R_{VIS}+R_{mIR}}$ (Hall et al. 2002). A cohort of optical-based snow products from MODIS (Hall et al. 2002), and snow algorithms from Advanced Very High Resolution Radiometer (AVHRR) satellite data record (Simpson et al. 1998) are available with a variety of spatial and temporal resolutions. Accuracy of such optical-based snow estimates can be compromised by cloud blockage. If such blockage is persistent, it can also hinder temporal continuity of the snow data.

Microwave radiation, on the contrary, can penetrate through clouds and offer a temporally continuous estimation of snow pack (Kongoli et al. 2012; Schanda et al. 1983). The MW approach renders a more significant advantage by allowing the estimation of SWE and SD, since microwaves can penetrate through snow as well (Durand et al. 2008). A great deal of research has been devoted to develop algorithms that can approximate SCA from MW data sets (Grody and Basist 1996). Interested readers are referred to Kongoli et al. (2007) for a detailed discussion on the empirical regression relationship between variations in observed SWE/SD and the difference in brightness temperature of two low frequency channels; and to Kunzi et al. (1982) and Goodison (1989) for static, as well as Foster et al. (2005) and Kelly et al. (2003) for dynamic empirical regression models to estimate SWE/SD.

Temporal frequency of MW-based snow estimates are, however, lower than that of optical-based products, since MW sensors are solely available onboard polar orbiting satellites that bear longer revisit times compared to the geostationary satellites that carry optical sensors. To overcome the limitations of each sensor, snow retrieval algorithms based on joint optical and MW data sets have been proposed Liang et al. (2008) and Foster et al. (2011).

A number of studies have incorporated snow information into drought analysis (Wiesnet 1981; Kongoli et al. 2012; Painter et al. 2013; Guan et al. 2013; Molotch and Margulis 2008), mainly focusing on assimilating remotely sensed snow data into hydrological models to improve runoff simulation and hydrological drought prediction (Dong et al. 2007; Andreadis and Lettenmaier 2005; Margulis et al. 2006; He et al. 2012). Snow derived drought indicators are still in their early stages of development, mainly due to the time lag between snow occurrence and change in the availability of streamflow and soil moisture. This time lag, varying from a few weeks (e.g., low elevation snow and in lower latitudes) to a few months (e.g., high elevation snow and in higher latitudes), warrants a unique opportunity for an early drought warning system. Estimation of the lag time between snowfall, snowmelt, and runoff is very complex, even in a specific location as it depends on seasonal temperatures and the timing of snow accumulation (McCabe and Clark 2005). This poses a significant challenge in deriving snow-based drought indices.

11.8 Multi-Index Multivariate Drought Monitoring

Numerous studies have argued that a more robust and integrated measure of drought should be adopted based on multiple variables/indicators (Hao and AghaKouchak 2013; Keyantash and Dracup 2004; Kao and Govindaraju 2010; Hao and Singh 2015), to capture a diverse range of vegetation response to drought across different biomes. Having a wide range of satellite observations, such as precipitation, soil moisture, relative humidity, temperature, total water storage anomalies, and vegetation health indices, offers the opportunity to investigate droughts from different viewpoints (Fig. 11.5).

A number of premier studies in this field promoted the Vegetation Drought Response Index (VegDRI) (Tadesse et al. 2005; Brown et al. 2008) to quantify vegetation drought stress using climate-based drought indices, satellite-based observations of vegetation conditions, and other biophysical information. VegDRI builds upon NDVI (Rouse et al. 1974), and incorporates climate-based data from SPI and the Palmer Drought Severity Index (PDSI) (Palmer 1968) to improve identification of the root causes of vegetation stress. NDVI-based information regarding vegetation health is distorted from a drought point of view by other factors, such as fire, land cover change, plant disease, pest infestation, biomass harvesting, and flooding.

In an effort to improve drought monitoring, Anderson et al. (2012b) successfully developed a composite soil moisture product using triple collocation analysis

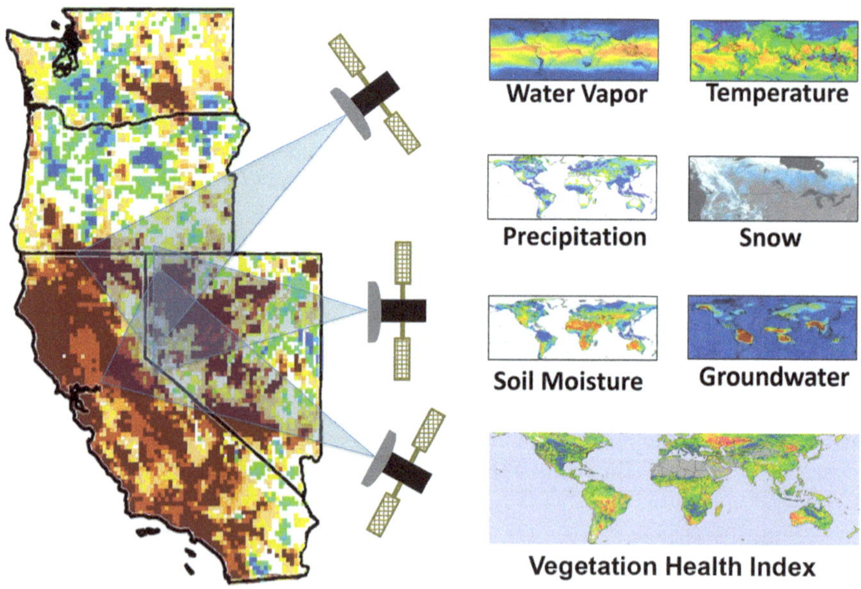

Fig. 11.5 Opportunities for multi-index multivariate drought monitoring using satellite observations (modified after AghaKouchak et al. 2015a)

(TCA) of different soil moisture products (i.e., microwave AMSR-E, thermal remote sensing using ALEXI, and physically based model simulations). The product shows promising results for monitoring of the 2010–2011 Horn of Africa drought.

The Microwave Integrated Drought Index (MIDI) by Zhang and Jia (2013), as an alternative composite model, incorporates in its formulation satellite-based precipitation data from the Tropical Rainfall Measuring Mission (TRMM), and soil moisture and land-surface temperature data from the Advanced Microwave Scanning Radiometer for EOS (AMSR-E) as $MIDI = \alpha PCI + \beta SMCI + (1 - \alpha - \beta)TCI$. In the recent equation, PCI signifies the precipitation condition index, SMCI denotes the soil moisture condition index, and TCI represents the temperature condition index. MIDI is developed for monitoring short-term droughts, specifically meteorological drought in semi-arid regions.

Following a similar scheme, the Scaled Drought Condition Index (SDCI) (Rhee et al. 2010) merges scaled TRMM-based precipitation data with LST and NDVI information to monitor agricultural drought. This index is defined as $SDCI = \alpha LST + \beta TRMM + \gamma NDVI$. It is worth mentioning that multiple indicators for drought monitoring should be carefully selected so as not to be fully correlated. In other words, multiple indices should render independent pieces of information, and redundancy should be avoided.

The composite approaches for drought monitoring discussed thus far solely depend on satellite obtained information in their analysis. Other approaches exist that incorporate information from physically based models along with remotely sensed data for drought analysis. An example of such an approach is the United States Drought Monitor (USDM) (Svoboda et al. 2002), which combines indicators from satellite observations (e.g., VegDRI, VHI, ESI, and GRACE TWS), in situ measurements, and expert opinion to deliver weekly drought monitoring information.

Finally, the Global Integrated Drought Monitoring and Prediction System (GIDMaPS) (Hao et al. 2014) utilizes the Multivariate Standardized Drought Index (MSDI) (Hao and AghaKouchak 2014) to deliver composite drought information. MSDI derives a composite drought index from precipitation and soil moisture, and provides a measure of agro-meteorological drought based on the joint state of precipitation and soil moisture (AghaKouchak 2015; Hao and AghaKouchak 2013). MSDI is a standardized indicator, hence comparable to other standardized indices, such as SPI and SSI, that can utilize satellite-based information as well as model simulations to compose a drought map. As an example, Fig. 11.6 displays SPI (top), SSI (middle), and MSDI (bottom) for Summer 1999–2001 based on NASA's Modern-Era Retrospective Analysis for Research and Applications (MERRA-Land) precipitation and soil moisture simulations.

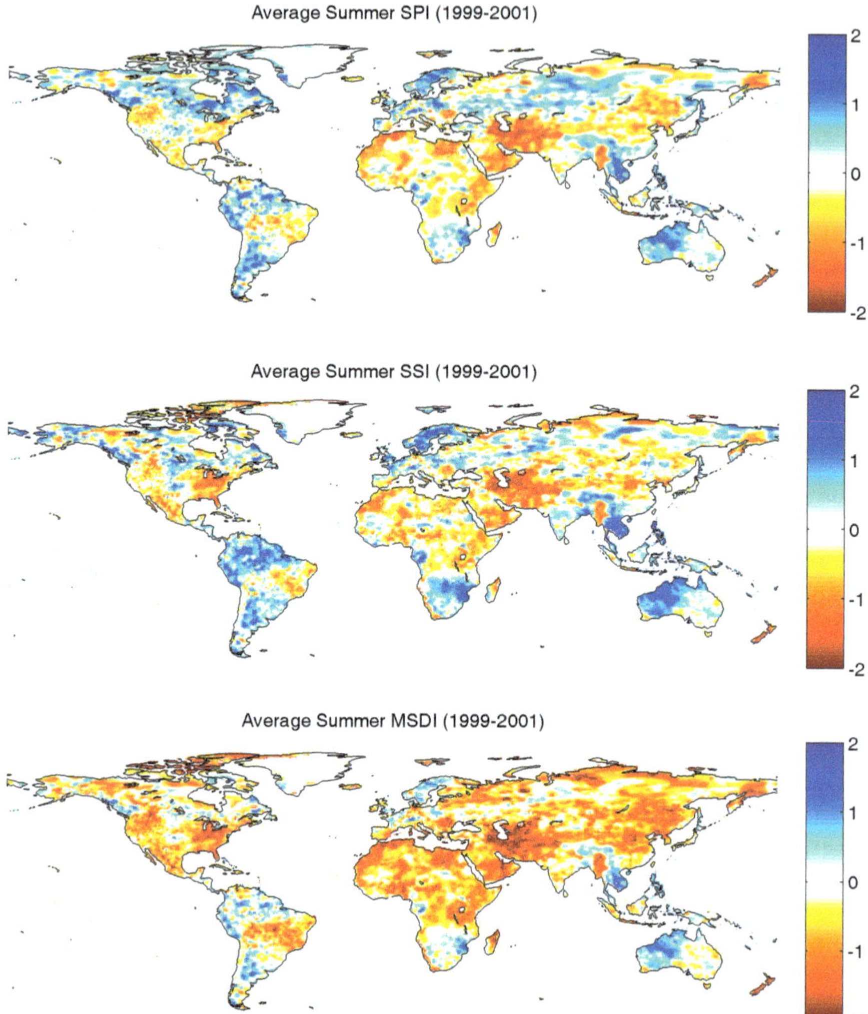

Fig. 11.6 (*top*) Standardized precipitation index (SPI), (*middle*) standardized soil moisture index (SSI), and (*bottom*) multivariate standardized drought index (MSDI) for summer 1999–2001. Input data are from modern-era retrospective analysis for research and applications (MERRA-Land) precipitation and soil moisture simulations

11.9 Ecosystem and Vegetation Health

There are numerous remotely sensed vegetation and ecosystem health models and indicators that can be combined with other satellite observations (Tucker and Choudhury 1987; Silleos et al. 2006; Nemani et al. 2009; Boken et al. 2005). The Normalized Difference Vegetation Index (NDVI) (Rouse et al. 1974; Karnieli et al. 2010; Tucker 1979; Funk and Budde 2009), for example, is a widely used vegetation greenness and health indicator that sheds light on the ecosystem response to water stress. Previous studies have shown that there exists a strong association between climatic variables (e.g., precipitation, soil moisture) and NDVI (Di et al. 1994; Adegoke and Carleton 2002; Richard and Poccard 1998; Hielkema et al. 1986; Farrar et al. 1994; Wang et al. 2001).

Based on the notion of NDVI, a series of indicators have been developed for different applications which employ visible satellite data, surface brightness temperature, and short wave infrared observations. These indicators include the following: Vegetation Condition Index (VCI) (Kogan and Sullivan 1993), Perpendicular Vegetation Index (PVI) (Wiegand et al. 1991), Transformed Vegetation Index (TVI) (Deering and Rouse 1975; Tucker 1979), Corrected Transformed Vegetation Index (CTVI) (Perry and Lautenschlager 1984), Standardized Vegetation Index (SVI) (Peters et al. 2002; Park et al. 2008), Soil-Adjusted Vegetation Index (SAVI) (Huete 1988), Normalized Ratio Vegetation Index (NRVI) (Baret and Guyot 1991), Distance Drought Index (DDI) (Qin et al. 2010), Perpendicular Drought Index (PDI) (Ghulam et al. 2007a), Modified Perpendicular Drought Index (MPDI) (Ghulam et al. 2007b), Normalized Difference Infrared Index (NDIIb6) (Hunt Jr and Rock 1989), Enhanced Vegetation Index (EVI) (Huete et al. 2002), Temperature Condition Index (TCI) (Kogan 1995), Normalized Difference Temperature Index (NDTI) (McVicar and Jupp 1999, 2002), Vegetation Temperature Condition Index (VTCI) (Wan et al. 2004), Normalized Difference Water Index (NDWI) (Gao 1996; Gu et al. 2008, 2007), Enhanced Thematic Mapper (ETM) (Jackson et al. 2004; Chen et al. 2005; Wang and Qu 2009), Normalized Difference Drought Index (NDDI) (Gu et al. 2007), Normalized Multi-band Drought Index (NMDI) (Wang and Qu 2007), and Shortwave infrared Drought Index (VSDI) (Zhang et al. 2013)—see AghaKouchak et al. (2015a), Payero et al. (2004), and Silleos et al. (2006) for a comprehensive list of indicators.

Because of the strong relationship between climatic conditions and vegetation response, NDVI and its derivatives have been widely used for large-scale drought monitoring (Tucker and Choudhury 1987; Roderick et al. 1999; Prince et al. 1998; Donohue et al. 2009; Nicholson et al. 1998; Ji and Peters 2003; Lu et al. 2003; McVicar and Jupp 1998). An exciting research direction would be the advancement of drought monitoring through effectively merging climate-based indicators with vegetation health indices. Understanding the lags between climate anomalies and the impact on vegetation can lead to significant improvement of drought monitoring and prediction.

11.10 Discussion

Satellite sensors and their associated products have become increasingly abundant since the 2000s, the number of which will only increase in the future. The most relevant available, or upcoming, missions for drought monitoring include the

Global Precipitation Mission (GPM), Geostationary Operational Environmental Satellites R series (GOES-R), GRACE Follow-On, SMAP, and SWOT missions. These satellites establish a comprehensive ground to study, monitor, and assess drought from different perspectives using multivariate analyses. However, the challenges of data continuity, unquantified uncertainty, sensor changes, community acceptability, and data maintenance remain for the community to resolve.

The continuity of data is essential to the development of reliable drought monitoring and assessment tools. The relatively short life span of satellites, mostly less than a decade of operation sometimes with a few years of extension, poses a challenge on the extent of reliability of the provided data sets for drought studies. Ideally, follow-up missions should extend the length of available data sets, but the planning, approval, and design of new satellites may take decades, not to mention the extensive monetary investment required. Examples of such follow-up missions include GPM, GOES-R, and GRACE Follow-On that are aimed to elude discontinuity of the current satellite-based precipitation and total water storage records. The Visible Infrared Imager Radiometer Suite (VIIRS) (Justice et al. 2013; Welsch et al. 2001; Vargas et al. 2013) is also a successor to MODIS and AVHRR. Acknowledging all these satellites, the availability of long-term multidecadal data for robust drought studies from a climatological perspective remains uncertain.

Managing the data volume and making them accessible to the public is another major challenge that requires continual investment in hardware infrastructure to store and serve the data, as well as training professionals to process the data and maintain the infrastructure.

Recent literature has proven that indicators based on multiple data sets enhance drought detection (Hao et al. 2014) and monitoring ability (Mu et al. 2013). Indeed, availability of multiple satellite data sets has empowered the development of numerous multi-index multi-sensor frameworks for drought analysis. This has refined our ability to describe the onset, development, and termination of drought (Keyantash and Dracup 2004; Kao and Govindaraju 2010; Tadesse et al. 2005; Hao and AghaKouchak 2013; Svoboda et al. 2002; Rajsekhar et al. 2014).

Snow is one of the least explored hydrological variables in drought monitoring studies. Incorporating snow data into seasonal precipitation and runoff forecasts can lead to major improvements in drought monitoring for regions that depend on snow melt, such as the western United States. Such data may be acquired from the Snow and Cold Land Processes (SCLP) mission (Rott et al. 2010). Snow information can be used in multivariate composite drought monitoring frameworks to address multiple aspects of drought simultaneously.

Limited and sparse ground observations of hydrological variables, such as total water storage, water vapor, soil moisture, snow, and streamflow, restrict the development of multi-index drought monitoring indicators. However, remotely sensed hydrological variables enable development of composite drought assessment frameworks, which have only recently been explored. Much research is still required to develop methodologies that deliver robust multivariate drought information.

What limits the application of satellite-based data sets for climatological drought assessment is the relatively short period of observation, with the exception of Landsat, GOES, and AVHRR-MODIS-VIIRS missions. Some missions, such as GRACE, provide only sightly more than a decade of data. Other satellite sensors are only research instruments, and their replacement is highly uncertain. Persistent long-term investments in satellites designed for Earth observation are essential to guarantee data continuity.

The absence of secure support and investment in this area is a major challenge for developing a consistent, long-term remotely sensed data record that is necessary to accurately detect anomalies against a historical record. It is worth mentioning that short-term records provide valuable drought assessment and monitoring information (Rodell 2012; Famiglietti and Rodell 2013), but are not enough for studies focused on a climatological perspective of drought. To diminish this limitation, some studies strive to construct longer inter-sensor records by combining data from multiple satellite missions, such as AVHRR and MODIS (Tucker et al. 2005). More effective models and frameworks are yet to be developed to merge different data sets and/or model simulations, and generate long-term data sets of hydrological and climatic variables (AghaKouchak and Nakhjiri 2012; Houborg et al. 2012; Andreadis and Lettenmaier 2005; Zaitchik et al. 2008)

Moreover, underlying uncertainties of satellite-based data sets due to retrieval and sensor errors are not yet fully quantified (Dorigo et al. 2010; Mehran and AghaKouchak 2014; Pinker et al. 2009). Several studies have tried to address this issue through the development of models and indicators for uncertainty quantification of remotely sensed data (Gebremichael 2010; AghaKouchak and Mehran 2013; Entekhabi et al. 2010b; Hossain and Huffman 2008); however, most satellite-based data sets still lack uncertainty estimates. Furthermore, land-surface and hydrological models that use such data as input and/or boundary conditions are also subject to model structural and parameter uncertainties (Li et al. 2012; Dong et al. 2007; Houborg et al. 2012). Model simulations, therefore, bear an amalgam of these uncertainties which is almost impossible to trace. The unavailability of uncertainty estimates for satellite-based data sets might avert usage of such data in decision-making and operational applications.

Remote sensing products should be carefully scrutinized, and their strengths and limitations should be communicated to decision-makers in a non-technical language. Moreover, to enhance the usage and improve the acceptability of current and future satellite data sets, further research should target the assessment and quantification of their underlying uncertainties.

Finally, it is critical to develop new products and tailor remotely sensed data into output formats amenable to drought monitoring systems. It is specifically important to communicate valuable information from emerging satellites to the drought community. This aspect of the data-to-information process is overlooked and requires more attention. It is critical to engage drought experts and decision-makers in this process to develop applicable tools and information from remote sensing data for operational drought monitoring and early warning systems.

Acknowledgements This study is supported by the National Aeronautics and Space Administration (NASA) award NNX15AC27G.

References

Adegoke JO, Carleton AM (2002) Relations between soil moisture and satellite vegetation indices in the us corn belt. J hydrometeorol 3(4):395–405

Adler R, Huffman G, Chang A, Ferraro R, Xie P, Janowiak J, Rudolf B, Schneider U, Curtis S, Bolvin D, Gruber A, Susskind J, Arkin P (2003) The version-2 global precipitation climatology project (GPCP) monthly precipitation analysis (1979–present). J hydrometeorol 4(6): 1147–1167

AghaKouchak A (2014) A baseline probabilistic drought forecasting framework using standardized soil moisture index: application to the 2012 united states drought. Hydrol Earth Syst Sci 18:2485–2492

AghaKouchak A (2015, A multivariate approach for persistence-based drought prediction: application to the 2010–2011 east Africa drought. J Hydrol 526:127–135. doi:10.1016/j.jhydrol.2014.09.063

AghaKouchak A, Mehran A (2013) Extended contingency table: performance metrics for satellite observations and climate model simulations. Water Resour Res 49:7144–7149. doi:10.1002/wrcr.20498

AghaKouchak A, Nakhjiri N (2012) A near real-time satellite-based global drought climate data record. Environ Res Lett 7(4):044,037. doi:10.1088/1748-9326/7/4/044037

AghaKouchak A, Behrangi A, Sorooshian S, Hsu K, Amitai E (2011) Evaluation of satellite-retrieved extreme precipitation rates across the central United States. J Geophys Res-Atmos 116:D02,115

AghaKouchak A, Mehran A, Norouzi H, Behrangi A (2012) Systematic and random error components in satellite precipitation data sets. Geophys Res Lett 39(9):L09,406

Aghakouchak A, Cheng L, Mazdiyasni O, Farahmand A (2014) Global warming and changes in risk of concurrent climate extremes: insights from the 2014 California drought. Geophys Res Lett 41. doi: 10.1002/2014GL062308

AghaKouchak A, Farahmand A, Melton F, Teixeira J, Anderson M, Wardlow B, Hain C (2015a) Remote sensing of drought: progress, challenges and opportunities. Rev Geophys 53 (2):452–480. doi:10.1002/2014RG000456

AghaKouchak A, Feldman D, Hoerling M, Huxman T, Lund J (2015b) Recognize anthropogenic drought. Nature 524(7566):409–4011. doi:10.1038/524409a

Allen RG, Tasumi M, Trezza R (2007) Satellite-based energy balance for mapping evapotranspiration with internalized calibration (metric)—model. J Irrig Drain Eng 133(4):380–394.

Ambaw GM (2013) Satellite based remote sensing of soil moisture for drought detection and monitoring in the horn of Africa, Ph.D. Thesis, Politecnico di Torino

Anagnostou EN, Maggioni V, Nikolopoulos EI, Meskele T, Hossain F, Papadopoulos A (2010) Benchmarking high-resolution global satellite rainfall products to radar and rain-gauge rainfall estimates. IEEE Trans Geosci Remote Sens 48(4):1667–1683

Anderson M, Kustas W (2008) Thermal remote sensing of drought and evapotranspiration. Eos Trans Am Geophys Union 89(26):233

Anderson L, Malhi Y, Aragao L, Saatchi S (2008) Spatial patterns of the canopy stress during 2005 drought in Amazonia. In: 2007 I.E. international geoscience and remote sensing symposium, IGARSS 2007, pp 2294–2297

Anderson M, Kustas W, Norman J, Hain C, Mecikalski J, Schultz L, Gonzalez-Dugo M, Cammalleri C, d'Urso G, Pimstein A, et al. (2011a) Mapping daily evapotranspiration at field to continental scales using geostationary and polar orbiting satellite imagery. Hydrol Earth Syst Sci 15(1):223–239

Anderson MC, Hain C, Wardlow B, Pimstein A, Mecikalski JR, Kustas WP (2011b) Evaluation of drought indices based on thermal remote sensing of evapotranspiration over the continental united states. J Climate 24(8):2025–2044

Anderson M, Hain C, Wardlow B, Pimstein A, Mecikalski J, Kustas W (2012a) A thermal-based evaporative stress index for monitoring surface moisture depletion. In: Remote sensing of drought: innovative monitoring approaches. CRC, pp 145–167

Anderson W, Zaitchik B, Hain C, Anderson M, Yilmaz M, Mecikalski J, Schultz L (2012b) Towards an integrated soil moisture drought monitor for east Africa. Hydrol Earth Syst Sci 16 (8):2893–2913

Anderson MC, Cammalleri C, Hain CR, Otkin J, Zhan X, Kustas W (2013a) Using a diagnostic soil-plant-atmosphere model for monitoring drought at field to continental scales. Procedia Environ Sci 19:47–56

Anderson MC, Hain C, Otkin J, Zhan X, Mo K, Svoboda M, Wardlow B, Pimstein A (2013b) An intercomparison of drought indicators based on thermal remote sensing and NLDAS-2 simulations with us drought monitor classifications. J Hydrometeorol 14(4):1035–1056

Andreadis KM, Lettenmaier D (2005) Assimilating passive microwave brightness temperature for snow water equivalent estimation. In: AMS annual meeting, 19th conference on hydrology

Arkin PA, Joyce R, Janowiak JE (1994) The estimation of global monthly mean rainfall using infrared satellite data: the goes precipitation index (GPI). Remote Sens Rev 11(1–4):107–124

Ashouri H, Hsu K-L, Sorooshian S, Braithwaite DK, Knapp KR, Cecil LD, Nelson BR, Prat OP (2015) Persiann-CDR: daily precipitation climate data record from multisatellite observations for hydrological and climate studies. Bull Am Meteorol Soc 96(1):69–83

Asner GP, Alencar A (2010) Drought impacts on the Amazon forest: the remote sensing perspective. New Phytol 187(3):569–578

Asrar G, Fuchs M, Kanemasu E, Hatfield J (1984) Estimating absorbed photosynthetic radiation and leaf area index from spectral reflectance in wheat. Agron J 76(2):300–306

Asrar G, Myneni RB, Kanemasu ET (1989) Estimation of plant-canopy attributes from spectral reflectance measurements

Bales RC, Molotch NP, Painter TH, Dettinger MD, Rice R, Dozier J (2006) Mountain hydrology of the western United States. Water Resour Res 42(8)

Baret F, Guyot G (1991) Potentials and limits of vegetation indices for LAI and APAR assessment. Remote Sens Environ 35(2):161–17

Barrett E, Herschy R (1989) Opportunities for satellite remote sensing in hydrology and water management. Geocarto Int 4(2):11–18

Boken VK, Cracknell AP, Heathcote RL, et al. (2005) Monitoring and predicting agricultural drought: a global study. Oxford University Press, Cary

Bolten JD, Crow WT, Zhan X, Jackson TJ, Reynolds CA (2010) Evaluating the utility of remotely sensed soil moisture retrievals for operational agricultural drought monitoring. IEEE J Sel Top Appl Earth Observations Remote Sens 3(1):57–66

Brown JF, Wardlow BD, Tadesse T, Hayes MJ, Reed BC (2008) The vegetation drought response index (VegDRI): a new integrated approach for monitoring drought stress in vegetation. Geosci Remote Sens. 45(1):16–46

Cashion J, Lakshmi V, Bosch D, Jackson TJ (2005) Microwave remote sensing of soil moisture: evaluation of the TRMM microwave imager (TMI) satellite for the little river watershed Tifton, Georgia. J Hydrol 307(1):242–253

Chappell A, Renzullo LJ, Raupach TH, Haylock M (2013) Evaluating geostatistical methods of blending satellite and gauge data to estimate near real-time daily rainfall for Australia. J Hydrol 493:105–114

Chen D, Huang J, Jackson TJ (2005) Vegetation water content estimation for corn and soybeans using spectral indices derived from modis near-and short-wave infrared bands. Remote Sens Environ. 98(2):225–236

Chen J, Wilson C, Tapley B, Yang Z, Niu G (2009) 2005 drought event in the Amazon river basin as measured by grace and estimated by climate models. J Geophys Res 114(B5):B05,404
Choi M, Jacobs J, Anderson M, Bosch D (2013) Evaluation of drought indices via remotely sensed data with hydrological variables. J Hydrol 476(7):265–273
Dalezios N, Blanta A, Spyropoulos N (2012) Assessment of remotely sensed drought features in vulnerable agriculture. Nat Hazards Earth Syst Sci 12(10):3139–3150
Damberg L, AghaKouchak A (2014) Global trends and patterns of drought from space. Theor Appl Climatol 117(3):441–448. doi:10.1007/s00704-013-1019-5
Deering D, Rouse J (1975) Measuring'forage production' of grazing units from landsat mss data. In: 10 th international symposium on remote sensing of environment, Ann Arbor, Mich, pp 1169–1178
Di L, Rundquist DC, Han L (1994) Modelling relationships between NDVI and precipitation during vegetative growth cycles. Int J Remote Sens 15(10):2121–2136
D'Odorico P, Caylor K, Okin GS, Scanlon TM (2007) On soil moisture–vegetation feedbacks and their possible effects on the dynamics of dryland ecosystems. J Geophys Res Biogeosci 112 (G4):2005–2012
Dong J, Walker JP, Houser PR, Sun C (2007) Scanning multichannel microwave radiometer snow water equivalent assimilation. J Geophys Res Atmos 112(D7) (1984–2012)
Donohue RJ, McVICAR T, Roderick ML (2009) Climate-related trends in Australian vegetation cover as inferred from satellite observations, 1981–2006. Glob Chang Biol 15(4):1025–1039
Donohue RJ, McVicar TR, Roderick ML (2010) Assessing the ability of potential evaporation formulations to capture the dynamics in evaporative demand within a changing climate. J Hydrol 386(1):186–197
Dorigo W, Scipal K, Parinussa R, Liu Y, Wagner W, De Jeu R, Naeimi V (2010) Error characterisation of global active and passive microwave soil moisture datasets. Hydrol Earth Syst Sci 14(12):2605–2616
Dozier J, Green RO, Nolin AW, Painter TH (2009) Interpretation of snow properties from imaging spectrometry. Remote Sens Environ 113:S25–S37
Durand M, Kim EJ, Margulis SA (2008) Quantifying uncertainty in modeling snow microwave radiance for a mountain snowpack at the point-scale, including stratigraphic effects. IEEE Trans Geosci Remote Sens 46(6):1753–1767
Easterling D (2013) Global data sets for analysis of climate extremes. In: Extremes in a changing climate. Springer, Berlin. doi:10.1007/978-94-007-4479-0 1
Ebert E, Janowiak J, Kidd C (2007) Comparison of near real time precipitation estimates from satellite observations and numerical models. Bull Am Meteorol Soci 88:47–64
Entekhabi D, Njoku EG, Houser P, Spencer M, Doiron T, Kim Y, Smith J, Girard R, Belair S, Crow W, et al. (2004) The hydrosphere state (hydros) satellite mission: an earth system pathfinder for global mapping of soil moisture and land freeze/thaw, IEEE Trans Geosci Remote Sens 42(10):2184–2195
Entekhabi D, Njoku EG, O'Neill PE, Kellogg KH, Crow WT, Edelstein WN, Entin JK, Goodman SD, Jackson TJ, Johnson J, et al. (2010a) The soil moisture active passive (SMAP) mission. Proc IEEE 98(5):704–716
Entekhabi D, Reichle RH, Koster RD, Crow WT (2010b) Performance metrics for soil moisture retrievals and application requirements. J Hydrometeorol 11(3):832–840
Famiglietti JS, Rodell M (2013) Water in the balance. Science 340(6138):1300–1301
Farahmand A, AghaKouchak A (2015) A generalized framework for deriving nonparametric standardized drought indicators. Adv Water Resour 76:140–145
Farahmand A, Teixeira J, AghaKouchak A (2015) A vantage from space can detect earlier drought onset: an approach using relative humidity. Sci Rep 5:8553. doi:10.1038/srep07093
Farrar T, Nicholson S, Lare A (1994) The influence of soil type on the relationships between NDVI, rainfall, and soil moisture in semiarid Botswana. ii. NDVI response to soil moisture. Remote Sens Environ 50(2):121–133

Fensholt R, Sandholt I, Stisen S, Tucker C (2006) Analysing NDVI for the African continent using the geostationary meteosat second generation SEVIRI sensor. Remote Sens Environ 101 (2):212–229

Foster JL, Sun C, Walker JP, Kelly R, Chang A, Dong J, Powell H (2005) Quantifying the uncertainty in passive microwave snow water equivalent observations. Remote Sens Environ 94(2):187–203

Foster JL, Hall DK, Eylander JB, Riggs GA, Nghiem SV, Tedesco M, Kim E, Montesano PM, Kelly RE, Casey KA, et al. (2011) A blended global snow product using visible, passive microwave and scatterometer satellite data. Int J Remote Sens 32(5):1371–1395

Funk C, Budde ME (2009) Phenologically-tuned modis NDVI-based production anomaly estimates for Zimbabwe. Remote Sens Environ 113(1):115–125

Gallagher J, Biscoe P, Hunter B (1976) Effects of drought on grain growth. Nature 264:541–542

Gao B-C (1996) NDWI—a normalized difference water index for remote sensing of vegetation liquid water from space. Remote Sens Environ 58(3):257–266

Gebremichael M (2010) Framework for satellite rainfall product evaluation. Geophysical monograph series, vol 191. Wiley Online Library, pp 265–275

Ghulam A, Qin Q, Zhan Z (2007a) Designing of the perpendicular drought index. Environ Geol 52 (6):1045–1052

Ghulam A, Qin Q, Teyip T, Li Z-L (2007b) Modified perpendicular drought index (MPDI): a real-time drought monitoring method. ISPRS J Photogramm Remote Sens 62(2):150–164

Godfray HCJ, Beddington JR, Crute IR, Haddad L, Lawrence D, Muir JF, Pretty J, Robinson S, Thomas SM, Toulmin C (2010) Food security: the challenge of feeding 9 billion people. Science 327(5967):812–818

Golian S, Mazdiyasni O, AghaKouchak A (2015) Trends in meteorological and agricultural droughts in Iran. Theor Appl Climatol 119:679–688. doi:10.1007/s00704-014-1139-6

Goodison BE (1989) Determination of areal snow water equivalent on the Canadian prairies using passive microwave satellite data. In: 1989 international geoscience and remote sensing symposium, 1989. IGARSS'89. 12th Canadian symposium on remote sensing, vol 3. IEEE, pp 1243–1246

Grasso VF, Singh A (2011) Early warning systems: state-of-art analysis and future directions, United Nations Environment Programme (UNEP)

Grody NC, Basist AN (1996) Global identification of snowcover using ssm/i measurements. IEEE Trans Geosci Remote Sens 34(1):237–249

Gruhier C, De Rosnay P, Hasenauer S, Holmes TR, De Jeu RA, Kerr YH, Mougin E, Njoku E, Timouk F, Wagner W, et al. (2010) Soil moisture active and passive microwave products: intercomparison and evaluation over a Sahelian site, HAL 00463919

Gu Y, Brown JF, Verdin JP, Wardlow B (2007) A five-year analysis of modis NDVI and NDWI for grassland drought assessment over the central great plains of the united states. Geophys Res Lett 34(6)

Gu Y, Hunt E, Wardlow B, Basara JB, Brown JF, Verdin JP (2008) Evaluation of modis NDVI and NDWI for vegetation drought monitoring using Oklahoma Mesonet soil moisture data. Geophys Rese Lett 35(22)

Guan B, Molotch NP, Waliser DE, Jepsen SM, Painter TH, Dozier J (2013) Snow water equivalent in the Sierra Nevada: blending snow sensor observations with snowmelt model simulations. Water Resour Res 49(8):5029–5046

Hall DK, Riggs GA, Salomonson VV, DiGirolamo NE, Bayr KJ (2002) Modis snow-cover products. Remote Sens Environ 83(1):181–194

Hao Z, AghaKouchak A (2013) Multivariate standardized drought index: a parametric multi-index model. Adv Water Res 57:12–18

Hao Z, AghaKouchak A (2014) A nonparametric multivariate multi-index drought monitoring framework. J Hydrometeorol 15:89–101

Hao Z, Singh VP (2015) Drought characterization from a multivariate perspective: a review. J Hydrol 527:668–678

Hao Z, AghaKouchak A, Nakhjiri N, Farahmand A (2014) Global integrated drought monitoring and prediction system. Sci Data 1:140,001. doi:10.1038/sdata.2014.1
Hatfield J, Asrar G, Kanemasu E (1984) Intercepted photosynthetically active radiation estimated by spectral reflectance. Remote Sens Environ 14(1):65–75
Hayes M, Svoboda M, Wilhite D, Vanyarkho O (1999) Monitoring the 1996 drought using the standardized precipitation index. Bull. Am. Meteor. Soc. 80:429–438
He M, Hogue T, Margulis S, Franz K (2012) An integrated uncertainty and ensemble-based data assimilation approach for improved operational streamflow predictions. Hydrol Earth Syst Sci 16(3):815–831
Heumann BW (2011) Satellite remote sensing of mangrove forests: recent advances and future opportunities. Progress Phys Geogr 35(1):87–108
Hidalgo HG (2004) Climate precursors of multidecadal drought variability in the western united states. Water Resour Res 40(12):W12,504
Hielkema J, Prince S, Astle W (1986) Rainfall and vegetation monitoring in the savanna zone of the Democratic Republic of Sudan using the NOAA advanced very high resolution radiometer. Int J Remote Sens 7(11):1499–1513
Hirpa FA, Gebremichael M, Hopson T (2010) Evaluation of high-resolution satellite precipitation products over very complex terrain in Ethiopia. J Appl Meteorol Climatol 49(5):1044–1051
Hobbins MT, Dai A, Roderick ML, Farquhar GD (2008) Revisiting the parameterization of potential evaporation as a driver of long-term water balance trends. Geophys Res Lett 35(12)
Hoerling M, Kumar A (2003) The perfect ocean for drought. Science 299(5607):691–694
Hoerling M., et al. (2013) An interpretation of the origins of the 2012 central great plains drought. Technical Report, Assessment Report, National Oceanic and Atmospheric Administration, Drought Task Force
Hong Y, Hsu K, Gao X, Sorooshian S (2004) Precipitation estimation from remotely sensed imagery using artificial neural network-cloud classification system. J Appl Meteorol Climatol 43:1834–1853
Hossain F, Anagnostou E (2005) Numerical investigation of the impact of uncertainties in satellite rainfall estimation and land surface model parameters on simulation of soil moisture. Adv Water Resour 28(12):1336–1350
Hossain F, Huffman G (2008) Investigating error metrics for satellite rainfall data at hydrologically relevant scales. J Hydrometeorol 9(3):563–575
Houborg R, Rodell M, Li B, Reichle R, Zaitchik BF (2012) Drought indicators based on model-assimilated gravity recovery and climate experiment (grace) terrestrial water storage observations. Water Resour Res 48(7)
Hsu K, Gao X, Sorooshian S, Gupta H (1997) Precipitation estimation from remotely sensed information using artificial neural networks. J Appl Meteorol 36:1176–1190
Huete AR (1988) A soil-adjusted vegetation index (SAVI). Remote Sens Environ 25(3):295–309
Huete A, Didan K, Miura T, Rodriguez EP, Gao X, Ferreira LG, (2002) Overview of the radiometric and biophysical performance of the modis vegetation indices. Remote Sens Environ 83(1):195–213
Huffman G, Adler R, Bolvin D, Gu G, Nelkin E, Bowman K, Stocker E, Wolff D (2007) The TRMM multi-satellite precipitation analysis: quasi-global, multiyear, combined-sensor precipitation estimates at fine scale. J hydrometeorol 8:38–55
Huffman GJ, Bolvin DT, Braithwaite D, Hsu K, Joyce R, Xie P, Yoo S-H (2013) NASA global precipitation measurement (GPM) integrated multi-satellite retrievals for GPM (IMERG). In: Algorithm Theoretical Basis Document, Version 4.1, NASA
Hunt Jr, ER, Rock BN (1989) Detection of changes in leaf water content using near-and middle-infrared reflectances. Remote Sens Environ 30(1):43–54
Idso S, Jackson R, Pinter P, Reginato R, Hatfield J (1981) Normalizing the stress-degree-day parameter for environmental variability. Agric Meteorol 24(1):45–55
Jackson TJ (1993) III. measuring surface soil moisture using passive microwave remote sensing. Hydrol Process 7(2):139–152

Jackson TJ (1997) Soil moisture estimation using special satellite microwave/imager satellite data over a grassland region. Water Resour Res 33(6):1475–1484
Jackson RD, Idso S, Reginato R, Pinter Jr P (1981) Canopy temperature as a crop water stress indicator. Water Resour Res 17(4):1133–1138
Jackson TJ, Chen D, Cosh M, Li F, Anderson M, Walthall C, Doriaswamy P, Hunt E (2004) Vegetation water content mapping using landsat data derived normalized difference water index for corn and soybeans. Remote Sens Environ 92(4):475–482
Ji L, Peters AJ (2003) Assessing vegetation response to drought in the northern great plains using vegetation and drought indices. Remote Sens Environ 87(1):85–98
Joyce R, Arkin PA (1997) Improved estimates of tropical and subtropical precipitation using the goes precipitation index. J Atmos Ocean Technol 14(5):997–1011
Joyce R, Janowiak J, Arkin P, Xie P (2004) Cmorph: a method that produces global precipitation estimates from passive microwave and infrared data at high spatial and temporal resolution. J Hydrometeorol 5:487–503
Justice CO, Román MO, Csiszar I, Vermote EF, Wolfe RE, Hook SJ, Friedl M, Wang Z, Schaaf CB, Miura T, et al. (2013) Land and cryosphere products from Suomi NPP VIIRS: overview and status. J Geophys Res Atmos 118(17):9753–9765
Kao S, Govindaraju R (2010) A copula-based joint deficit index for droughts. J Hydrol 380 (1–2):121–134
Karnieli A, Agam N, Pinker RT, Anderson M, Imhoff ML, Gutman GG, Panov N, Goldberg A (2010) Use of NDVI and land surface temperature for drought assessment: merits and limitations. J Climate 23(3):618–633
Kelly RE, Chang AT, Tsang L, Foster JL (2003) A prototype AMSR-E global snow area and snow depth algorithm. IEEE Trans Geosci Remote Sens 41(2):230–242
Keyantash J, Dracup J (2004) An aggregate drought index: assessing drought severity based on fluctuations in the hydrologic cycle and surface water storage. Water Resour Res 40(9): W09,304. doi:10.1029/2003WR002610
Kidd C (2001) Satellite rainfall climatology: a review. Int J Climatol 21:1041–1066
Kim J, Hogue TS (2012) Improving spatial soil moisture representation through integration of AMSR-E and modis products. IEEE Trans Geosci Remote Sens 50(2):446–460
Kogan F (1995) Application of vegetation index and brightness temperature for drought detection. Adv Space Res 15(11):91–100
Kogan F, Sullivan J (1993) Development of global drought-watch system using NOAA/AVHRR data. Adv Space Res 13(5):219–222
Kongoli C, Dean CA, Helfrich SR, Ferraro RR (2007) Evaluating the potential of a blended passive microwave-interactive multi-sensor product for improved mapping of snow cover and estimations of snow water equivalent. Hydrol Process 21(12):1597–1607
Kongoli C, Romanov P, Ferraro R (2012) Snow cover monitoring from remote sensing satellites. In: Remote sensing of drought: innovative monitoring approaches. CRC, Boca Raton, pp 359–386
Koster RD, Suarez MJ, Ducharne A, Stieglitz M, Kumar P (2000) A catchment-based approach to modeling land surface processes in a general circulation model: 1. model structure. J Geophys Res Atmos (1984–2012) 105(D20):24,809–24,822
Krajewski WF, Anderson MC, Eichinger WE, Entekhabi D, Hornbuckle BK, Houser PR, Katul GG, Kustas WP, Norman JM, Peters-Lidard C, et al. (2006) A remote sensing observatory for hydrologic sciences: a genesis for scaling to continental hydrology. Water Resour Res 42(7)
Kubota T, Shige S, Hashizume H, Aonashi K, Takahashi N, Seto S, Takayabu YN, Ushio T, Nakagawa K, Iwanami K, et al. (2007) Global precipitation map using satellite-borne microwave radiometers by the GSMaP project: production and validation. IEEE Trans Geosci Remote Sens 45(7):2259–2275
Kummerow C, Olson WS, Giglio L (1996) A simplified scheme for obtaining precipitation and vertical hydrometeor profiles from passive microwave sensors. IEEE Trans Geosci Remote Sens 34(5):1213–1232

Kummerow C, Hong Y, Olson W, Yang S, Adler R, McCollum J, Ferraro R, Petty G, Shin D-B, Wilheit T (2001) The evolution of the Goddard profiling algorithm (GPROF) for rainfall estimation from passive microwave sensors. J Appl Meteorol 40(11):1801–1820

Kunzi KF, Patil S, Rott H (1982) Snow-cover parameters retrieved from nimbus-7 scanning multichannel microwave radiometer (SMMR) data. IEEE Trans Geosci Remote Sens GE-20 (4):452–467

Leblanc MJ, Tregoning P, Ramillien G, Tweed SO, Fakes A (2009) Basin-scale, integrated observations of the early 21st century multiyear drought in southeast Australia. Water Resour Res 45(4)

Levizzani V (2008) Satellite clouds and precipitation observations for meteorology and climate. In: Sorooshian S, Hsu K-L, Coppola E, Tomassetti B, Verdecchia M, Visconti G (eds) Hydrological modeling and the water cycle. Springer, Berlin, pp 49–68

Levizzani V, Bauer P, Turk F (2007) Measurement of Precipitation from space: EURAINSAT and future. Springer, Dordrecht

Li B, Rodell M, Zaitchik BF, Reichle RH, Koster RD, van Dam TM (2012) Assimilation of grace terrestrial water storage into a land surface model: evaluation and potential value for drought monitoring in western and central Europe. J Hydrol 446:103–115

Liang T, Zhang X, Xie H, Wu C, Feng Q, Huang X, Chen Q (2008) Toward improved daily snow cover mapping with advanced combination of modis and AMSR-E measurements. Remote Sens Environ 112(10):3750–3761

Liu Y, Parinussa R, Dorigo W, Jeu RD, Wagner W, Dijk AV, McCabe M, Evans J (2011) Developing an improved soil moisture dataset by blending passive and active microwave satellite-based retrievals. Hydrol Earth Syst Sci 15(2):425–436

Liu X, Zhu X, Pan Y, Li S, Liu Y, Ma Y (2016) Agricultural drought monitoring: progress, challenges, and prospects. J Geogr Sci 26(6):750–767

Long D, Scanlon BR, Longuevergne L, Sun A-Y, Fernando DN, Himanshu S (2013) Grace satellites monitor large depletion in water storage in response to the 2011 drought in Texas. Geophys Res Lett 40(13):3395–3401

Lorenz C, Tourian MJ, Devaraju B, Sneeuw N, Kunstmann H (2015) Basin-scale runoff prediction: An ensemble Kalman filter framework based on global hydrometeorological data sets. Water Resour Res 51(10):8450–8475

Lu H, Raupach MR, McVicar TR, Barrett DJ (2003) Decomposition of vegetation cover into woody and herbaceous components using AVHRR NDVI time series. Remote Sens Environ 86 (1):1–18

Maggioni V, Sapiano MR, Adler RF (2016) Estimating uncertainties in high-resolution satellite precipitation products: systematic or random error? J Hydrometeorol 17(4):1119–1129

Margulis SA, Wood EF, Troch PA (2006) The terrestrial water cycle: modeling and data assimilation across catchment scales. J Hydrometeorol 7(3):309–311

McCabe GJ, Clark MP (2005) Trends and variability in snowmelt runoff in the western united states. J Hydrometeorol 6(4):476–482

McKee T, Doesken N, Kleist J (1993) The relationship of drought frequency and duration to time scales. In: In proceedings of the 8th conference of applied climatology, 17–22 January 1993, Anaheim, CA. American Meteorological Society, pp 179–184

McVicar TR, Jupp DL (1998) The current and potential operational uses of remote sensing to aid decisions on drought exceptional circumstances in Australia: a review. Agric Syst 57 (3):399–468

McVicar TR, Jupp DL (1999) Estimating one-time-of-day meteorological data from standard daily data as inputs to thermal remote sensing based energy balance models. Agric Forest Meteorol 96(4):219–238

McVicar TR, Jupp DL (2002) Using covariates to spatially interpolate moisture availability in the Murray–darling basin: a novel use of remotely sensed data. Remote Sens Environ 79 (2):199–212

McVicar TR, Roderick ML, Donohue RJ, Li LT, Van Niel TG, Thomas A, Grieser J, Jhajharia D, Himri Y, Mahowald NM, et al. (2012) Global review and synthesis of trends in observed terrestrial near-surface wind speeds: implications for evaporation. J Hydrol 416:182–205
Mehran A, AghaKouchak A (2014) Capabilities of satellite precipitation datasets to estimate heavy precipitation rates at different temporal accumulations. Hydrol Process 28:2262–2270. doi:10.1002/hyp.9779
Mehran A, Mazdiyasni O, AghaKouchak A (2015) A hybrid framework for assessing socioeconomic drought: linking climate variability, local resilience, and demand. J Geophys Res 121. doi:10.1002/2015JD023147
Mo KC (2011) Drought onset and recovery over the United States. J Geophys Res Atmos 116: D20,106. doi:10.1029/2011JD016168
Molotch NP, Bales RC (2006) Comparison of ground-based and airborne snow surface albedo parameterizations in an alpine watershed: impact on snowpack mass balance. Water Resour Res 42(5)
Molotch NP, Margulis SA (2008) Estimating the distribution of snow water equivalent using remotely sensed snow cover data and a spatially distributed snowmelt model: a multi-resolution, multi-sensor comparison. Adv Water Resour 31(11):1503–1514
Moran M, Clarke T, Inoue Y, Vidal A (1994) Estimating crop water deficit using the relation between surface-air temperature and spectral vegetation index. Remote Sens Environ 49 (3):246–263
Morgan J (1989) Satellite remote sensing in meteorology and climatology-status, perspectives and challenges. In: Deutsche Meteorologen-Tagung ueber Atmosphaere, Ozeane, Kontinente, Kiel, Federal Republic of Germany, May 16–19. Annalen der Meteorologie, vol 26, pp 39–43
Mu Q, Heinsch FA, Zhao M, Running SW (2007) Development of a global evapotranspiration algorithm based on modis and global meteorology data. Remote Sens Environ 111(4):519–536
Mu Q, Jones LA, Kimball JS, McDonald KC, Running SW (2009) Satellite assessment of land surface evapotranspiration for the pan-arctic domain. Water Resour Res 45(9)
Mu Q, Zhao M, Running SW (2011) Improvements to a modis global terrestrial evapotranspiration algorithm. Remote Sens Environ 115(8):1781–1800
Mu Q, Zhao M, Kimball JS, McDowell NG, Running SW (2013) A remotely sensed global terrestrial drought severity index. Bull Am Meteorol Soc 94:83–98
NASA (2010) 2010 science plan for NASA's science mission directorate. Technical Report, National Aeronautics and Space Administration, Washington, DC
Nemani R, Hashimoto H, Votava P, Melton F, Wang W, Michaelis A, Mutch L, Milesi C, Hiatt S, White M (2009) Monitoring and forecasting ecosystem dynamics using the terrestrial observation and prediction system (tops). Remote Sens Environ 113(7):1497–1509
Nguyen P, Thorstensen A, Sorooshian S, Hsu K, AghaKouchak A (2015) Flood forecasting and inundation mapping using HiResFlood-UCI and near real-time satellite precipitation data: the 2008 Iowa flood. J Hydrometeorol (16):1171–1183
Nicholson SE, Tucker CJ, Ba M (1998) Desertification, drought, and surface vegetation: an example from the west African Sahel. Bull Am Meteorol Soc 79(5):815–829
Njoku EG, Entekhabi D (1996) Passive microwave remote sensing of soil moisture. J Hydrol 184 (1):101–129
Njoku EG, Jackson TJ, Lakshmi V, Chan TK, Nghiem SV (2003) Soil moisture retrieval from AMSR-E. IEEE Trans Geosci Remote Sens 41(2):215–229
Otkin JA, Anderson MC, Hain C, Svoboda M (2014) Examining the relationship between drought development and rapid changes in the evaporative stress index. J Hydrometeorol 15 (3):938–956
Painter TH, Seidel FC, Bryant AC, McKenzie Skiles S, Rittger K (2013) Imaging spectroscopy of albedo and radiative forcing by light-absorbing impurities in mountain snow. J Geophys Res Atmos 118(17):9511–9523
Palmer W (1965) Meteorological drought. Technical Report, Weather Bureau Res. Paper 45, U.S. Dept. of Commerce, 58 pp

Palmer WC (1968) Keeping track of crop moisture conditions, nationwide: the new crop moisture index. Weatherwise 21(4)

Palmer WC, Havens AV (1958) A graphical technique for determining evapotranspiration by the Thornthwaite method. Mon Weather Rev 86(4):123–128

Paridal BR, Collado WB, Borah R, Hazarika MK, Samarakoon L (2008) Detecting drought-prone areas of rice agriculture using a modis-derived soil moisture index. GISci Remote Sens 45 (1):109–129

Park J-S, Kim K-T, Choi Y-S (2008) Application of vegetation condition index and standardized vegetation index for assessment of spring drought in south Korea. In: IEEE international geoscience and remote sensing symposium, 2008 (IGARSS 2008), vol 3. IEEE, pp III–774

Payero J, Neale C, Wright J (2004) Comparison of eleven vegetation indices for estimating plant height of alfalfa and grass. Appl Eng Agric 20:385–393

Perry CR, Lautenschlager LF (1984) Functional equivalence of spectral vegetation indices. Remote Sens Environ 14(1):169–182

Peters AJ, Walter-Shea EA, Ji L, Vina A, Hayes M, Svoboda MD (2002) Drought monitoring with NDVI-based standardized vegetation index. Photogramm Eng Remote Sens 68(1):71–75

Pinker RT, Sun D, Hung M-P, Li C, Basara JB (2009) Evaluation of satellite estimates of land surface temperature from GOES over the United States. J Appl Meteorol Climatol 48 (1):167–180

Prakash S, Mitra AK, AghaKouchak A, Pai D (2015) Error characterization of TRMM multisatellite precipitation analysis (TMPA-3b42) products over India for different seasons. J Hydrol 529:1302–1312

Prakash S, Mitra AK, Pai D, AghaKouchak A (2016a) From TRMM to GPM: how well can heavy rainfall be detected from space? Adv Water Resour 88:1–7

Prakash S, Mitra AK, AghaKouchak A, Liu Z, Norouzi H, Pai D (2016b) A preliminary assessment of GPM-based multi-satellite precipitation estimates over a monsoon dominated region. J Hydrol. doi:10.1016/j.jhydrol.2016.01.029

Price JC (1982) Estimation of regional scale evapotranspiration through analysis of satellite thermal-infrared data. IEEE Trans Geosci Remote Sens GE-20(3):286–292

Prince SD, Colstoun D, Brown E, Kravitz L (1998) Evidence from rain-use efficiencies does not indicate extensive Sahelian desertification. Glob Chang Biol 4(4):359–374

Qin Q, Jin C, Zhang N, Yang X (2010) An two-dimensional spectral space based model for drought monitoring and its re-examination. In: 2010 I.E. international geoscience and remote sensing symposium (IGARSS). IEEE, pp 3869–3872

Rajsekhar D, Singh VP, Mishra AK (2014) Multivariate drought index: an information theory based approach for integrated drought assessment. J Hydrol. doi:10.1016/j.jhydrol.2014.11.031

Rasmusson EM, Wallace JM, et al. (1983) Meteorological aspects of the el nino/southern oscillation. Science 222(4629):1195–1202

Reager JT, Thomas AC, Sproles EA, Rodell M, Beaudoing HK, Li B, Famiglietti JS (2015) Assimilation of grace terrestrial water storage observations into a land surface model for the assessment of regional flood potential. Remote Sens 7(11):14,663. doi:10.3390/rs71114663

Reichle RH, Koster RD, Dong J, Berg AA (2004) Global soil moisture from satellite observations, land surface models, and ground data: implications for data assimilation. J Hydrometeorol 5 (3):430–442

Rhee J, Im J, Carbone GJ (2010) Monitoring agricultural drought for arid and humid regions using multi-sensor remote sensing data. Remote Sens Environ 114(12):2875–2887

Richard Y, Poccard I (1998) A statistical study of NDVI sensitivity to seasonal and interannual rainfall variations in southern Africa. Int J Remote Sens 19(15):2907–2920

Rodell M (2012) Satellite gravimetry applied to drought monitoring. In: Remote Sensing of Drought: Innovative Monitoring Approaches. CRC, p 261

Rodell M, Famiglietti J (2002) The potential for satellite-based monitoring of groundwater storage changes using grace: the high plains aquifer, central US. J Hydrol 263(1):245–256

Rodell M, Chen J, Kato H, Famiglietti JS, Nigro J, Wilson CR (2007) Estimating groundwater storage changes in the Mississippi river basin (USA) using grace. Hydrogeol J 15(1):159–166
Roderick ML, Noble IR, Cridland SW (1999) Estimating woody and herbaceous vegetation cover from time series satellite observations. Glob Ecol Biogeogr 8(6):501–508
Romanov P, Gutman G, Csiszar I (2000) Automated monitoring of snow cover over north America with multispectral satellite data. J Appl Meteorol 39(11):1866–1880
Rott H, Yueh SH, Cline DW, Duguay C, Essery R, Haas C, Heliere F, Kern M, Macelloni G, Malnes E, et al. (2010) Cold regions hydrology high-resolution observatory for snow and cold land processes. Proc IEEE 98(5):752–765
Rouse J, Haas R, Schell J, Deering D, Harlan J (1974) Monitoring the vernal advancement and retrogradation (greenwave effect) of natural vegetation. Texas A & M University, Remote Sensing Center, College Station
Running SW, Nemani RR, Peterson DL, Band LE, Potts DF, Pierce LL, Spanner MA (1989) Mapping regional forest evapotranspiration and photosynthesis by coupling satellite data with ecosystem simulation. Ecology 70:(4)1090–1101
Running SW, Nemani RR, Heinsch FA, Zhao M, Reeves M, Hashimoto H (2004) A continuous satellite-derived measure of global terrestrial primary production. Bioscience 54(6):547–560
Schanda E, Matzler C, Kunzi K (1983) Microwave remote sensing of snow cover. Int J Remote Sens 4(1):149–158
Senay GB, Bohms S, Verdin JP (2012) Remote sensing of evapotranspiration for operational drought monitoring using principles of water and energy balance. In: Wardlow BD, Anderson MC, Verdin JP (eds) Remote sensing of drought: innovative monitoring approaches. CRC, pp 123–144
Sheffield J, Goteti G, Wen F, Wood E (2004) A simulated soil moisture based drought analysis for the United States. J Geophys Res Atmos 109:D24
Sheffield J, Goteti G, Wood E (2006) Development of a 50-yr, high resolution global dataset of meteorological forcings for land surface modeling. J. Climate 13:3088–3111
Sheffield J, Wood E, Roderick M (2012) Little change in global drought over the past 60 years. Nature 491(7424):435–438
Shen HW, Tabios GQ (1996) Modeling of precipitation-based drought characteristics over California. Centers for Water and Wildland Resources, Davis
Silleos NG, Alexandridis TK, Gitas IZ, Perakis K (2006) Vegetation indices: advances made in biomass estimation and vegetation monitoring in the last 30 years. Geocarto Int 21(4):21–28
Simpson J, Stitt J, Sienko M (1998) Improved estimates of the areal extent of snow cover from AVHRR data. J Hydrol 204(1):1–23
Sorooshian S, Hsu K, Gao X, Gupta H, Imam B, Braithwaite D (2000) Evolution of the Persiann system satellite-based estimates of tropical rainfall. Bull Am Meteorol Soc 81(9):2035–2046
Sorooshian S, AghaKouchak A, Arkin P, Eylander J, Foufoula-Georgiou E, Harmon R, Hendrickx JMH, Imam B, Kuligowski R, Skahill B, Skofronick-Jackson G (2011) Advanced concepts on remote sensing of precipitation at multiple scales. Bull Am Meteorol Soc 92(10):1353–1357
Sorooshian S, Nguyen P, Sellars S, Braithwaite D, AghaKouchak A, Hsu K (2014) Satellite-based remote sensing estimation of precipitation for early warning systems. Extreme Nat Hazards Disaster Risks Soc Implications 1:99–112
Svoboda M, LeComte D, Hayes M, Heim R, Gleason K, Angel J, Rippey B, Tinker R, Palecki M, Stooksbury D, Miskus D, Stephens S (2002) The drought monitor. Bull Am Meteorol Soc 83 (8):1181–1190
Tadesse T, Brown J, Hayes M (2005) A new approach for predicting drought-related vegetation stress: integrating satellite, climate, and biophysical data over the US central plains. ISPRS J Photogramm Remote Sens 59(4):244–253; 30th international symposium on remote sensing of environment, Honolulu, HI, 2003
Takada M, Mishima Y, Natsume S (2009) Estimation of surface soil properties in peatland using ALOS/PALSAR. Landsc Ecol Eng 5(1):45–58

Tapley BD, Bettadpur S, Watkins M, Reigber C (2004) The gravity recovery and climate experiment: mission overview and early results. Geophys Res Lett 31(9)
Thomas AC, Reager JT, Famiglietti JS, Rodell M (2014) A grace-based water storage deficit approach for hydrological drought characterization. Geophys Res Lett 41(5):1537–1545
Tian Y, Peters-Lidard C, Eylander J, Joyce R, Huffman G, Adler R, Hsu K, Turk F, Garcia M, Zeng J (2009) Component analysis of errors in satellite-based precipitation estimates. J Geophys Res 114:D24101
Tourian M, Elmi O, Chen Q, Devaraju B, Roohi S, Sneeuw N (2015) A spaceborne multisensor approach to monitor the desiccation of lake Urmia in Iran. Remote Sens Environ 156:349–360
Tsakiris G, Vangelis H (2005) Establishing a drought index incorporating evapotranspiration. Eur Water 9(10):3–11
Tsakiris G, Pangalou D, Vangelis H (2007) Regional drought assessment based on the reconnaissance drought index (RDI). Water Resour Manag 21(5):821–833
Tucker CJ (1979) Red and photographic infrared linear combinations for monitoring vegetation. Remote Sens Environ 8(2):127–150
Tucker CJ, Choudhury BJ (1987) Satellite remote sensing of drought conditions. Remote Sens Environ 23(2):243–251
Tucker CJ, Pinzon JE, Brown ME, Slayback DA, Pak EW, Mahoney R, Vermote EF, El Saleous N (2005) An extended AVHRR 8-km NDVI dataset compatible with modis and spot vegetation NDVI data. Int J Remote Sens 26(20):4485–4498
Turk FJ, Rohaly GD, Hawkins J, Smith EA, Marzano FS, Mugnai A, Levizzani V (1999) Meteorological applications of precipitation estimation from combined ssm/i, TRMM and infrared geostationary satellite data. In: Microwave Radiometry and Remote Sensing of the Earth's Surface and Atmosphere. VSP International Science Publishers, pp 353–363
UNESCO (1979) Map of the world distribution of arid regions. Technical Report, The United Nations Educational, Scientific and Cultural Organization (UNESCO), Paris, France
van Dijk A, Renzullo L, Rodell M (2011) Use of gravity recovery and climate experiment terrestrial water storage retrievals to evaluate model estimates by the Australian water resources assessment system. Water Resour Res 47(11)
van Dijk AI, Beck HE, Crosbie RS, Jeu RA, Liu YY, Podger GM, Timbal B, Viney NR (2013) The millennium drought in southeast Australia (2001–2009): natural and human causes and implications for water resources, ecosystems, economy, and society. Water Resour Res 49 (2):1040–1057
Van Loon AF (2015) Hydrological drought explained. Wiley Interdiscip Rev Water 2(4):359–392
Van Loon AF, Gleeson T, Clark J, Van Dijk AI, Stahl K, Hannaford J, Di Baldassarre G, Teuling AJ, Tallaksen LM, Uijlenhoet R, et al. (2016) Drought in the anthropocene. Nat Geosci 9(2):89–91
Vargas M, Miura T, Shabanov N, Kato A (2013) An initial assessment of Suomi NPP VIIRS vegetation index EDR. J Geophys Res Atmos 118(22):12–301
Wagner W, Noll J, Borgeaud M, Rott H (1999) Monitoring soil moisture over the Canadian prairies with the ERS scatterometer. IEEE Trans Geosci Remote Sens 37(1):206–216
Wagner W, Dorigo W, de Jeu R, Fernandez D, Benveniste J, Haas E, Ertl M (2012) Fusion of active and passive microwave observations to create an essential climate variable data record on soil moisture. In: XXII ISPRS congress, Melbourne, Australia
Wan Z, Wang P, Li X (2004) Using modis land surface temperature and normalized difference vegetation index products for monitoring drought in the southern great plains, USA. Int J Remote Sens 25(1):61–72
Wang K, Dickinson RE (2012) A review of global terrestrial evapotranspiration: observation, modeling, climatology, and climatic variability. Rev Geophys 50(2)
Wang L, Qu JJ (2007) NMDI: a normalized multi-band drought index for monitoring soil and vegetation moisture with satellite remote sensing. Geophys Res Lett 34(20)
Wang L, Qu JJ (2009) Satellite remote sensing applications for surface soil moisture monitoring: a review. Front Earth Sci Chin 3(2):237–247

Wang A, Bohn TJ, Mahanama SP, Koster RD, Lettenmaier DP (2009) Multimodel ensemble reconstruction of drought over the continental United States. J Climate 22(10):2694–2712
Wang J, Price K, Rich P (2001) Spatial patterns of NDVI in response to precipitation and temperature in the central great plains. Int J Remote Sens 22(18):3827–3844
Wardlow B, Anderson M, Verdin J (2012) Remote sensing of drought. CRC, Boca Raton
Wegren SK (2011) Food security and Russia's 2010 drought. Eurasian Geogr Econ 52(1):140–156
Welsch C, Swenson H, Cota SA, DeLuccia F, Haas JM, Schueler C, Durham RM, Clement JE, Ardanuy PE (2001) VIIRS (visible infrared imager radiometer suite): a next-generation operational environmental sensor for NPOESS. In: IEEE 2001 international geoscience and remote sensing symposium, 2001. IGARSS'01, vol. 3. IEEE, pp 1020–1022
Werick W, Willeke G, Guttman N, Hosking J, Wallis J (1994) National drought atlas developed. Eos Trans Am Geophys Union 75:89
Whitcraft AK, Becker-Reshef I, Justice CO (2015) A framework for defining spatially explicit earth observation requirements for a global agricultural monitoring initiative (GEOGLAM). Remote Sens 7(2):1461–1481
Wiegand C, Richardson A, Escobar D, Gerbermann A (1991) Vegetation indices in crop assessments. Remote Sens Environ 35(2):105–119
Wiesnet D (1981) Winter snow drought. Eos Trans Am Geophys Union 62(14):137–137
Wilhite DA (2005) Drought and water crises: science, technology, and management issues, vol 86. CRC, Boca Raton
Wilson WJ, Yueh SH, Dinardo SJ, Chazanoff SL, Kitiyakara A, Li FK, Rahmat-Samii Y (2001) Passive active l-and s-band (pals) microwave sensor for ocean salinity and soil moisture measurements. IEEE Trans Geosci Remote Sens 39(5):1039–1048
Wiscombe WJ, Warren SG (1980) A model for the spectral albedo of snow. i: Pure snow. J Atmos Sci 37(12):2712–2733
Wood A, Lettenmaier D (2008) An ensemble approach for attribution of hydrologic prediction uncertainty. Geophys Res Lett 35(14)
Yao Y, Liang S, Qin Q, Wang K (2010) Monitoring drought over the conterminous united states using modis and NCEP reanalysis-2 data. J Appl Meteorol Climatol 49(8):1665–1680
Yao Y, Liang S, Qin Q, Wang K, Zhao S (2011) Monitoring global land surface drought based on a hybrid evapotranspiration model. Int J Appl Earth Obs Geoinf 13(3):447–457
Yin D, Roderick ML, Leech G, Sun F, Huang Y (2014) The contribution of reduction in evaporative cooling to higher surface air temperatures during drought. Geophys Res Lett 41 (22):7891–7897
Yirdaw SZ, Snelgrove KR, Agboma CO (2008) Grace satellite observations of terrestrial moisture changes for drought characterization in the Canadian prairie. J Hydrol 356(1):84–92
Zaitchik BF, Rodell M, Reichle RH (2008) Assimilation of grace terrestrial water storage data into a land surface model: results for the Mississippi river basin. J Hydrometeorol 9(3):535–548
Zhang A, Jia G (2013) Monitoring meteorological drought in semiarid regions using multi-sensor microwave remote sensing data. Remote Sens Environ 134:12–23
Zhang N, Hong Y, Qin Q, Liu L (2013) VSDI: a visible and shortwave infrared drought index for monitoring soil and vegetation moisture based on optical remote sensing. Int J Remote Sens 34(13):4585–4609

ERRATUM TO

Chapter 3: The NASA Global Flood Mapping System

F. Policelli, Dan Slayback, Bob Brakenridge, Joe Nigro, Alfred Hubbard, Ben Zaitchik, Mark Carroll, and Hahn Jung

V. Lakshmi (ed.), *Remote Sensing of Hydrological Extremes*, Springer Remote Sensing/Photogrammetry, DOI 10.1007/978-3-319-43744-6_3

DOI 10.1007/978-3-319-43744-6_3

The original version of this chapter was published with an error in the title. The title should read **The NASA Global Flood Mapping System**

The updated original online version for this chapter can be found at DOI 10.1007/978-3-319-43744-6_3

V. Lakshmi (ed.), *Remote Sensing of Hydrological Extremes*, Springer Remote Sensing/Photogrammetry, DOI 10.1007/978-3-319-43744-6_12

Index

V. Lakshmi (ed.), *Remote Sensing of Hydrological Extremes*, Springer Remote Sensing/Photogrammetry, DOI 10.1007/978-3-319-43744-6

FSC
www.fsc.org
MIX
Papier | Fördert
gute Waldnutzung
FSC® C083411

Kolibri 360 GmbH
Ferdinand-Jühlke-Straße 7
99095 Erfurt, Deutschland
produktsicherheit@kolibri360.de